Annals of Mathematics Studies
Number 185

Degenerate Diffusion Operators Arising in Population Biology

Charles L. Epstein
and
Rafe Mazzeo

PRINCETON UNIVERSITY PRESS
PRINCETON AND OXFORD
2013

Published by Princeton University Press, 41 William Street,
Princeton, New Jersey 08540

In the United Kingdom: Princeton University Press, 6 Oxford Street,
Woodstock, Oxfordshire OX20 1TW

press.princeton.edu

Library of Congress Cataloging-in-Publication Data

Epstein, Charles L.
Degenerate diffusion operators arising in population biology / Charles L. Epstein and Rafe Mazzeo.
pages cm. – (Annals of mathematics studies ; number 185)
Includes bibliographical references and index.
ISBN 978-0-691-15712-2 (hardcover : alk. paper) –
ISBN 978-0-691-15715-3 (pbk. : alk. paper)
1. Elliptic operators. 2. Markov processes. 3. Population biology–Mathematical models. I. Mazzeo, Rafe. II. Title.
QA329.42.E67 2013
577.8′801519233–dc23
2012022328

British Library Cataloging-in-Publication Data is available

This book has been composed in LaTeX.

The publisher would like to acknowledge the authors of this volume for providing the camera-ready copy from which this book was printed.

Printed on acid-free paper ∞

Printed in the United States of America

10 9 8 7 6 5 4 3 2 1

Out here in the perimeter there are no stars
Out here we is stone
Immaculate.

Jim Morrison and the Doors, “The Wasp (Texas Radio and the Big Beat)” (*L.A. Woman*)

Contents

Preface

This *lange megillah* is concerned with establishing properties of a mathematical model in population genetics that some might regard, in light of what is being modeled, as entirely obvious. Once written down, however, a mathematical model has a life of its own; it must be addressed in its own terms, and understood without reference to its origins.

The models we consider are phrased as partial differential equations, which arise as limits of finite Markov chains. The existence of solutions to these partial differential equations and their properties are suggested by the physical, economic, or biological systems under consideration, but logically speaking, are entirely independent of them. What is far from obvious are the regularity properties of these solutions, and, as is so often the case, the existence of solutions actually hinges on these very subtle properties. Using a Schauder method, we *prove* the existence of solutions to a class of degenerate parabolic and elliptic equations that arise in population genetics and mathematical finance.

The archetypes for these equations arise as infinite population limits of the Wright-Fisher models in population genetics. These describe the prevalence of a mutant allele, in a population of fixed size, under the effects of genetic drift, mutation, migration and selection. The formal generator of the infinite population limit acts on functions defined on [0, 1] (the space of frequencies) and is given by

$$L_{WF} = x(1-x)\partial_x^2 + [b_0(1-x) - b_1 x + sx(1-x)]\partial_x. \tag{1}$$

Processes defined by such operators were studied by Feller in the early 1950s and used to great effect by Kimura et al. in the 1960s and '70s to give quantitative answers to a wide range of questions in population genetics. Notwithstanding, a modern appreciation of the analytic properties of these equations is only now coming into focus.

In this monograph we provide analytic foundations for equations of this type and their natural higher dimensional generalizations. We call these operators generalized Kimura diffusions. They act on functions defined on generalizations of convex polyhedra, which are called manifolds with corners. We provide the basic Hölder space-type estimates for operators in this class with which we establish the existence of solutions. These operators satisfy a strong form of the maximum principle, which implies uniqueness.

The partial differential operators we consider are degenerate and the underlying manifolds with corners are themselves singular. This inevitably produces significant technical challenges in the analysis of such equations, and explains, in part, the length of this text. The Markov processes defined by these operators provide fundamental models for many Biological and Economic situations, and it is for this reason that we feel that these operators merit such a detailed and laborious treatment.

A large portion of this book is devoted to a careful exploration of operators of the form

$$L_{\boldsymbol{b},m} = \sum_{i=1}^{n} [x_i \partial_{x_i}^2 + b_i \partial_{x_i}] + \sum_{l=1}^{m} \partial_{y_l}^2, \tag{2}$$

acting on functions defined in $\mathbb{R}_+^n \times \mathbb{R}^m$. Here the coefficients $\{b_i\}$ are non-negative constants. These operators are interesting in their own right, arising as models in population genetics, but for us, they are largely building blocks for the analysis of general Kimura diffusions. These are the analogues, in the present context, of the "constant coefficient elliptic operators" in classical elliptic theory. A notable feature of this family is that, because the coefficient of $\partial_{x_i}^2$ vanishes at $x_i = 0$, we need to retain the first order transverse term. The value of $\boldsymbol{b} = (b_1, \ldots, b_n)$ has a pervasive effect on the behavior of the solution. Much of our analysis rests upon explicit formulæ for the solutions of the initial value problems:

$$\partial_t v - L_{\boldsymbol{b},m} v = 0 \text{ with } v(\boldsymbol{x}, \boldsymbol{y}, 0) = f(\boldsymbol{x}, \boldsymbol{y}). \tag{3}$$

Using these models, we have succeeded in developing a rather complete existence and regularity theory for general Kimura diffusions with Hölder continuous data. This in turn suffices to prove the existence of a C_0-semi-group acting on continuous functions, and shows that the associated martingale problem has a unique solution. The existence of a strong Markov process, which in applications to genetics describes the statistical behavior of individual populations, follows from this.

In special situations, such results have been established by other authors, but without either the precise control on the regularity of solutions, or the generality considered herein. As long as it is, this text just begins to scratch the surface of this very rich field. We have largely restricted our attention to the solutions with the best possible regularity properties, which leads to considerable simplifications. For applications it will be important to consider solutions with more complicated boundary behavior; we hope that this text will provide a solid foundation for these investigations.

ACKNOWLEDGMENTS

We would like to acknowledge the generous financial support and unflagging personal support provided by Ben Mann and the DARPA FunBio project. It is certainly the case that without Ben's encouragement, we would never have undertaken this project. We would like to thank our FunBio colleagues[1] who provided us with the motivation

[1]Phil Benfey, Michael Deem, Richard Lenski, Jack Morava, Lior Pachter, Herbert Edelsbrunner, John Harer, Jim Damon, Peter Bates, Joshua Weitz, Konstantin Mischaikow, Gunnar Carlsson, Bernd Sturmfels, Tim Buchman, Ary Goldberger, Jonathan Eisen, Olivier Porquie, Thomas Fink, Ned Wingreen, Jonathan Dushoff, Peter Nara, *inter alios*, and research staff: Mark Filipkowski, Shauna Koppel, Rachel Scholz, Matthew Clement, Traci Kiesling, Traci Pals, *inter alios*.

and knowledge base to pursue this project, and Simon Levin for his leadership and inspiration. CLE would like to thank Josh Plotkin, Warren Ewens, Todd Parsons, and Ricky Der, from whom he has learned what little he knows about population genetics.

We would both like to thank Charlie Fefferman for showing us an explicit formula for $k_t^0(x, y)$, which set us off in the very fruitful direction pursued herein. We would also like to thank Dan Stroock for his help with connections to Probability Theory. We would like to thank Camelia Pop for providing a long list of corrections.

CLE would like to acknowledge the financial support of DARPA through grants HR0011-05-1-0057 and HR0011-09-1-0055, the support of NSF through grant DMS-0603973, and the hospitality and support provided by Leslie Greengard and the Courant Institute, where this text was completed. He would also like to thank his wife Jane for doing the hard work to get us there and back.

RM would like to acknowledge the financial support of DARPA through grants HR0011-05-1-0057 and HR0011-09-1-0055, and the support of NSF through grants DMS-0805529 and DMS-1105050. He is also immensely grateful to his wife Laurence for her general forbearance and support, and to his daughters Sabine and Sophie for making it all worthwhile.

Chapter One

Introduction

In population genetics one frequently replaces a discrete Markov chain model, which describes the random processes of genetic drift, with or without selection, and mutation with a limiting, continuous time and space, stochastic process. If there are $N+1$ possible types, then the configuration space for the resultant continuous Markov process is typically the N-simplex

$$\mathcal{S}_N = \{(x_1, \ldots, x_N) : x_j \geq 0 \text{ and } x_1 + \cdots + x_N \leq 1\}. \tag{1.1}$$

If a different scaling is used to define the limiting process, different domains might also arise. As a geometrical object the simplex is quite complicated. Its boundary is not a smooth manifold, but has a stratified structure with strata of codimensions 1 through N. The codimension 1 strata are

$$\Sigma_{1,l} = \{x_l = 0\} \cap \mathcal{S}_N \text{ for } l = 1, \ldots, N, \tag{1.2}$$

along with

$$\Sigma_{1,0} = \{x_1 + \cdots + x_N = 1\} \cap \mathcal{S}_N. \tag{1.3}$$

Components of the stratum of codimension $1 < l \leq N$ arise by choosing integers $0 \leq i_1 < \cdots < i_l \leq N$ and forming the intersection:

$$\Sigma_{1,i_1} \cap \cdots \cap \Sigma_{1,i_l}. \tag{1.4}$$

The simplex is an example of a *manifold with corners*. The singularity of its boundary significantly complicates the analysis of differential operators acting functions defined in $\mathcal{S}_N$.

In the simplest case, without mutation or selection, the limiting operator of the Wright-Fisher process is the Kimura diffusion operator, with formal generator:

$$L_{\text{Kim}} = \sum_{i,j=1}^{N} x_i(\delta_{ij} - x_j)\partial_{x_i}\partial_{x_j}. \tag{1.5}$$

This is the "backward" Kolmogorov operator for the limiting Markov process. This operator is elliptic in the interior of $\mathcal{S}_N$ but the coefficient of the second order normal derivative tends to zero as one approaches a boundary. We can introduce local coordinates $(x_1, y_1, \ldots, y_{N-1})$ near the interior of a point on one of the faces of $\Sigma_{1,l}$, so that the boundary is given locally by the equation $x_1 = 0$, and the operator then takes the form

$$x_1\partial_{x_1}^2 + \sum_{l=1}^{N-1} x_1 c_{1l}\partial_{x_1}\partial_{y_l} + \sum_{l,m=1}^{N-1} c_{lm}\partial_{y_m}\partial_{y_l}, \tag{1.6}$$

where the matrix c_{lm} is positive definite. To include the effects of mutation, migration and selection, one typically adds a vector field:

$$V = \sum_{i=1}^{N} b_i(x)\partial_{x_i}, \tag{1.7}$$

where V is inward pointing along the boundary of $\mathcal{S}_N$. In the classical models, if only the effect of mutation and migration are included, then the coefficients $\{b_i(x)\}$ can be taken to be linear polynomials, whereas selection requires at least quadratic terms.

The most significant feature is that the coefficient of $\partial_{x_1}^2$ vanishes exactly to order 1. This places L_{Kim} outside the classes of degenerate elliptic operators that have already been analyzed in detail. For applications to Markov processes the difficulty that presents itself is that it is not possible to introduce a square root of the coefficient of the second order terms that is Lipschitz continuous up to the boundary. Indeed the best one can hope for is Hölder-$\frac{1}{2}$. The uniqueness of the solutions to either the forward Kolmogorov equation, or the associated stochastic differential equation, cannot then be concluded using standard methods.

Even in the presence of mutation and migration, the solutions of the heat equation for this operator in 1-dimension was studied by Kimura, using the fact that $L_{\text{Kim}} + V$ preserves polynomials of degree d for each d. In higher dimensions it was done by Karlin and Shimakura by showing the existence of a complete basis of polynomial eigenfunctions for this operator. This in turn leads to a proof of the existence of a polynomial solution to the initial value problem for $[\partial_t - (L_{\text{Kim}} + V)]v = 0$ with polynomial initial data. Using the maximum principle, this suffices to establish the existence of a strongly continuous semi-group acting on $\mathcal{C}^0$, and establish many of its basic properties, see [38]. This general approach has been further developed by Barbour, Etheridge, Ethier, and Griffiths, see [17, 2, 16, 25].

As noted, if selection is also included, then the coefficients of V are at least quadratic polynomials, and can be quite complicated, see [9]. So long as the second order part remains L_{Kim}, then a result of Ethier, using the Trotter product formula, makes it possible to again define a strongly continuous semi-group, see [18]. Various extensions of these results, using a variety of functional analytic frameworks, were made by Athreya, Barlow, Bass, Perkins, Sato, Cerrai, Clément, and others, see [1, 4, 6, 7, 8].

For example Cerrai and Clément constructed a semi-group acting on $\mathcal{C}^0([0, 1]^N)$, with the coefficient a_{ij} of $\partial_{x_i}\partial_{x_j}$ assumed to be of the form

$$a_{ij}(x) = m(x)A_{ij}(x_i, x_j). \tag{1.8}$$

Here $m(x)$ is strictly positive. In [1, 4, 3], Bass and Perkins along with several collaborators, study a class of equations, similar to that defined below. Their work has many points of contact with our own, and we discuss it in greater detail at the end of Section 1.5.

We have not yet said anything about boundary conditions, which would seem to be a serious omission for a PDE on a domain with a boundary. Indeed, one would expect that there would be an infinite dimensional space of solutions to the homogeneous equation. It is possible to formulate local boundary conditions that assure uniqueness, but, in some sense, this is not necessary. As a result of the degeneracy of the principal part, uniqueness for these types of equations can also be obtained as a consequence of regularity alone! We illustrate this in the simplest 1-dimensional case, which is the equation, with $b(0) \geq 0,\ b(1) \leq 0$,

$$\partial_t v - [x(1-x)\partial_x^2 + b(x)\partial_x]v = 0 \text{ and } v(x,0) = f(x). \tag{1.9}$$

If we assume that $\partial_x v(x,t)$ extends continuously to $[0,1] \times (0,\infty)$ and

$$\lim_{x\to 0^+} x(1-x)\partial_x^2 v(x,t) = \lim_{x\to 1^-} x(1-x)\partial_x^2 v(x,t) = 0, \tag{1.10}$$

then a simple maximum principle argument shows that the solution is unique. In our approach, such regularity conditions naturally lead to uniqueness, and little effort is expended in the consideration of boundary conditions. In Chapter 3 we prove a generalization of the Hopf boundary point maximum principle that demonstrates, in the general case, how regularity implies uniqueness.

1.1 GENERALIZED KIMURA DIFFUSIONS

In his seminal work, Feller analyzed the most general closed extensions of operators, like those in (1.9), which generate Feller semi-groups in 1-dimension, see [20]. He analyzes the resolvent kernel, using methods largely restricted to ordinary differential equations. In [10], Chen and Stroock use probabilistic methods to prove estimates on the fundamental solution of the parabolic equation, $\partial_t u = x(1-x)\partial_x^2 u$.

Up to now very little is known, in higher dimensions, about the analytic properties of the solution to the initial value problem for the heat equation

$$\partial_t v - (L_{\text{Kim}} + V)v = 0 \text{ in } (0,\infty) \times \mathcal{S}_N \text{ with } v(0,x) = f(x). \tag{1.11}$$

Indeed, if we replace L_{Kim} with a qualitatively similar second order part, which does not take one of the forms described above, then even the existence of a solution is not known. In this monograph we introduce a very flexible analytic framework for studying a large class of equations, which includes all standard models, of this type appearing in population genetics, as well as the SIR model for epidemics, see [19, 38], and models that arise in Mathematical Finance, see [21]. Our approach is to introduce non-isotropic Hölder spaces with respect to which we establish sharp existence and regularity results for the solutions to heat equations of this type, as well as the corresponding elliptic problems. Using the Lumer-Phillips theorem we conclude that the $\mathscr{C}^0$-graph closure of this operator generates a strongly continuous semi-group.

The approach here is an extension of our work on the 1-d case in [15], which allows us to prove existence, uniqueness and regularity results for a class of higher dimensional, degenerate diffusion operators. While our methods also lead to a precise description of the heat kernel in the 1-dimensional case, this has proved considerably more challenging in higher dimensions. It is hoped that a combination of the analytic techniques used here, and the probabilistic techniques from [10] will lead to good descriptions of the heat kernels in higher dimensions.

Our analysis applies to a class of operators that we call *generalized Kimura diffusions*, which act on functions defined on *manifolds with corners*. Such spaces generalize the notion of a regular convex polyhedron in $\mathbb{R}^N$, e.g., the simplex. Working in this more general context allows for a great deal of flexibility, which proves indispensable in the proof of our basic existence result.

Locally a manifold with corners, P, can be described as a subset of $\mathbb{R}^N$ defined by inequalities. Let $\{p_k(x) : k = 1, \ldots, K\}$ be smooth functions in the unit ball $B_1(0) \subset \mathbb{R}^N$, vanishing at 0, with $\{dp_k(0) : k = 1, \ldots, K\}$ linearly independent; clearly $K \leq N$. Locally P is diffeomorphic to

$$\bigcap_{k=1}^{K} \{x \in B_1(0) : p_k(x) \geq 0\}. \tag{1.12}$$

We let $\Sigma_k = P \cap \{x : p_k(x) = 0\}$; suppose that Σ_k contains a non-empty, open $(N-1)$-dimensional hypersurface and that dp_k is non-vanishing in a neighborhood of Σ_k. The boundary of P is a stratified space, where the strata of codimension n locally consists of points where the boundary is defined by the vanishing of n functions with independent gradients. The components of the codimension 1 part of the bP are called *faces*. As in (1.4), the codimension-n stratum of bP is formed from intersections of n faces.

The *formal* generator is a degenerate elliptic operator of the form

$$L = \sum_{i,j=1}^{N} A_{ij}(x)\partial_{x_i}\partial_{x_j} + \sum_{j=1}^{N} b_j(x)\partial_{x_j}. \tag{1.13}$$

Here $A_{ij}(x)$ is a smooth, symmetric matrix valued function in P. The second order term is positive definite in the interior of P and degenerates along the hypersurface boundary components in a specific way. For each k

$$\sum_{i,j=1}^{N} a_{ij}(x)\partial_{x_i} p_k(x)\partial_{x_j} p_k(x) \propto p_k(x) \text{ as } x \text{ approaches } \Sigma_k. \tag{1.14}$$

On the other hand,

$$\sum_{i,j=1}^{N} a_{ij}(x)v_i v_j > 0 \text{ for } x \in \operatorname{int}\Sigma_k \text{ and } v \neq 0 \in T_x\Sigma_k. \tag{1.15}$$

The first order part of L is an inward pointing vector field

$$Vp_k(x) = \sum_{j=1}^{N} b_j(x)\partial_{x_j} p_k(x) \geq 0 \text{ for } x \in \Sigma_k. \tag{1.16}$$

We call a second order partial differential operator defined on P, which is non-degenerate elliptic in int P, with this local description near any boundary point a *generalized Kimura diffusion* .

If p is a point on the stratum of bP of codimension n, then locally there are coordinates $(x_1, \ldots, x_n; y_1, \ldots, y_m)$ so that p corresponds to $(\mathbf{0}; \mathbf{0})$, and a neighborhood, U, of p is given by

$$U = \{(\boldsymbol{x}; \boldsymbol{y}) \in [0, 1)^n \times (-1, 1)^m\}. \tag{1.17}$$

In these local coordinates a generalized Kimura diffusion, L, takes the form

$$L = \sum_{i=1}^{n} [a_{ii}(\boldsymbol{x}; \boldsymbol{y}) x_i \partial_{x_i}^2 + \widetilde{b}_i(\boldsymbol{x}; \boldsymbol{y})\partial_{x_i}] + \sum_{1 \leq i \neq j \leq n} x_i x_j a_{ij}(\boldsymbol{x}; \boldsymbol{y})\partial_{x_i x_j}^2 + \\ \sum_{i=1}^{n}\sum_{k=1}^{m} x_i b_{ik}(\boldsymbol{x}; \boldsymbol{y})\partial_{x_i y_k}^2 + \sum_{k,l=1}^{m} c_{kl}(\boldsymbol{x}; \boldsymbol{y})\partial_{y_k y_l}^2 + \sum_{k=1}^{m} d_k(\boldsymbol{x}; \boldsymbol{y})\partial_{y_k}; \tag{1.18}$$

(a_{ij}) and (c_{kl}) are symmetric matrices, the matrices (a_{ii}) and (c_{kl}) are strictly positive definite. The coefficients $\{\widetilde{b}(\boldsymbol{x}; \boldsymbol{y})\}$ are non-negative along $bP \cap U$, so that first order part is inward pointing.

Let P be a compact manifold with corners and L a generalized Kimura diffusion defined on P. Broadly speaking, our goal is to prove the existence, uniqueness and regularity of solutions to the equation

$$\begin{aligned} &(\partial_t - L)u = g \text{ in } P \times (0, \infty) \\ &\text{with } u(p, 0) = f(p), \end{aligned} \tag{1.19}$$

with certain boundary behavior along $bP \times [0, \infty)$, for data g and f satisfying appropriate regularity conditions. These results in turn can be used to prove the existence of a strongly continuous semi-group acting on $\mathscr{C}^0(P)$, with formal generator L. This is the "backward Kolmogorov equation." The solution to the "forward Kolmogorov equation," $(\partial_t - L^*)m = \nu$, is then given by the adjoint semi-group, canonically defined on a (non-dense) domain in $[\mathscr{C}^0(P)]' = \mathcal{M}(P)$, the space of finite Borel measures on P.

1.2 MODEL PROBLEMS

The problem of proving the existence of solutions to a class of PDEs is essentially a matter of finding a good class of model problems, for which existence and regularity can be established, more or less directly, and then finding a functional analytic setting in which to do a perturbative analysis of the equations of interest. The model operators for Kimura diffusions are the differential operators, defined on $\mathbb{R}_+^n \times \mathbb{R}^m$, by

$$L_{\boldsymbol{b},m} = \sum_{j=1}^{n} [x_j \partial_{x_j}^2 + b_j \partial_{x_j}] + \sum_{k=1}^{m} \partial_{y_k}^2. \tag{1.20}$$

Here $\boldsymbol{b} = (b_1, \ldots, b_n)$ is a non-negative vector.

We have not been too explicit about the boundary conditions that we impose along $b\mathbb{R}_+^n \times \mathbb{R}^m$. This condition can be defined by a local Robin-type formula involving the value of the solution and its normal derivative along each hypersurface boundary component of bP. For $b > 0$, the 1-dimensional model operator, $L_b = x\partial_x^2 + b\partial_x$, has two indicial roots

$$\beta_0 = 0, \beta_1 = 1 - b, \tag{1.21}$$

that is

$$L_b x^{\beta_0} = L_b x^{\beta_1} = 0. \tag{1.22}$$

If $b \neq 1$, then the boundary condition,

$$\lim_{x \to 0+} [\partial_x (x^b u(x,t)) - b x^{b-1} u(x,t)] = 0, \tag{1.23}$$

excludes the appearance of terms like x^{1-b} in the asymptotic expansion of solutions along $x = 0$. In fact, this condition insures that u is as smooth as possible along the boundary: if $g = 0$ and f has m derivatives then the solution to (1.19), satisfying (1.23), does as well. This boundary condition can be encoded as a regularity condition, that is $u(\cdot, t) \in \mathcal{C}^1([0, \infty)) \cap \mathcal{C}^2((0, \infty))$, with

$$\lim_{x \to 0+} x\partial_x^2 u(x,t) = 0 \tag{1.24}$$

for $t > 0$. We call the unique solution to a generalized Kimura diffusion, satisfying this condition, or its higher dimensional analogue, the *regular* solution. The majority of this monograph is devoted to the study of regular solutions.

In applications to probability one often seeks solutions to equations of the form $Lw = g$, where w satisfies a Dirichlet boundary condition: $w \restriction_P = h$. Our uniqueness results often imply that these equations *cannot* have a regular solution, for example, when $g \geq 0$. In the classical case the solutions to these problems can sometimes be written down explicitly, and are seen to involve the non-zero indicial roots. Usually these satisfy the other natural boundary condition, a la [20]. In 1-dimension, when $b \neq 0, 1$ it is:

$$\lim_{x \to 0+} [\partial_x (x u(x,t)) - (2 - b) u] = 0, \tag{1.25}$$

and allows for solutions that are $O(x^{1-b})$ as $x \to 0^+$. These are not smooth up to the boundary, even if the data is. The adjoint of L is naturally defined as an operator on $\mathcal{M}(P)$, the space finite Borel measures on P. It is more common to study this operator using techniques from probability theory, see [40].

For a generalized Kimura diffusion in dimensions greater than 1, the coefficient of the normal first derivative can vary as one moves along the boundary. For example, in 2-dimensions one might consider the operator

$$L = x\partial_x^2 + \partial_y^2 + b(y)\partial_x. \tag{1.26}$$

If $b(y)$ is not constant, then, with the boundary condition

$$\lim_{x\to 0^+} [\partial_x(xu(x, y, t)) - (2 - b(y))u(x, y, t)] = 0, \tag{1.27}$$

one would be faced with the very thorny issue of a varying indicial root on the outgoing face of the heat or resolvent kernel. As it is, we get a varying indicial root on the incoming face, a fact which already places the analysis of this problem beyond what has been achieved using the detailed kernel methods familiar in geometric microlocal analysis. The natural boundary condition for the adjoint operator includes the condition:

$$\lim_{x\to 0^+} [\partial_x(xu(x, y, t)) - b(y)u(x, y, t)] = 0, \tag{1.28}$$

allowing for solutions that behave like $x^{b(y)-1}$, as $x \to 0^+$.

The solution operators for the 1-dimensional model problems are given by simple explicit formulæ. If $b > 0$, then the heat kernel is

$$k_t^b(x, y)dy = \left(\frac{y}{t}\right)^b e^{-\frac{x+y}{t}} \psi_b\left(\frac{xy}{t^2}\right)\frac{dy}{y}, \tag{1.29}$$

where

$$\psi_b(z) = \sum_{j=0}^{\infty} \frac{z^j}{j!\Gamma(j+b)}. \tag{1.30}$$

If $b = 0$ then

$$k_t^0(x, y) = e^{-\frac{x}{t}}\delta_0(y) + \left(\frac{x}{t}\right) e^{-\frac{x+y}{t}} \psi_2\left(\frac{xy}{t^2}\right)\frac{dy}{t}. \tag{1.31}$$

In either case k_t^b is smooth as $x \to 0^+$ and displays a y^{b-1} singularity as $y \to 0^+$. It is notable that the character of the kernel changes dramatically as $b \to 0$, nonetheless the regular solutions to these heat equations satisfy uniform estimates even as $b \to 0^+$. This fact is quite essential for the success of our approach.

The structure of these operators suggests that a natural functional analytic setting in which to do the analysis might be that provided by the anisotropic Hölder spaces defined by the singular, but incomplete metric on $\mathbb{R}_+^n \times \mathbb{R}^m$:

$$ds_{WF}^2 = \sum_{j=1}^{n} \frac{dx_j^2}{x_j} + \sum_{k=1}^{m} dy_m^2. \tag{1.32}$$

Similar spaces have been introduced by other authors for problems with similar degeneracies, see [11, 12, 28, 32, 23, 24]. In [22] Goulaouic and Shimakura proved a priori estimates in a Hölder space of this general sort in a case where the operator has this type of degeneracy, but the boundary is smooth. As was the case in these earlier works, we introduce two families of anisotropic Hölder spaces, which we denote by $\mathscr{C}^{k,\gamma}_{WF}(P)$, and $\mathscr{C}^{k,2+\gamma}_{WF}(P)$, for $k \in \mathbb{N}_0$, and $0 < \gamma < 1$. In this context, heuristically an operator A is "elliptic of second order" if $A^{-1} : \mathscr{C}^{k,\gamma}_{WF}(P) \to \mathscr{C}^{k,2+\gamma}_{WF}(P)$. Note that $\mathscr{C}^{k,2+\gamma}_{WF}(P) \subseteq \mathscr{C}^{k+1,\gamma}_{WF}(P)$, but $\mathscr{C}^{k+2,\gamma}_{WF}(P) \nsubseteq \mathscr{C}^{k,2+\gamma}_{WF}(P)$, which explains the need for two families of spaces.

In this monograph we consider the problem in (1.19) for f and g belonging to these Hölder spaces. The results obtained suffice to prove the existence of a semi-group on the space $\mathscr{C}^0(P)$, but establishing the refined regularity properties of this semi-group and its adjoint require the usage of a priori estimates. These are of a rather different character from the analysis presented here; we will return to this question in a subsequent publication.

In [23, 24], Graham studies model operators of the forms $x(\partial_x^2 + \Delta_{\mathbb{R}^n}) + (1-\lambda)\partial_x$, and $x\partial_x^2 + \Delta_{\mathbb{R}^n} + (1-\lambda)\partial_x$, acting on functions defined on the half-space $[0, \infty) \times \mathbb{R}^n$. Using kernel methods, he proves sharp estimates for the solution of the inhomogeneous Dirichlet problem, in both isotropic and an-isotropic Hölder spaces. Graham works directly with the solution kernels for the elliptic problems, whereas we prove estimates for the parabolic problem and obtain the elliptic case via the Laplace transform.

As manifolds with corners have non-smooth boundaries, and the Kimura diffusions are degenerate elliptic operators, the analysis of (1.19) can be expected to be rather challenging. We have already indicated a variety of problems that arise:

1. The principal part of L degenerates at the boundary.

2. The boundary of P is not smooth.

3. The "indicial roots" vary with the location of the point on bP.

4. The character of the solution operator is quite different at points where the vector field is tangent to bP.

Along the boundary $\{x_j = 0\}$, the first and second order terms in (1.20), $b_j\partial_{x_j}$ and $x_j\partial_{x_j}^2$, respectively, are of comparable "strength." It is a notable and non-trivial fact that estimates for the solutions of these equations can be proved in these Hölder spaces, without regard for the value of $\boldsymbol{b} \geq \boldsymbol{0}$. As there is an explicit formula for the fundamental solution, the analysis of these model operators, while tedious and time consuming, is elementary. Indeed the solution of the homogeneous Cauchy problem,

$$\begin{aligned}(\partial_t - L_{\boldsymbol{b},m})u &= 0 \text{ in } P \times (0, \infty)\\ \text{with } u(p, 0) &= f(p),\end{aligned} \tag{1.33}$$

has an analytic extension to $\operatorname{Re} t > 0$, which satisfies many useful estimates.

To obtain a gain of derivatives where $\operatorname{Re} t > 0$, in a manner that can be extended beyond the model problems, one must address the inhomogeneous problem, which has somewhat simpler analytic properties. By this device, one can also estimate the Laplace transform of the heat semi-group, which is the resolvent operator:

$$(\mu - L_{\boldsymbol{b},m})^{-1} = \int_0^\infty e^{tL_{\boldsymbol{b},m}} e^{-\mu t} dt. \tag{1.34}$$

The estimates for the inhomogeneous problem show that, in an appropriate sense, $(\mu - L_{\boldsymbol{b},m})^{-1}$ gains two derivatives and is analytic in the complement of $(-\infty, 0]$. Finally one can re-synthesize the heat operator from the resolvent, via contour integration:

$$e^{tL_{\boldsymbol{b},m}} = \frac{1}{2\pi i} \int_C (\mu - L_{\boldsymbol{b},m})^{-1} e^{\mu t} d\mu, \tag{1.35}$$

where C is of the form $|\arg \mu| = \frac{\pi}{2} + \alpha$, for an $0 < \alpha < \frac{\pi}{2}$. This shows that, for t with positive real part, $e^{tL_{\boldsymbol{b},m}}$ also gains two derivatives.

Remark. In probability theory it is more common to consider the operators obtained from $L_{\boldsymbol{b},m}$ under the change of variables $x_j = \frac{w_j^2}{2}$. In this coordinate system, the model operators become:

$$L_{\boldsymbol{b},m} = \frac{1}{2} \sum_{j=1}^{n} \left[\partial_{w_j}^2 + \frac{2b_j - 1}{w_j} \partial_{w_j} \right] + \sum_{k=1}^{m} \partial_{y_k}^2. \tag{1.36}$$

The processes these operators generate are called Bessel processes. For a recent paper on this subject see, for example, [5].

1.3 PERTURBATION THEORY

The next step is to use these estimates for the model problems in a perturbative argument to prove existence and regularity for a generalized Kimura diffusion operator L on a manifold with corners, P. The boundary of a manifold with corners is a stratified space, which produces a new set of difficulties. To overcome this we use an induction on the maximal codimension of the strata of bP.

The induction starts with the simplest case where bP is just a manifold, and P is then a manifold with boundary. In this case, we can use the model operators to build a parametrix for the solution operator to the heat equation in a neighborhood of the boundary, $\widehat{Q}_b^t$. It is a classical fact that there is an exact solution operator, $\widehat{Q}_i^t$, defined in the complement of a neighborhood of the boundary, for, in any such subset of P, L is a non-degenerate elliptic operator. Using a partition of unity these operators can be "glued together" to define a parametrix, $\widehat{Q}^t$, for the solution operator. The Laplace transform

$$\widehat{R}(\mu) = \int_0^\infty e^{\mu t} \widehat{Q}^t dt \tag{1.37}$$

is then a right parametrix for $(\mu - L)^{-1}$. Using the estimates and analyticity for the model problems, and the properties of the interior solution operator, we can show that

$$(\mu - L)\widehat{R}(\mu) = \mathrm{Id} + E(\mu), \tag{1.38}$$

where $E(\mu)$ is analytic in $\mathbb{C} \setminus (-\infty, 0]$, and the Neumann series for $(\mathrm{Id} + E(\mu))^{-1}$ converges in the operator norm topology for μ in sectors $|\arg \mu| \leq \pi - \alpha$, for any $\alpha > 0$, if $|\mu|$ is sufficiently large. This allows us to show that

$$(\mu - L)^{-1} = \widehat{R}(\mu)(\mathrm{Id} + E(\mu))^{-1} \tag{1.39}$$

is analytic and satisfies certain estimates in

$$T_{\alpha,R} = \{\mu : |\arg \mu| < \pi - \alpha, \quad |\mu| > R\}, \tag{1.40}$$

for any $0 < \alpha$, and R depending on α.

For t in the right half plane we can now reconstruct the heat semi-group acting on the Hölder spaces:

$$e^{tL} = \frac{1}{2\pi i} \int\limits_{bT_{\alpha,R}} (\mu - L)^{-1} e^{\mu t} d\mu \tag{1.41}$$

for an appropriate choice of α. This allows us to verify that e^{tL} has an analytic continuation to $\mathrm{Re}\, t > 0$, which satisfies the desired estimates with respect to the anisotropic Hölder spaces defined above. The proof for the general case now proceeds by induction on the maximal codimension of the strata of bP. In all cases we use the model operators to construct a boundary parametrix $\widehat{Q}_b^t$ near the maximal codimensional part of bP. The induction hypothesis provides an exact solution operator in the "interior," $\widehat{Q}_i^t$, with certain properties, which we once again glue together to get $\widehat{Q}^t$. A key step in the argument is to verify that the heat operator we finally obtain satisfies the induction hypotheses. The representation of e^{tL} in (1.41) is a critical part of this argument.

1.4 MAIN RESULTS

With these preliminaries we can state our main results. The sharp estimates for operators e^{tL} and $(\mu - L)^{-1}$ are phrased in terms of two families of Hölder spaces. For $k \in \mathbb{N}_0$ and $0 < \gamma < 1$, we define the spaces $\mathscr{C}^{k,\gamma}_{WF}(P)$, $\mathscr{C}^{k,2+\gamma}_{WF}(P)$, and their "heat-space" analogues, $\mathscr{C}^{k,\gamma}_{WF}(P \times [0, T])$, $\mathscr{C}^{k,2+\gamma}_{WF}(P \times [0, T])$, see Chapter 5. For example: in the 1-dimensional case $f \in \mathscr{C}^{0,\gamma}_{WF}([0, \infty))$ if f is continuous and

$$\sup_{x \neq y} \frac{|f(x) - f(y)|}{|\sqrt{x} - \sqrt{y}|^{\gamma}} < \infty; \tag{1.42}$$

it belongs to $\mathscr{C}^{0,2+\gamma}_{WF}([0, \infty))$ if f, $\partial_x f$, and $x\partial_x^2 f$ all belong to $\mathscr{C}^{0,\gamma}_{WF}([0, \infty))$, with

$$\lim_{x \to 0^+, \infty} x\partial_x^2 f(x) = 0.$$

For $k \in \mathbb{N}$, we say that $f \in \mathscr{C}^{k,\gamma}_{WF}([0,\infty))$, if $f \in \mathscr{C}^k([0,\infty))$, and $\partial_x^k f \in \mathscr{C}^{0,\gamma}_{WF}([0,\infty))$. A function $g \in \mathscr{C}^{0,\gamma}_{WF}([0,\infty)\times[0,\infty))$, if $g \in \mathscr{C}^0([0,\infty)\times[0,\infty))$, and

$$\sup_{(x,t)\neq(y,s)} \frac{|g(x,t)-g(y,s)|}{[|\sqrt{x}-\sqrt{y}|+\sqrt{|t-s|}]^{\gamma}} < \infty, \tag{1.43}$$

etc.

Much of this monograph is concerned with proving detailed estimates for the model problems with respect to these Hölder spaces and then using perturbative arguments to obtain analogous results for a general Kimura diffusion on an arbitrary compact manifold with corners.

To describe the uniqueness properties for solutions to these equations, we need to consider the geometric structure of the boundary of P, and its relationship to L. As noted bP is a stratified space, with hypersurface boundary components $\{\Sigma_{1,j} : j = 1,\dots,N_1\}$. A boundary component of codimension n is a component of an intersection

$$\Sigma_{1,i_1} \cap \dots \cap \Sigma_{1,i_n}, \tag{1.44}$$

where $1 \le i_1 < \dots < i_n \le N_1$. A component of bP is *minimal* if it is an isolated point or a positive dimensional manifold without boundary. We denote the set of minimal components by $bP_{\min}$. Fix a generalized Kimura diffusion operator L. Let $\{\rho_j : j = 1,\dots,N_1\}$ be defining functions for the hypersurface boundary components. We say that L is *tangent* to $\Sigma_{1,j}$ if $L\rho_j \restriction_{\Sigma_{1,j}} = 0$, and *transverse* if there is a $c > 0$ so that

$$L\rho_j \restriction_{\Sigma_{1,j}} > c. \tag{1.45}$$

DEFINITION 1.4.1. *The terminal boundary of P relative to L, $bP_{\text{ter}}(L)$, consists of elements of $bP_{\min}$ to which L is tangent, along with boundary strata, Σ of P to which L is tangent, and such that $L_\Sigma = L \restriction_\Sigma$ is transverse to all components of $b\Sigma$.*

For the model space we say that

$$f \in \mathscr{D}^2_{WF}(\mathbb{R}^n_+ \times \mathbb{R}^m) \subset \dot{\mathscr{C}}^1(\mathbb{R}^n_+ \times \mathbb{R}^m) \cap \mathscr{C}^2((0,\infty)^n \times \mathbb{R}^m) \tag{1.46}$$

if the scaled second derivatives

$$x_i\partial^2_{x_i} f(\boldsymbol{x};\boldsymbol{y}),\ x_ix_j\partial^2_{x_ix_j} f(\boldsymbol{x},\boldsymbol{y}),\ x_i\partial^2_{x_iy_l} f(\boldsymbol{x},\boldsymbol{y}),\ \partial^2_{y_ly_k} f(\boldsymbol{x},\boldsymbol{y}) \tag{1.47}$$

extend continuously to $\mathbb{R}^n_+ \times \mathbb{R}^m$. We also assume that $x_ix_j\partial^2_{x_ix_j} f(\boldsymbol{x},\boldsymbol{y})$ tends to zero if either x_i or x_j goes to zero, and $x_i\partial^2_{x_i} f(\boldsymbol{x};\boldsymbol{y})$ and $x_i\partial^2_{x_iy_l} f(\boldsymbol{x},\boldsymbol{y})$ go to zero as x_i goes to zero. A function $f \in \mathscr{C}^1(P) \cap \mathscr{C}^2(\text{int}\, P)$ belongs to $\mathscr{D}^2_{WF}(P)$ if it belongs to these local spaces in each local coordinate chart. Using a variant of the Hopf maximum principle, we can prove

THEOREM 1.4.2. *Let P be a compact manifold with corners, and L a generalized Kimura diffusion defined on P. Suppose that L is either tangent or transverse to every hypersurface boundary component of bP, and let $bP_{\text{ter}}(L)$ denote the set of terminal components of the boundary stratification relative to L. The cardinality of the* *set* $bP_{\text{ter}}(L)$ *equals the dimension of the null-space of* L *acting on* $\mathfrak{D}^2_{WF}(P)$*, which is also the dimension of* $\ker \overline{L}^*$*. The null-space of* L *is represented by smooth non-negative functions; the null-space of* $\overline{L}^*$ *by non-negative measures supported on the components of* $bP_{\text{ter}}(L)$*.*

The existence and regularity results for the heat equation defined by a general Kimura diffusion, L, on a manifold with corners, P, are summarized in the next two results:

THEOREM 1.4.3. *Let* P *be a compact manifold with corners,* L *a generalized Kimura diffusion on* P*,* $k \in \mathbb{N}_0$ *and* $0 < \gamma < 1$*. If* $f \in \mathscr{C}^{k,\gamma}_{WF}(P)$*, then there is a unique solution*

$$v \in \mathscr{C}^{k,\gamma}_{WF}(P \times [0,\infty)) \cap \mathscr{C}^{\infty}(P \times (0,\infty)),$$

to the initial value problem

$$(\partial_t - L)v = 0 \text{ with } v(p,0) = f(p). \tag{1.48}$$

This solution has an analytic continuation to t *with* $\operatorname{Re} t > 0$*.*

We have a similar result for the inhomogeneous problem:

THEOREM 1.4.4. *Let* P *be a compact manifold with corners,* L *a generalized Kimura diffusion on* P*,* $k \in \mathbb{N}_0$*,* $0 < \gamma < 1$*, and* $T > 0$*. If* $g \in \mathscr{C}^{k,\gamma}_{WF}(P \times [0,T])$*, then there is a unique solution*

$$u \in \mathscr{C}^{k,2+\gamma}_{WF}(P \times [0,T])$$

to

$$(\partial_t - L)u = g \text{ with } u(p,0) = 0, \tag{1.49}$$

which satisfies estimates of the form

$$\|u\|_{WF,k,2+\gamma,T} \leq M_{k,\gamma} \exp(C_{k,\gamma} T) \|g\|_{WF,k,\gamma,T}. \tag{1.50}$$

We also have a result for the resolvent of L acting on the spaces $\mathscr{C}^{k,2+\gamma}_{WF}(P)$, showing that $(\mu - L)^{-1}$ is an elliptic operator with respect to our scales of Banach spaces.

THEOREM 1.4.5. *Let* P *be a compact manifold with corners,* L *a generalized Kimura diffusion on* P*,* $k \in \mathbb{N}_0$*,* $0 < \gamma < 1$*. The spectrum,* E*, of the unbounded, closed operator* L*, with domain*

$$\mathscr{C}^{k,2+\gamma}_{WF}(P) \subset \mathscr{C}^{k,\gamma}_{WF}(P),$$

is independent of k*, and* γ*. It is a discrete set lying in a conic neighborhood of* $(-\infty, 0]$*. The eigenfunctions belong to* $\mathscr{C}^{\infty}(P)$*.*

Remark. Note that $\mathscr{C}^{k,2+\gamma}_{WF}(P)$ is *not* a dense subspace of $\mathscr{C}^{k,\gamma}_{WF}(P)$.

1.5 APPLICATIONS IN PROBABILITY THEORY

The principal sources for operators of the type studied here are infinite population limits of Markov chains in population genetics, and certain classes of "linear" models in mathematical finance. In this context the operator L, acting on a dense domain in $\mathscr{C}^0(P)$ is called the backward Kolmogorov operator. Its formal adjoint, which acts on the dual space, $\mathscr{M}(P)$, of finite signed Borel measures on P, is the forward Kolmogorov operator. The standard way to address the adjoint operator is to study the martingale problem associated with L on $\mathscr{C}^0([0,\infty); P)$. Letting $\omega \in \mathscr{C}^0([0,\infty); P)$, for each $t \in [0,\infty)$, we define

$$x(t) : \mathscr{C}^0([0,\infty); P) \to P,$$

by $x(t)[\omega] = \omega(t)$. We let $\mathscr{F}_t$ denote the σ-field generated by $\{x(s) : 0 \leq s \leq t\}$ and $\mathscr{F}$ the σ-field generated by $\{x(s) : s \geq 0\}$. For each $q \in P$, a probability measure $\mathbb{P}_q$ on $(\mathscr{C}^0([0,\infty); P), \mathscr{F})$ is a solution of the martingale problem associated with L and starting from $q \in P$ at time $t = 0$, if

$$\begin{gathered}\mathbb{P}_q(x(0) = q) = 1 \text{ and} \\ \left\{ f(x(t)) - \int_0^t Lf(x(s))ds \right\}_{t\geq 0}\end{gathered} \tag{1.51}$$

is a $\mathbb{P}_q$-martingale with respect to $\{\mathscr{F}_t\}_{t\geq 0}$, see [39].

The existence results in Theorems 1.4.3 or 1.4.5 suffice to prove that the associated martingale problem has a unique solution. A standard argument then shows that the paths for associated strong Markov process remain, almost surely, within P. From this we can deduce a wide variety of results about the forward Kolmogorov equation, and the solutions of the associated stochastic differential equation. The precise nature of these results depends on the behavior of the vector field V along bP. As this analysis requires techniques quite distinct from those employed here, we defer these questions to a future, joint publication with Daniel Stroock.

Using the Lumer-Phillips theorem, these results also suffice to prove that the $\mathscr{C}^0(P)$-graph closure, $\overline{L}$, of L acting on $\mathscr{C}^3(P)$ is the generator of a strongly continuous contraction semi-group. At present we have not succeeded in showing that the resolvent of $\overline{L}$ is compact, and will return to this question in a later publication. We have nonetheless been able to characterize the null-space of adjoint operator $\overline{L}^*$, under a natural clean intersection condition for the vector field V. This allows for an analysis of the asymptotic behavior of the solution to $\partial_t \nu - L^*\nu = 0$, as $t \to \infty$, see formula (12.63), and (12.62), for the asymptotics of $e^{tL} f$.

In [1, 4] Bass, Perkins, et al. have employed methods, similar to our own, to study operators of the form

$$L_{BP} = \sum_{i,j=1}^n \sqrt{x_i x_j} a_{ij}(\boldsymbol{x}) \partial^2_{x_i x_j} + \sum_{i=1}^n b_i(\boldsymbol{x}) \partial_{x_i} \tag{1.52}$$

acting on functions in $\mathscr{C}_b^2(\mathbb{R}_+^n)$. Here $b_i(\boldsymbol{x}) \geq 0$ along $b\mathbb{R}_+^n$. They have also considered other degenerate operators of this general type. Their main goal is to show the uniqueness of the solution to the martingale problem defined by L_{BP}. To that end they introduce *weighted* Hölder spaces, which take the place of our anisotropic spaces. In the 1-dimensional case, the weighted γ-semi-norm is defined by

$$[f]_{BP,\gamma} = \sup_{x\in\mathbb{R}_+;\, h>0} \left[\frac{|f(x) - f(x+h)|}{h^\gamma} x^{\frac{\gamma}{2}} \right]. \tag{1.53}$$

They prove estimates for the heat kernels of model operators, equivalent to $L_{\boldsymbol{b},0}$, with respect to these Hölder spaces. Under a smallness assumption on the off-diagonal elements of the coefficient matrix $(a_{ij}(\boldsymbol{x}))$, they are able to control the error terms introduced by replacing L_{BP} by the model operator

$$L_{BP,0} = \sum_{i=1}^{n} [x_i a_{ii}(\mathbf{0})\partial_{x_i}^2 + b_i(\mathbf{0})\partial_{x_i}] \tag{1.54}$$

well enough to construct a resolvent operator $(L_{BP} - \mu)^{-1}$ for $\mu > 0$. This suffices for their applications to the martingale problem defined by L_{BP}. Notice that with this approach, only "pure corner" models are used, and no consideration is given to operators of the form $L_{\boldsymbol{b},m}$ with $m > 0$. For domains much more general than $\mathbb{R}_+^n$ it is difficult to see how to make such an approach viable.

The operators we treat are somewhat more restricted, in that we take the coefficients of the off-diagonal terms to have the form $x_i x_j a_{ij}(\boldsymbol{x}; \boldsymbol{y})$. Our method could equally well be applied to operators of the form considered by Bass and Perkins, i.e., with $x_i x_j$ replaced by $\sqrt{x_i x_j}$, if we were to append smallness hypotheses for the off-diagonal terms, similar to those they employ. We briefly consider this more general class of operators in Section 11.3. After slightly modifying the definitions of the higher order Hölder norms to include certain increasing weights, many of our results could be generalized to include certain non-compact cases.

Our aims were of a more analytic character, and take us far beyond what is needed to show the uniqueness of the solution to the martingale problem. This leads us to consider such things as the higher order regularity of solutions with smoother initial data, the analytic extension of the semi-group in time, and the higher order mapping properties of the resolvent operator. We also show the ellipticity of the resolvent, with a gain of 2 derivatives with respect to the anisotropic Hölder norms. While this does not appear explicitly in [4], a similar result, with respect to the weighted Hölder norms, should follow from what they have proved.

1.6 ALTERNATE APPROACHES

We wish to mention a few other approaches which have been successfully employed to study similar degenerate elliptic and parabolic equations. As outlined above, the approach here is a hybrid one: we use the explicit kernels for the solution operators of the model problems to derive a priori estimates in adapted Hölder spaces, and then, using these a priori estimates, carry out a perturbation argument to pass from the model problem to the actual problem. A guiding principle throughout is to frame definitions and constructions based on the geometry of the underlying WF metrics. It is also viable to study such problems using only a priori estimates, or through a purely parametrix based approach.

We have already mentioned the papers [11], [12] and [28]. Each of these treats very similar classes of degenerate operators on a manifold with boundary, with the defining function for the boundary multiplying either just the second normal derivative or else all second derivative terms, respectively. Motivation for these papers comes out of the study of the porous medium equation. The work of Daskalopoulos and Hamilton is based on a priori estimates derived through the maximum principle. Koch's paper, by contrast, uses some powerful techniques involving singular integral operators to derive similar estimates.

It is not immediately apparent, but the model operators

$$L_1 := x\partial_x^2 + \sum_{j=1}^{m} \partial_{y_j}^2 + b\partial_x, \quad \text{and}$$

$$L_2 := x(\partial_x^2 + \sum_{j=1}^{m} \partial_{y_j}^2) + b\partial_x$$

are essentially equivalent. Indeed, replacing L_1 by xL_1 and changing variables by setting $\xi = \sqrt{x}$ gives

$$\frac{1}{4}\xi^2\partial_\xi^2 + \frac{1}{2}(b - \frac{1}{2})\xi\partial_\xi + \xi^2 \sum_{j=1}^{m} \partial_{y_j}^2,$$

which has the same structure as xL_2. In this representation the indicial roots of both xL_1 and xL_2 are $\alpha \in \{0, 1-b\}$. The functions $\xi^{2\alpha}$ (x^α) are solutions of the normal equation

$$\left[\frac{1}{4}\xi^2\partial_\xi^2 + \frac{1}{2}(b - \frac{1}{2})\xi\partial_\xi + \lambda\xi^2\right]\xi^\alpha = O(\xi^{\alpha+1}), \tag{1.55}$$

which evidently do not depend on λ.

The operators xL_1 and xL_2 are *uniformly degenerate operators* on a manifold with boundary. By definition, any such operator is one which can be written as sums of products of the basic vector fields $x\partial_x$ and $x\partial_{y_j}$. Thus, a model prototype for second order "elliptic" differential operators has the form

$$(x\partial_x)^2 + \sum_{j=1}^{m} (x\partial_{y_j})^2 + bx\partial_x. \tag{1.56}$$

In other words, the linear operators considered in [11], [12], [28] and [23, 24] are subsumed into the more general framework of uniformly degenerate operators.

Another instance of a much-studied operator of this form is the Laplacian on hyperbolic space, $\Delta_{\mathbb{H}^n}$, which has the form (1.56) with $b = 1 - n$. A key difference between the usual theory developed for that operator, and indeed also for the linearizations of the porous medium equations in [11], [23, 24] and [28], and our study of the generalized Kimura diffusions is that the implicit boundary condition (1.23) used here is of Neumann type (at least when $b < 1$). In other words, for the hyperbolic Laplacian one typically picks out the solution determined by a simple growth condition, e.g., $\Delta_{\mathbb{H}^n} u = f$ with $u \in L^2$. Neumann conditions are well-known, even in the non-degenerate setting, to have a more global nature. On the other hand, a very important feature of our class of generalized Kimura operators L is the fact that the indicial root 0 is constant, and in particular is independent of both the first order coefficient b as well as the spectral parameter λ when we study the resolvent of L. This is in marked contrast with $\Delta_{\mathbb{H}^n} - \lambda$ where the indicial roots depend on the spectral parameter. While this property of Kimura diffusions is not inconsistent with the general theory of elliptic uniformly degenerate operators, it indicates a special feature of these operators that makes tractable much of the analysis here.

A comprehensive study of uniformly degenerate elliptic operators on manifolds with boundary, and of the slightly more general class of edge operators, was undertaken in [32]. The emphasis in that paper is on mapping properties on weighted Sobolev and Hölder spaces, and on the construction of parametrices using the methods of geometric microlocal analysis. The focus is on identifying the precise behavior of the Schwartz kernels of the various operators (the resolvent operator, heat kernel, etc.) along the boundaries and near to the intersection of the diagonal with the boundaries. These techniques give very detailed information, but have the disadvantage that they require an elaborate formalism.

The geometric-microlocal methods and those employed in [11, 28] have been developed *only* for operators on manifolds with boundary, but not corners. It has proved challenging to generalize the microlocal approach to situations where the boundary is stratified, but see [33, 34]. It is likely that the amount of effort required to do this would be at least what has been done in the present monograph. Successful completion of such a program would give a precise description of the resolvent kernel itself, which would more than suffice to prove local regularity results. In this monograph we do not give precise descriptions of the resolvent or heat kernels, nor have we established local regularity results.

1.7 OUTLINE OF TEXT

The book is divided in three parts:

I. **Wright-Fisher Geometry and the Maximum Principle: Chapters 2.1-3.** Chapter 2.1 introduces the geometric preliminaries needed to analyze generalized Kimura diffusions. In Chapter 2.2 we show that coordinates

$$(x_1, \ldots, x_M; y_1, \ldots, y_{N-M})$$

can be introduced in the neighborhood of a boundary point of codimension M so that the boundary is locally given by $\{x_1 = \cdots = x_M = 0\}$ and the second order purely normal part of L takes the form

$$\sum_{j=1}^{M} x_i \partial_{x_i}^2. \tag{1.57}$$

This generalizes a 1-dimensional result in [20]. In Chapter 3 we prove maximum principles for the parabolic and elliptic equations,

$$(\partial_t - L)u = g \text{ and } (\mu - L)w = f, \text{ respectively,} \tag{1.58}$$

from which the uniqueness results follow easily. Of particular note is an analogue of the Hopf boundary point maximum principle, which allows very detailed analyses of the $\ker L$ and $\ker \overline{L}^*$.

II. **Analysis of Model Problems: Chapters 4–9.** In Chapter 4 we introduce the model problems and the solution operator for the associated heat equations. These operators,

$$L_{\boldsymbol{b},m} = \sum_{j=1}^{m} [x_j \partial_{x_j}^2 + b_j \partial_{x_j}] + \sum_{l=1}^{m} \partial_{y_l}^2, \tag{1.59}$$

act on functions defined on $S_{n,m} = \mathbb{R}_+^n \times \mathbb{R}^m$, where $n+m = N$, and give a good approximation for the behavior of the heat kernel $(\partial_t - L)^{-1}$ in neighborhoods of different types of boundary points. We state and prove elementary features of these operators, that generalize results proved in [15], and show that the model heat operators have an analytic continuation to the right half plane:

$$H_+ = \{t : \operatorname{Re} t > 0\}. \tag{1.60}$$

In Chapter 5 we introduce the degenerate Hölder spaces on the spaces $S_{n,m}$, and their heat-space counterparts on $S_{n,m} \times [0, T]$. These are, in essence, Hölder spaces defined by the incomplete metric on $S_{n,m}$ given by

$$ds_{WF}^2 = \sum_{j=1}^{n} \frac{dx_i^2}{x_i} + \sum_{l=1}^{m} dy_l^2. \tag{1.61}$$

We also establish the basic properties of these spaces.

Chapters 6–9 are devoted to analyzing the heat and resolvent operators for the model problems acting on the Hölder spaces defined in Chapter 5. This is a very long and tedious process because many cases need to be considered and, in each case, many estimates are required. Conceptually, however, these results are elementary. The estimates are pointwise estimates done in Hölder spaces, which means one can vary a single variable at a time. As the model heat kernels are products of 1-dimensional heat kernels, this reduces essentially every question one might want to answer to one of proving estimates for the 1-dimensional kernels. We call this the *1-variable-at-a-time* method. In higher dimensions, the resolvent kernel, which is the Laplace transform of the heat kernel, is *not* a product of 1-dimensional kernels. This makes it far more difficult to deduce the mapping properties of the resolvent from its kernel, and explains why we use the representation as a Laplace transform.

The proof of the estimates on the 1-dimensional heat kernels, defined by the operators $x\partial_x^2 + b\partial_x$ are given in Appendix A. Analogous results for the Euclidean heat kernel are stated in Chapter 8. The proofs of these lemmas, which are elementary, are left to the reader. A notable feature of the estimates for the degenerate model problem is the fact that the constants remain uniformly bounded as $b \to 0$. This despite the fact that the character of the heat kernel changes quite dramatically at $b = 0$, see (1.29) and (1.31). This is also in sharp contrast to the analysis of similar problems in [11], where a positive lower bound is assumed for the coefficient of the analogous vector field.

III. **Analysis of Generalized Kimura Diffusions: Chapters 10–12.** This part of the book represents the culmination of all the work done up to this point. We consider a generalized Kimura diffusion operator L defined on a compact manifold with corners P. In Chapter 10 we prove the existence of solutions to the heat equation

$$(\partial_t - L)u = g \text{ in } P \times (0, T] \text{ with } u(p, 0) = f(p), \tag{1.62}$$

with data in $(g, f) \in \mathscr{C}_{WF}^{k,\gamma}(P \times [0, T]) \times \mathscr{C}_{WF}^{k,2+\gamma}(P)$. We show that the solution belongs to $\mathscr{C}_{WF}^{k,2+\gamma}(P \times [0, T])$ (Theorems 10.0.2 and 10.5.2). The case $g = 0$ provides a solution to the Cauchy problem, but it is not optimal as regards either the regularity of the solution, or the domain of the time variable, defects that are corrected in Chapter 11. The proof of these results is an intricate induction argument, where we induct over the maximal codimension of bP. This argument allows us to handle one stratum at a time. The underlying geometric fact is a "doubling theorem," which shows that any neighborhood, *complementary* to the highest codimension stratum of bP, can be embedded into a manifold with corners $\widetilde{P}$ where the maximum codimension of $b\widetilde{P}$ is one less than that of bP, (Theorem 10.2.1). This explains why we need to consider domains well beyond those that can be easily embedded into Euclidean space.

We first treat the lowest differentiability case ($k = 0$) and then use an extension of the contraction mapping theorem to towers of Banach spaces (Theorem 10.8.1), to obtain the mapping results for $k > 0$. These results (even in the $k = 0$ case) suffice to prove that the graph closure of L acting on $\mathscr{C}^3(P)$ is the generator of strongly continuous semi-group in $\mathscr{C}^0(P)$.

We next consider the operators L_γ, defined as L acting on the domain $\mathscr{C}_{WF}^{0,2+\gamma}(P)$. As a map from $\mathscr{C}_{WF}^{0,2+\gamma}(P)$ to $\mathscr{C}_{WF}^{0,\gamma}(P)$, L_γ is a Fredholm operator of index 0. In

Chapter 11 we use essentially the same parametrix construction as used to prove Theorem 10.0.2 to prove the existence of the resolvent operator

$$(\mu - L_\gamma)^{-1} : \mathscr{C}^{k,\gamma}_{WF}(P) \longrightarrow \mathscr{C}^{k,2+\gamma}_{WF}(P),$$

for μ in the complement of discrete set lying in a conic neighborhood of $(-\infty, 0]$. These are the expected "elliptic" estimates for operators with this type of degeneracy. In fact the spectrum of L acting on $\mathscr{C}^{k,\gamma}_{WF}(P)$ does not depend on k or γ, as the resolvent is compact and all eigenfunctions belong to $\mathscr{C}^\infty(P)$. Using the analyticity properties of the resolvent, we give an alternate construction, using a contour integral, for the semi-group, acting on $\mathscr{C}^{0,\gamma}_{WF}(P)$:

$$e^{tL} = \frac{1}{2\pi i}\int_{\Gamma_\alpha} e^{t\mu}(\mu - L)^{-1} d\mu, \tag{1.63}$$

where Γ_α bounds a region of the form $|\arg\mu| > \pi - \alpha$, and $|\mu| > R_\alpha$, (Theorem 11.2.1). As we can take α to be any positive number, this shows that the semi-group is holomorphic in the right half plane.

Finally in Chapter 12 we give a good description of the null-space of L_γ, and show that the non-zero spectrum of L_γ lies in a half plane $\operatorname{Re}\mu < \eta < 0$. We also deduce various properties of the semi-group, defined by the graph closure, $\overline{L}$, of L, acting on $\mathscr{C}^0(P)$. The adjoint operator, $\overline{L}^*$, is defined on a domain $\operatorname{Dom}(\overline{L}^*) \subset \mathscr{M}(P)$, which is not dense. We describe the subspace $\mathscr{M}^{\odot}(P)$, defined by Lumer and Phillips, on which $\overline{L}^*$ defines a C_0-semi-group. Although we have not yet proved the compactness of the resolvent of $\overline{L}$, we obtain a rather complete description of the null-space of $\overline{L}^*$. Using this we give the long time asymptotics for $e^{tL}f$, assuming that $f \in \mathscr{C}^{0,\gamma}_{WF}(P)$, for any $0 < \gamma$, as well as those for $e^{t\overline{L}^*}\nu$, for ν in $\mathscr{M}^{\odot}(P)$. Finally we consider the problem of finding "irregular" solutions to the equation $Lw = f$, when the maximum principle precludes the existence of a regular solution.

IV. **Proof of 1-dimensional Estimates: Appendix A.** In the Appendix we give careful proofs of the estimates for the degenerate, 1-dimensional heat kernels used in the perturbation theory. These arguments are complicated by the fact that the heat kernel displays both the additive and multiplicative group structures on $\mathbb{R}_+$:

$$k_t^b(x, y)dy = \left(\frac{y}{t}\right)^b e^{-\frac{x+y}{t}} \psi_b\left(\frac{xy}{t^2}\right)\frac{dy}{y}. \tag{1.64}$$

The arguments involve Taylor's theorem, the asymptotic expansion of the heat kernel where $\frac{|\sqrt{x}-\sqrt{y}|}{\sqrt{t}}$ tends to infinity and Laplace's method. We obtain mapping properties for $b > 0$, with uniform constants as b tends to zero. Using compactness of the embeddings, $\mathscr{C}^{k,\gamma}_{WF} \hookrightarrow \mathscr{C}^{k,\tilde{\gamma}}_{WF}$, for $\tilde{\gamma} < \gamma$, e.g., we then extend these results to $b = 0$.

1.8 NOTATIONAL CONVENTIONS

- We use $\mathbb{R}_+ = [0, \infty)$, hence $\mathscr{C}_c^\infty(\mathbb{R}_+)$ consists of smooth functions on $\mathbb{R}_+$, supported in finite intervals $[0, R]$.
- The right half plane is denoted

$$H_+ = \{t \in \mathbb{C} : \operatorname{Re} t > 0\}. \tag{1.65}$$

- We let

$$S_{n,m} = \mathbb{R}_+^n \times \mathbb{R}^m. \tag{1.66}$$

- For $\boldsymbol{b} = (b_1, \dots, b_n) \in \mathbb{R}_+^n$, the model operator is:

$$L_{\boldsymbol{b},m} = \sum_{j=1}^m [x_j \partial_{x_j}^2 + b_j \partial_{x_j}] + \sum_{l=1}^m \partial_{y_l}^2. \tag{1.67}$$

- If $a(t)$ is an operator valued function, acting on functions in a space $\mathscr{X}$, and $g : [0, T] \to \mathscr{X}$ is measurable, then the notation $A^t g$ usually means

$$A^t g = \int_0^t a(t-s) g(s) ds. \tag{1.68}$$

- For $\phi \in [0, \frac{\pi}{2})$ we define the sector

$$S_\phi = \{t \in \mathbb{C} : |\arg t| < \frac{\pi}{2} - \phi\}. \tag{1.69}$$

  We also let $S_{\frac{\pi}{2}} = [0, \infty)$.

- For $\boldsymbol{x}, \boldsymbol{x}' \in \mathbb{R}_+^n$ we let

$$\rho_s(\boldsymbol{x}, \boldsymbol{x}') = \sum_{j=1}^n |\sqrt{x_j} - \sqrt{x_j'}|; \tag{1.70}$$

  for $\boldsymbol{y}, \boldsymbol{y}' \in \mathbb{R}^m$ we let

$$\rho_e(\boldsymbol{y}, \boldsymbol{y}') = \sum_{k=1}^m |y_k - y_k'|. \tag{1.71}$$

  We then let

$$\rho((\boldsymbol{x}, \boldsymbol{y}), (\boldsymbol{x}', \boldsymbol{y}')) = \rho_s(\boldsymbol{x}, \boldsymbol{x}') + \rho_e(\boldsymbol{y}, \boldsymbol{y}'). \tag{1.72}$$

  We also use

$$d_{WF}((\boldsymbol{x}, \boldsymbol{y}), (\boldsymbol{x}', \boldsymbol{y}')) = \rho((\boldsymbol{x}, \boldsymbol{y}), (\boldsymbol{x}', \boldsymbol{y}')). \tag{1.73}$$

  When there is also a time variable,

$$d_{WF}((\boldsymbol{x}, \boldsymbol{y}, t), (\boldsymbol{x}', \boldsymbol{y}', t')) = \rho((\boldsymbol{x}, \boldsymbol{y}), (\boldsymbol{x}', \boldsymbol{y}')) + \sqrt{|t - t'|}. \tag{1.74}$$

- If $\boldsymbol{a}$ and $\boldsymbol{b}$ are vectors in $\mathbb{R}^n$, then $\boldsymbol{a} < \boldsymbol{b}$, or $\boldsymbol{a} \leq \boldsymbol{b}$ means that

$$a_j < b_j (\text{ or } a_j \leq b_j) \text{ for } j = 1, \ldots, n. \tag{1.75}$$

We let $\mathbf{0} = (0, \ldots, 0)$, and $\mathbf{1} = (1, \ldots, 1)$; the dimension will be clear from the context.

Part I

Wright-Fisher Geometry and the Maximum Principle

Chapter Two

Wright-Fisher Geometry

We begin by describing the class of spaces we work on and then the structure of the operators we consider.

2.1 POLYHEDRA AND MANIFOLDS WITH CORNERS

The natural domains of definition for generalized Kimura diffusions are polyhedra in Euclidean space or, more generally, abstract manifolds with corners. In order to set notation and fix ideas, we review this class of objects here and discuss the main properties about them that will be needed below. A more complete discussion can be found in [35, 36].

The standard N-dimensional Euclidean space is denoted $\mathbb{R}^N$. For any $n = 1, \dots, N$, let us set $m = N - n$ and define the positive n-orthant in $\mathbb{R}^N$ as the subset

$$S_{n,m} := \mathbb{R}_+^n \times \mathbb{R}^m = \{(\boldsymbol{x}, \boldsymbol{y}) \in \mathbb{R}^n \times \mathbb{R}^m : x_j \geq 0 \ j = 1, \dots, n\}. \tag{2.1}$$

Recall the standard definition of a smooth manifold: A paracompact, Hausdorff topological space, M, is called an N-dimensional smooth manifold if every point $p \in M$ has a neighborhood $\mathcal{U}_p$ which is identified homeomorphically with an open set $\mathcal{V}_p$ around the origin in $\mathbb{R}^N$. Here p mapped to the origin, and such that the identifications between these various subsets of $\mathbb{R}^N$ are diffeomorphisms. More specifically, if $\psi_p : \mathcal{U}_p \to \mathcal{V}_p$ is the homeomorphism, then for $p \neq q$:

$$\psi_p \circ \psi_q^{-1} : \psi_q(\mathcal{U}_q \cap \mathcal{U}_p) \longrightarrow \psi_p(\mathcal{U}_q \cap \mathcal{U}_p) \tag{2.2}$$

is a diffeomorphism. The mappings ψ_p are sometimes called charts, and the compositions $\psi_p \circ \psi_q^{-1}$ are called transition functions.

Generalizing this, we say that P is an N-dimensional manifold with corners up to codimension n if for every $p \in P$, there is a neighborhood $\mathcal{U}_p$ and a homeomorphism ψ_p from $\mathcal{U}_p$ to a neighborhood of 0 in $\mathbb{R}_+^\ell \times \mathbb{R}^{N-\ell}$ for some $\ell \in \{0, \dots, n\}$, with $\psi_p(p) = 0$, and such that the overlap maps, defined exactly as in (2.2), are diffeomorphisms. (Recall that a mapping between two relatively open sets in $\mathbb{R}_+^n \times \mathbb{R}^{N-n}$ is a diffeomorphism if it is the restriction of a diffeomorphism between two absolute open sets in $\mathbb{R}^N$.) If such a map exists, we say that a point p lies on a corner of codimension ℓ. The fact that the codimension associated to any point is well-defined is a basic fact from differential topology known as the invariance of domain lemma.

Using these local charts, we can meaningfully define all the usual flora and fauna of differential geometry in this setting. For example, we can discuss smooth functions, vector fields, differential forms, etc., simply by identifying these objects using the charts to the familiar objects of each type on the orthants in $\mathbb{R}^N$. The fact that such objects are well-defined follows from the fact that the transition functions between the charts are diffeomorphisms and each of these classes of objects (smooth functions, etc.) are preserved by diffeomorphisms.

It follows directly from this definition that the set of points p lying on a corner of codimension $0 < \ell \leq N$ constitute a possibly open, and possibly disconnected, smooth manifold, Σ_ℓ, of dimension $N - \ell$. If ℓ is strictly less than the maximal codimension, n, then Σ_ℓ is open and

$$\overline{\Sigma_\ell} = \cup_{j=\ell}^{n} \Sigma_j, \tag{2.3}$$

where the union here is only over the union of corners Σ_j such that $\Sigma_j \cap \overline{\Sigma_\ell} \neq \emptyset$. Each component of Σ_ℓ is called a corner of P of codimension ℓ. The corners of codimension one are the boundary hypersurfaces, which we sometimes also call the faces, of P. We henceforth make the global hypothesis that the closure of each connected corner of P, of any codimension ℓ, is itself an embedded manifold with corners at most up to codimension $N - \ell$. The important part of this hypothesis is the embeddedness of this closure. In the sequel we consider components of the boundary stratification to be these closed manifolds with corners.

DEFINITION 2.1.1. *The stratum of bP of codimension ℓ consists of the closures, in P, of the connected components of Σ_ℓ. We call these connected subsets the components of the stratum of bP of codimension ℓ, or more briefly, components of bP.*

We now prove several useful facts about this class of objects. In the following, fix any manifold with corners P, and denote by $\{H_1, \ldots, H_A\}$ its set of boundary hypersurfaces. Note that every corner of P of codimension ℓ arises as a component of an intersection $H_{i_1} \cap \ldots \cap H_{i_\ell}$. For simplicity we usually assume that P is compact, though the results below extend easily to the noncompact setting (sometimes with a few extra hypotheses).

LEMMA 2.1.2. *If H is any boundary hypersurface of P, then there is a smooth vector field V defined in a neighborhood of the closure of H which is inward pointing, nowhere vanishing, transverse to H, and which is also tangent to all other boundary faces and corners at bH.*

PROOF. It is easy to construct such a vector field near the origin in $\mathbb{R}_+^n \times \mathbb{R}^{N-n}$: using the coordinates $(\boldsymbol{x}, \boldsymbol{y})$, normalized so that H is locally $\{x_1 = 0\}$. In this chart we let V be the vector field ∂_{x_1}. Now choose a finite number of coordinate charts $\mathcal{U}_\alpha$, which provide an open cover of H and such that each $\mathcal{U}_\alpha$ is mapped by a chart to a relative open ball in some orthant of $\mathbb{R}^N$. Let $\{\chi_\alpha\}$ be a partition of unity subordinate to this open cover. For each α, let V_α be the coordinate vector field in $\mathcal{U}_\alpha$ defined above, and set $V = \sum \chi_\alpha V_\alpha$. This vector field clearly satisfies the conclusions of the lemma. □

LEMMA 2.1.3. *For any boundary hypersurface H, there is a neighborhood $\mathcal{U}$ of H in P, which is diffeomorphic to a product $H \times [0, 1)$.*

PROOF. Let V be the vector field defined above relative to the boundary hypersurface H. Assuming that H is compact, there exists some $\epsilon > 0$ such that the flow by the one-parameter family of diffeomorphisms Φ_t associated to the vector field V is defined on all of H for $0 \leq t < \epsilon$. This gives a diffeomorphism between $H \times [0, \epsilon)$ and some neighborhood $\mathcal{U}$ of H. A rescaling in t gives a diffeomorphism from $H \times [0, 1)$. □

LEMMA 2.1.4. *Let K be a corner of codimension ℓ in P. Let $B^{\ell}_{+} = \{x : |x| < 1\} \cap \mathbb{R}^{\ell}_{+}$. Then there is a diffeomorphism between $K \times B^{\ell}_{+}$ and a neighborhood $\mathcal{U}$ of K in P.*

PROOF. Let $H_1, \ldots, H_\ell$ be the boundary hypersurfaces which intersect along K. Let V_j be an inward pointing vector field transversal to H_j, as constructed above. For any $x \in B^{\ell}_{+}$ with $|x| < \epsilon$, let $V_x = \sum x_j V_j$ and Φ_x be the time one flow of the one-parameter family of diffeomorphisms associated to V_x. Then

$$K \times \epsilon B^{\ell}_{+} \ni (q, x) \longmapsto \Phi_x(q)$$

is the desired mapping. □

LEMMA 2.1.5. *For each boundary face H of P there is a function ρ_H which is everywhere positive in $P \setminus H$ and which vanishes on H and has non-vanishing differential there. Any such function is called a boundary defining function for H.*

PROOF. This may be constructed using a partition of unity exactly as in the first lemma. Alternately, if $\mathcal{U}$ is a neighborhood of H, which has been identified diffeomorphically with a product $H \times [0, 1)$, then we can set ρ_H to equal the projection onto the second coordinate in this neighborhood. This function is then extended to the rest of P as a strictly positive function using a partition of unity. □

There is one other construction which plays an important role below. We present it in a sequence of two lemmas.

LEMMA 2.1.6. *For each boundary face H, there is a new manifold with corners $\widetilde{P}$ which is obtained by doubling P across H.*

PROOF. Let $\widetilde{P}$ be the disjoint union of two copies of P identified by the identity mapping along H. If we wish to work within the setting of oriented manifolds, then one copy should be P itself and the other $-P$, i.e., P with the opposite orientation. We give this space the structure of a manifold with corners by specifying a collection of coordinate charts. First, use all charts of P which are disjoint from H. Then, near any point $p \in \overline{H}$, choose a chart $\psi : \mathcal{U} \to (\mathbb{R}_+)^{\ell} \times \mathbb{R}^{N-\ell-1}$ for H as a manifold with corners, and define the extended chart $\widetilde{\psi} : \mathcal{U} \times (-\epsilon, \epsilon) \to (\mathbb{R}_+)^{\ell} \times \mathbb{R}^{N-\ell-1} \times \mathbb{R}$ by $\widetilde{\psi}(q, t) = (\psi(q), t)$. This provides a chart around all points of the image of H in $\widetilde{P}$, and it is clear that the transition functions are smooth. □

LEMMA 2.1.7. *Let P be a compact manifold with corners. Let K be the corner of maximal codimension n. Then there is a new space $\widetilde{P}$, which is a manifold with corners only up to codimension $n - 1$, obtained by* doubling *P across K, such that $P \setminus K$ is identified with an open subset of $\widetilde{P}$.*

Remark. In the proof of the main existence theorem we use a related though somewhat different doubling construction, see Theorem 10.2.1.

PROOF. Notice that K is a closed, and possibly disconnected, manifold of dimension $N-n$. For simplicity we assume that K is connected, but removing this only complicates the notation slightly. We first define the radial blowup of P along the submanifold K. Let B be the unit ball around 0 in $\mathbb{R}^n_+$, which we describe via the polar coordinates (r,θ) where $0 \le r < 1$ and $\theta \in S^n_+ = \{\theta \in \mathbb{R}^{n+1} : |\theta| = 1,\ \theta_j \ge 0\ \forall\, j\}$. This coordinate system is degenerate at $r=0$ since the entire spherical orthant $\{0\}\times S^n_+$ is collapsed to a point. We define the radial blowup of B at 0, denoted $\widehat{B} = [B, \{0\}]$, by replacing the origin by a copy of S^n_+; in other words, $\widehat{B}$ is simply a copy of the cylinder $[0,1)\times S^n_+$.

Next, define the radial blowup of P along K, denoted $\widehat{P} = [P; K]$. In terms of the identification of the tubular neighborhood $\mathcal{U}$ of K in P as $K \times B$, we replace each B by $\widehat{B}$. This space is still a manifold with corners up to codimension n; the codimension n corners are now the products of the vertices of S^n_+ with K. There is a new, possibly disconnected, hypersurface boundary, $H_0 = K \times S^n_+$ (at $r=0$).

The final step is to define $\widetilde{P}$ to be the double of $\widehat{P}$ across the face H_0 in the sense of the previous lemma. This space no longer has any corners of codimension n. From the construction it is clear that $P \setminus K$ embeds in $\widetilde{P}$ as an open set. □

An important class of manifolds with corners is provided by the regular polyhedra $P \subset \mathbb{R}^N$. Recall first that a polyhedron is a domain in $\mathbb{R}^N$ whose boundary lies in a union of hyperplanes and is a finite union of regions $\{Q_i\}$, each itself a polyhedron in a hyperplane $H_i \subset \mathbb{R}^N$. Of particular interest to us here are the convex polyhedra, which are by definition determined by a finite number of affine inequalities:

$$P = \{z \in \mathbb{R}^N : z\cdot\omega_j \ge c_j,\ j = 1,\dots,A\},$$

where $\{\omega_1,\dots,\omega_A\} \subset \mathbb{R}^N$ is a finite set of vectors, and the c_j are real numbers. The various faces and corners of P are the subsets determined by replacing any subcollection of these inequalities by the corresponding equalities.

Amongst the convex polyhedra we distinguish the subclass of regular convex polyhedra P. By definition, P is a regular convex polyhedron if it is convex and if near any corner, P is the intersection of no more than N half-spaces with corresponding normal vectors ω_j linearly independent. It is clear from these definitions that any regular convex polyhedron is a manifold with corners. Namely, if $p \in P$ lies in a corner of codimension ℓ, hence is an element of ℓ independent hyperplanes $\{z\cdot\omega_j = c_j\}$, then there is an affine change of variables which carries a neighborhood of p in P to a neighborhood of 0 in $\mathbb{R}^\ell_+ \times \mathbb{R}^{N-\ell}$. A polyhedron that is not regular or non-convex, when endowed with the smooth structure given by the embedding into Euclidean space, does not satisfy one or more of the defining properties of manifolds with corners. It should be noted that unlike convex polyhedra, which are always contractible, manifolds with corners can have very complicated topologies.

2.2 NORMAL FORMS AND WRIGHT-FISHER GEOMETRY

We are now in a position to define the general class of elliptic Kimura operators on a manifold with corners P. These, and their associated heat operators, are our main objects of study. The definition we give is coordinate-dependent, but we indicate how to formulate this in a coordinate-independent way. Our goal in this chapter is to show that there is a local normal form for any operator L in this class which shows that it can be regarded as a perturbation in a small enough neighborhood of one of the model Kimura operators $L_{\boldsymbol{b},m}$ introduced in the Introduction. The reduction to this normal form is assisted by use of geometric constructions with respect to a singular Riemannian metric on P. This metric (or any one equivalent to it) is also instrumental in the formulation of the correct function spaces on which we let L act; this is the topic of Chapter 5. The normal form in this multidimensional setting generalizes the normal form, originally introduced by Feller, which is fundamental in the analysis of the 1-dimensional case in [15].

DEFINITION 2.2.1. *Let P be a manifold with corners. A second order differential operator L defined on P is called a generalized Kimura diffusion operator if it satisfies the following set of conditions:*

i) L is elliptic in the interior of P.

ii) If q is a boundary point of P which lies in the interior of a corner of codimension n, then there are local coordinates $(\boldsymbol{x}, \boldsymbol{y})$, $\boldsymbol{x} = (x_1, \dots, x_n)$, $\boldsymbol{y} = (y_1, \dots, y_m)$ so that in the neighborhood

$$\mathcal{U} = \{0 \geq x_j < 1 \,\forall\, j \leq n,\ |y_k| < 1 \,\forall\, k \leq m\}$$

the operator takes the form

$$L = \sum_{j=1}^{n} a_{ii} x_i \partial_{x_i}^2 + \sum_{1\leq i\neq j\leq n} x_i x_j a_{ij} \partial_{x_i x_j}^2 + \sum_{i=1}^{n}\sum_{k=1}^{m} x_i b_{ik} \partial_{x_i y_k}^2 + \sum_{k,l=1}^{m} c_{kl} \partial_{y_k y_l}^2 + V. \quad (2.4)$$

For simplicity we assume that all coefficients lie in $\mathscr{C}^\infty(P)$. We also assume that (a_{ij}) and (c_{kl}) are symmetric matrices.

iii) The vector field V is inward pointing at all boundaries and corners of P.

iv) The matrices (a_{ii}) and (c_{kl}) are strictly positive definite.

The distinguishing features are the simple vanishing of the coefficients of the second order terms normal to the boundary and the ellipticity in all other directions. This leads to a coordinate-invariant definition. Recall that any second order operator L in the variables $\boldsymbol{x}, \boldsymbol{y}$ has a principal symbol

$$\sigma_2(L)(\boldsymbol{x}, \boldsymbol{y}, \boldsymbol{\xi}, \boldsymbol{\eta}) = \sum_{i,j=1}^{n} \hat{a}_{ij}(\boldsymbol{x}, \boldsymbol{y})\xi_i\xi_j + \sum_{i=1}^{n}\sum_{k=1}^{m} \hat{b}_{ik}(\boldsymbol{x}, \boldsymbol{y})\xi_i\eta_k + \sum_{k,l=1}^{m} \hat{c}_{kl}(\boldsymbol{x}, \boldsymbol{y})\eta_k\eta_l.$$

Here $\boldsymbol{\xi}$ and $\boldsymbol{\eta}$ are the dual (cotangent) variables associated to $\boldsymbol{x}$ and $\boldsymbol{y}$. We require then that $\sigma_2(L)(\boldsymbol{x}, \boldsymbol{y}, \boldsymbol{\xi}, \boldsymbol{\eta})$ is non-negative for all $(\boldsymbol{\xi}, \boldsymbol{\eta})$, and strictly positive when all $x_j > 0$ and $(\boldsymbol{\xi}, \boldsymbol{\eta}) \neq (\mathbf{0}, \mathbf{0})$. Furthermore its characteristic set

$$\text{Char}(L) = \{(\boldsymbol{x}, \boldsymbol{y}, \boldsymbol{\xi}, \boldsymbol{\eta}) : \sigma_2(L)(\boldsymbol{x}, \boldsymbol{y}, \boldsymbol{\xi}, \boldsymbol{\eta}) = 0\}$$

is equal to the set of all conormal vectors to bP, or more precisely to the set of all points (q, ν) where $q \in bP$ and $\nu \in T_q^*P$ vanishes on the tangent spaces of all boundary hypersurfaces which contain q. Finally, we require that $\sigma_2(L)$ vanishes precisely to first order at this set.

It is immediate to see that any operator which satisfies the first definition satisfies all the coordinate-invariant conditions above. For the converse, observe that in any local coordinate system, at a point q in the interior of the face where $x_j = 0$, the conormal is spanned by the covector $\nu = dx_j$, i.e., $\xi_j = 0$ and all other ξ_i and η_k vanish. Hence if we write $\sigma_2(L)(\boldsymbol{x}, \boldsymbol{y}, \boldsymbol{\xi}, \boldsymbol{\eta})$ as a quadratic form as above, then for this (q, ν), $\hat{a}_{ij}(q) = 0$ for all $i \leq n$, and this vanishing is simple. This gives $\hat{a}_{jj}(\boldsymbol{x}, \boldsymbol{y}) = x_j a_{jj}(\boldsymbol{x}, \boldsymbol{y})$ and for $i \neq j, \hat{a}_{ij}(\boldsymbol{x}, \boldsymbol{y}) = x_i x_j a_{ij}(\boldsymbol{x}, \boldsymbol{y})$. The rest of the verification is straightforward.

As noted in the Introduction, we could enlarge our family somewhat by allowing terms of the form

$$\sqrt{x_i x_j} a_{ij} \partial^2_{x_i x_j} \text{ and } \sqrt{x_i} b_{ik} \partial^2_{x_i y_k}, \tag{2.5}$$

if we append a smallness hypothesis on the coefficients $\{a_{ij}(\boldsymbol{x}; \boldsymbol{y}), b_{ik}(\boldsymbol{x}; \boldsymbol{y})\}$. Operators of this type were analyzed by Bass and Perkins. Indeed we can consider operators of this more general type where the coefficients are smooth functions of the variables $(\sqrt{x_1}, \dots, \sqrt{x_n}; y_1, \dots, y_m)$. We discuss this in more detail in Section 11.3, but leave a more thorough discussion of these generalizations to the interested reader.

Let H be a boundary hypersurface of P, with ρ a $\mathcal{C}^2$-function, vanishing on H. The first order part V of L is tangent to H if and only if

$$V\rho \restriction_{\rho=0} = 0. \tag{2.6}$$

From the form of the operator it is easy to see that this is true if and only if

$$L\rho \restriction_{\rho=0} = 0. \tag{2.7}$$

In this case we say that L *is tangent to* H. If L is tangent to H, then there is a naturally induced operator L_H acting on $\mathcal{C}^\infty(H)$.

DEFINITION 2.2.2. *If L is tangent to H, then the restriction of L to H, L_H, is given by the prescription*

$$L_H u := (L\widetilde{u}) \restriction_H \ \forall u \in \mathscr{C}^\infty(H).$$

Here $\widetilde{u}$ is any smooth extension of u to a neighborhood of H in P. The operator L_H is a generalized Kimura diffusion operator on the manifold with corners H.

As we have already mentioned, it is very helpful to consider a singular Riemannian metric on P such that the second order terms of L agree with those in the Laplacian for g. The change of variables which brings L into a normal form is then simply a Fermi coordinate system for this metric. To do this, we regard the principal symbol of L as a dual metric on the cotangent bundle; we write this as

$$g^{-1} = \begin{pmatrix} XA_d & 0 \\ 0 & C \end{pmatrix} + \begin{pmatrix} XA_oX & XB \\ B^tX & 0 \end{pmatrix}. \tag{2.8}$$

Here, B is an $n \times m$ matrix, C is a positive definite $m \times m$ matrix, A_o is off-diagonal (and has all diagonal entries equal to zero) and X and A_d are the diagonal matrices given by

$$X = \begin{pmatrix} x_1 & 0 & \dots & 0 \\ 0 & x_2 & \dots & 0 \\ & & \ddots & \\ 0 & 0 & \dots & x_n \end{pmatrix}, \quad A_d = \begin{pmatrix} a_{11} & 0 & \dots & 0 \\ 0 & a_{22} & \dots & 0 \\ & & \ddots & \\ 0 & 0 & \dots & a_{nn} \end{pmatrix}. \tag{2.9}$$

We then compute that the inverse of this matrix, i.e., the metric tensor itself, takes the form:

$$g = \begin{pmatrix} X^{-1}\widetilde{A}_d + \widetilde{A}_o & \widetilde{B} \\ \widetilde{B}^t & \widetilde{C} \end{pmatrix}. \tag{2.10}$$

The block submatrices here are all smooth, with $\widetilde{A}_d$ positive definite and diagonal, $\widetilde{A}_o$ having vanishing diagonal entries and $\widetilde{C}$ positive definite.

The metric g is singular when any x_j vanishes, but it can be desingularized by changing variables via

$$d\zeta_j = \frac{dx_j}{\sqrt{x_j}}, \quad \text{i.e.,}\ \zeta_j = 2\sqrt{x_j}. \tag{2.11}$$

In these coordinates, the metric takes the form

$$g = \sum_{i=1}^{n} \widetilde{a}_{ii}\, d\zeta_i^2 + \sum_{i,j=1}^{n} \widetilde{a}_{ij}\zeta_i\zeta_j\, d\zeta_i d\zeta_j + \sum_{i=1}^{n}\sum_{k=1}^{m} \widetilde{b}_{ik}\zeta_i\, d\zeta_i dy_k + \sum_{k,l=1}^{m} \widetilde{c}_{kl}\, dy_k dy_l. \tag{2.12}$$

The coefficients $\tilde{a}_{ij}$, $\tilde{b}_{ik}$ and $\tilde{c}_{kl}$ are smooth functions of $(\zeta_1^2, \dots, \zeta_n^2; y_1, \dots, y_m)$. This implies that the metric g can be extended by even reflection across each hyperplane $\zeta_j = 0$ to define a smooth, non-degenerate metric on a full neighborhood $\mathcal{V}$ of the origin in $\mathbb{R}^{n+m}$. This means that each boundary hypersurface $S_i = \{\zeta_i = 0\}$ is the fixed point set of the locally defined isometry $\zeta_i \to -\zeta_i$, and hence S_i, and any intersection $S_J = \bigcap_{j\in J} S_{i_j}$, is totally geodesic. Let S be the intersection of all the S_i, i.e., the corner of maximal codimension which intersects $\mathcal{V}$; for simplicity, assume that its codimension is n.

Assuming that $\mathcal{V}$ is sufficiently small, then for each $p \in \mathcal{V}$ there is a unique closest point $\Pi_J(p)$ to p in S_J; if $S_J = S$, the maximal codimension corner, then we write this projection as Π_S. The signed distance function $\rho_i(p) = \text{s-dist}_g(p, S_i)$ to each hypersurface is smooth. Abusing notation slightly, let $\mathbf{y} = (y_1, \dots, y_m)$ be the composition of the original set of local coordinates restricted to S with the projection Π_S. Then $(\rho_1, \dots, \rho_n, y_1, \dots, y_m)$ is a new set of smooth local coordinates in $\mathcal{V}$.

Let us compute the metric coefficients of g with respect to these coordinates. In fact, we first compute the coefficients for the corresponding co-metric, i.e., the entries of the matrix

$$g^{-1} = \begin{pmatrix} \langle d\rho_i, d\rho_j\rangle & \langle d\rho_i, dy_l\rangle \\ \langle dy_k, d\rho_j\rangle & \langle dy_k, dy_l\rangle \end{pmatrix}. \tag{2.13}$$

From general considerations, since each of the ρ_i are distance functions, we have

$$\langle d\rho_i, d\rho_i\rangle = 1, \quad i = 1, \dots, n,$$

in the entire neighborhood $\mathcal{V}$. Now, because of the reflectional symmetries, $d\rho_j$ is clearly orthogonal to $d\rho_i$ when either ρ_i or ρ_j equal 0, and similarly, $d\rho_i$ is orthogonal to dy_k when $\rho_i = 0$; this means that there are smooth functions $\widetilde{\alpha}_{ij}$ and $\widetilde{\beta}_{il}$ such that

$$\langle d\rho_i, d\rho_j\rangle = \rho_i\rho_j\widetilde{\alpha}_{ij}, \quad \text{and } \langle d\rho_i, dy_k\rangle = \rho_i\widetilde{\beta}_k.$$

Inserting these expressions into the matrix above and changing coordinates by setting $\rho_j = 2\sqrt{x_j}$, we obtain the matrix of coefficients for this co-metric. This has the form (2.8) with $A_d = \text{Id}_n$. Finally, taking the inverse of this matrix gives the normal form we are seeking. We summarize this in the

PROPOSITION 2.2.3. *Let $q \in bP$ lie in a boundary face of codimension n. Then there is a neighborhood $\mathcal{U}$ of q and smooth local coordinates $(\mathbf{x}, \mathbf{y})$ in this neighborhood, with q corresponding to $(\mathbf{0}; \mathbf{0})$, in terms of which L takes the form*

$$L = \sum_{i=1}^{n} x_i\partial_{x_i}^2 + \sum_{1\le k,l\le m} c'_{kl}\partial_{y_k}\partial_{y_l} + \sum_{1\le i\neq j\le n} x_ix_ja'_{ij}\partial_{x_i}\partial_{x_j} + \sum_{i=1}^{n}\sum_{l=1}^{m} x_ib'_{il}\partial_{x_i}\partial_{y_l} + V. \tag{2.14}$$

Here V is an inward pointing vector field, and $(c'_{kl}(\mathbf{x}, \mathbf{y}))$ is a smooth family of positive definite matrices.

DEFINITION 2.2.4. *The coordinates introduced in this proposition are called* adapted local coordinates *centered at q.*

We have written this expression in such a way as to emphasize that the first two sums on the right are in some sense the principal parts of L, and the other second order terms should be regarded as lower order perturbations. The body of our work below is devoted to showing that this is exactly the case. On the other hand, the first order term V (or at least the part of it which is not tangent to bP) is definitely not a lower order perturbation.

The key point in this normal form is that all of the coefficients of the leading terms $x_i\partial_{x_i}^2$ are simultaneously equal to 1. By a linear change of the $\mathbf{y}$ coordinates we can also make $c'_{kl}(\mathbf{0},\mathbf{0}) = \delta_{kl}$. Hence restricting to an even smaller neighborhood, and writing the normal part of V at $(\mathbf{0},\mathbf{0})$ as $\sum_{i=1}^n b_i\partial_{x_i}$, then we see that

$$L_{\mathbf{b},m} := \sum_{i=1}^{n}\left(x_i\partial_{x_i}^2 + b_i\partial_{x_i}\right) + \sum_{k=1}^{m}\partial_{y_k}^2 \tag{2.15}$$

should provide a good model for L. Note that since V is inward pointed, we have $b_i \geq 0$ for each i.

Chapter Three

Maximum Principles and Uniqueness Theorems

One of the most important features associated with any scalar parabolic or elliptic problem is the use of the maximum principle. Although this seems to give only qualitative properties of solutions, it can actually be used to deduce many quantitative results, including even the parabolic Schauder estimates. On a more basic level, it is the key ingredient in proving uniqueness of solutions to such an equation. We now develop the maximum principle and its main consequences, both for the model operators $\partial_t - L_{b,m}$ on an open orthant, and for the general Kimura diffusion operators $\partial_t - L$ on a compact manifold with corners, as well as their elliptic analogues. This generalizes the results for the 1-dimensional case in [15]. Of particular note in this regard is a generalization of the Hopf boundary point maximum principle, given in Lemma 3.2.6. This result allows us to precisely describe the null-space of L and its adjoint L^*.

3.1 MODEL PROBLEMS

We begin with maximum principles for the model operators.

PROPOSITION 3.1.1. *Suppose that u is a subsolution of the model Kimura diffusion equation $\partial_t u \leq L_{\boldsymbol{b},m} u$ on $[0, T] \times S_{n,m}$ such that*

$$u \in \mathscr{C}^0([0, T] \times S_{n,m}) \cap \mathscr{C}^1((0, T] \times S_{n,m}), \tag{3.1}$$

and $u \in \mathscr{C}^2$ away from the boundaries of $S_{n,m}$ for $t > 0$. Suppose also that $x_j \partial_{x_j}^2 u \to 0$ as $x_j \to 0$ for each $j \leq n$. Suppose finally that

$$|u(\boldsymbol{x}, \boldsymbol{y}, t)| \leq A e^{a(|\boldsymbol{x}|+|\boldsymbol{y}|^2)} \tag{3.2}$$

for some $a, A > 0$. Then

$$\sup_{[0,T]\times S_{n,m}} u(\boldsymbol{x}, \boldsymbol{y}, t) = \sup_{S_{n,m}} u(\boldsymbol{x}, \boldsymbol{y}, 0). \tag{3.3}$$

PROOF. We show that if $u(\boldsymbol{x}, \boldsymbol{y}, 0) \leq 0$, then $u(\boldsymbol{x}, \boldsymbol{y}, t) < 0$ for all $0 < t \leq T$.

It is straightforward to check that the function $e^{x/(\tau - t)}$ on $[0, \tau)_t \times \mathbb{R}_+$ is a solution of the equation $(\partial_t - L_0)u = 0$. Using the relation $\partial_x^k L_0 = L_k \partial_x^k$, we obtain that

$$\partial_x^k e^{x/(\tau-t)} = \frac{1}{(\tau - t)^k} e^{x/(\tau-t)}$$

is a solution of $(\partial_t - L_k)u = 0$, and hence also, if $\mathbf{k}$ is the multi-index $(k_1, \dots, k_n) \in \mathbb{N}^n$, then

$$U_{1,\tau,\mathbf{k}}(x,t) = \sum_{j=1}^{n} (\tau - t)^{-k_j - 1} e^{x_j/(\tau - t)}$$

solves $(\partial_t - L_{\mathbf{k}})U_{1,\tau,\mathbf{k}} = 0$. Suppose that $k_j > b_j$ for each j. Then

$$(\partial_t - L_{\boldsymbol{b},m})U_{1,\tau,\mathbf{k}} = \sum_{j=1}^{n} (k_j + 1 - b_j) x_j \partial_{x_j} U_{1,\tau,\mathbf{k}} > 0, \ 0 \leq t < \tau.$$

Next define

$$U_{2,\tau}(y,t) = \frac{1}{(\tau - t)^{m/2}} e^{|y|^2/2(\tau - t)}.$$

Finally, given the subsolution u in the statement of the proposition, set

$$v(\boldsymbol{x}, \boldsymbol{y}, t) = u(\boldsymbol{x}, \boldsymbol{y}, t) - \epsilon_1 U_{1,\tau,\mathbf{k}}(\boldsymbol{x}, t) - \epsilon_2 U_{2,\tau}(\boldsymbol{y}, t) + \epsilon_3 \frac{1}{1+t}.$$

We see that

$$(\partial_t - L_{\boldsymbol{b},m})v < 0,$$

so that v is a strict subsolution of this equation. Let

$$D_R = [0, \tau'] \times B_R(0) \text{ and } D_R' = \{0\} \times B_R(0) \cup [0, \tau'] \times bB_R(0),$$

$B_R(0)$ is the ball of radius R around the origin in $\mathbb{R}_+^n \times \mathbb{R}^m$ and $0 < \tau' < \tau$. We claim that the supremum of v on D_R is attained on D_R' but not at any one of the boundaries where any $x_j = 0$. The fact that this supremum must occur on D_R' follows from the standard maximum principle. If the supremum were to occur when $x_j = 0$, then using that $u \in \mathscr{C}^1$ up to this boundary, we see that the corresponding derivative $\partial_{x_j} v \geq 0$ at this point, which is impossible. This proves the claim. Finally, since u grows no faster than $e^{a(|x|+|y|^2)}$, we can choose $1/\tau < a$ and R sufficiently large to ensure that v is as negative as we wish on the entire side boundary $(0, \tau'] \times bB_R(0)$.

We conclude that

$$u(\boldsymbol{x}, \boldsymbol{y}, t) \leq \epsilon_1 U_{1,\tau,k}(\boldsymbol{x}, 0) + \epsilon_2 U_{2,\tau}(\boldsymbol{y}, 0) - \epsilon_3,$$

for any $\epsilon_1, \epsilon_2, \epsilon_3 > 0$. Now let these parameters tend to zero to see that $u \leq 0$ when $t > 0$, as desired. □

3.2 KIMURA DIFFUSION OPERATORS ON MANIFOLDS WITH CORNERS

The corresponding result for a general variable coefficient Kimura diffusion, L on a manifold with corners, P, requires slightly stronger hypotheses. We let $\mathscr{D}^2_{WF}(P)$ denote a certain subspace of $\mathscr{C}^1(P) \cap \mathscr{C}^2(\text{int}\, P)$ adapted to the degeneracies of L. In a neighborhood, U, of a boundary point p_0 of codimension M we introduce local coordinates

$$(x_1, \dots, x_M, y_1, \dots, y_{N-M}),$$

so that the stratum of the boundary through p_0 is locally given by

$$\Sigma \cap U = \{(\mathbf{0}, \mathbf{y}) : |\mathbf{y}| < 1\}. \tag{3.4}$$

A function $f \in \mathscr{C}^1(\overline{U}) \cap \mathscr{C}^2(U)$ belongs to $\mathscr{D}^2_{WF}(U)$ if, for $1 \leq i, j \leq M$, and $1 \leq l, m \leq N - M$ the functions

$$x_i \partial^2_{x_i} f,\ x_i x_j \partial_{x_i} \partial_{x_j} f,\ x_i \partial_{x_i} \partial_{y_l} f,\ \partial_{y_l} \partial_{y_m} f \tag{3.5}$$

extend continuously to $bP \cap U$, with the first three types of expressions vanishing whenever x_i or x_j vanishes. These conditions are clearly coordinate invariant.

DEFINITION 3.2.1. *A function in $\mathscr{C}^1(P) \cap \mathscr{C}^2(\text{int } P)$ belongs to $\mathscr{D}^2_{WF}(P)$ if its restrictions to neighborhoods of boundary points belong to each of these local $\mathscr{D}^2_{WF}$-spaces.*

Our first result shows that on $\mathscr{D}^2_{WF}(P)$, a Kimura diffusion is a dissipative operator.

LEMMA 3.2.2. *Let $w \in \mathscr{D}^2_{WF}(P)$, and suppose that w assumes a local maximum at $p_0 \in P$, then $Lw(p_0) \leq 0$.*

PROOF. If $p_0 \in \text{int } P$, then this is obvious, as L is strongly elliptic in the interior, and annihilates the constant function. Suppose that p_0 is a boundary point of codimension n, and $(x_1, \dots, x_n, y_1, \dots, y_m)$ are adapted local coordinates. We normalize so that p_0 corresponds to $\mathbf{0}$; the stratum bP through p_0 is locally given by $\Sigma = \{x_1 = \cdots = x_n = 0\}$. The regularity assumptions show that w restricted to Σ is locally $\mathscr{C}^2$ and the local form for L given in (2.14) shows that

$$Lw(p_0) = \sum_{k,l} c'_{kl} \partial_{y_k} \partial_{y_l} w(p_0) + V w(p_0). \tag{3.6}$$

Since $V(p_0)$ is inward pointing and p_0 is a local maximum, it is clear that

$$V w(p_0) \leq 0. \tag{3.7}$$

Since $w \restriction_\Sigma$ is locally $\mathscr{C}^2$, the second order part of Lw at p_0 is also non-positive, thus proving the lemma. □

In order to refine this result, we must describe in more detail the structure of bP, and the relationship of L to the various components of the stratification of bP. First, let $\{\Sigma_{1,j} : j = 1, \dots, N_1\}$ denote the connected hypersurface boundary components of P and $\{\rho_j\}$ their respective defining functions. If Σ is a component of bP of codimension n, then there are n hypersurface boundary components $\{\Sigma_{1,j_i} : i = 1, \dots, n\}$ so that Σ is a connected component of the intersection

$$\Sigma_{1,j_1} \cap \cdots \cap \Sigma_{1,j_n}. \tag{3.8}$$

The first order part V is tangent to Σ near p_0, if and only if $V\rho_{j_i} = 0$ for $i = 1, \dots, n$; this is evidently equivalent to the condition $L\rho_{j_i} = 0$ for this collection of indices. If this holds at all points of Σ, then we say that L *is tangent to* Σ. If, on the other hand, there is a $c > 0$ so that

$$L\rho_{j_i} \restriction_{\Sigma} > c, \text{ for } i = 1, \dots, n, \tag{3.9}$$

then we say that L is *transverse to* Σ. These conditions are independent of the choice of defining function. We write $bP^T(L)$ for the union of boundary components to which L is tangent, and $bP^{\pitchfork}(L)$ for the union of boundary components to which L is transverse.

The following non-degeneracy assumption about L simplifies many of the global results.

DEFINITION 3.2.3. *We say that* L meets bP cleanly, *if for each* $1 \leq j \leq N_1$, *either* $L\rho_j \restriction_{\{\rho_j=0\}} \equiv 0$, *or there exists a* $c_j > 0$, *so that*

$$L\rho_j \restriction_{\{\rho_j=0\}} > c_j. \tag{3.10}$$

Briefly, for each j, the vector field is either tangent or transverse to $\Sigma_{1,j}$, so cleanness prevents such behavior as V lying tangent to some $\Sigma_{1,j}$ only along a proper closed subset (possibly in $b\Sigma_{1,j}$). A boundary component belongs to $bP^T(L)$ if and only if it is a component of the intersection of a collection of boundary faces to which L is tangent. A boundary component belongs to $bP^{\pitchfork}(L)$ if and only if it is a component of the intersection of a collection of boundary faces to which L is transverse. There may be boundary components that belong to neither of these extremes.

Any boundary component Σ is itself a manifold with corners. If L is tangent to Σ, then we write L_Σ for the generalized Kimura diffusion defined by restriction of L to Σ. It is clear that if U is any neighborhood of a point p_0 in the interior of Σ, if $w \in \mathscr{D}^2_{WF}(U)$, and if L is tangent to Σ, then

$$[Lw] \restriction_{\Sigma \cap U} = L_\Sigma [w \restriction_{\Sigma \cap U}]. \tag{3.11}$$

The first basic result is the following.

LEMMA 3.2.4. *Let* $w \in \mathscr{D}^2_{WF}(P)$, *and suppose that* w *is a subsolution of* L, *i.e.,* $Lw \geq 0$. *Then* w *cannot assume a local maximum in the interior of* P, *or in the interior of any component* $\Sigma \in bP^T(L)$, *unless* w *is constant on* P, *or on that component, respectively.*

PROOF. Since L is a non-degenerate elliptic operator in int P, it follows from the standard strong maximum principle that w does not assume a local maximum in int P unless w is constant. The regularity hypothesis, and the assumption that L is tangent to Σ, show that $w \restriction_\Sigma \in \mathscr{D}^2_{WF}(\Sigma)$, and

$$L_\Sigma [w \restriction_\Sigma] \geq 0. \tag{3.12}$$

Hence the first part of this proof applies to show that w cannot assume its maximum in int Σ, unless $w \restriction_\Sigma$ is constant. □

In order to apply this result to determine the null-space of L, we need to discuss two further types of boundary components. First, amongst the collection of all components of the stratification of bP, certain ones are minimal in the sense that they themselves have no boundary; these components are either points or closed manifolds. We denote by $bP_{\min}$ the union of all such components; the different components of this set may have different dimensions. These minimal components are the minimal elements in the maximal well-ordered chains of boundary components, where the ordering is given by containment in the closure.

LEMMA 3.2.5. *Every component of bP is either minimal or else contains elements of $bP_{\min}$ in its closure.*

PROOF. This follows directly by induction on the maximal codimension of corners in P. If $\Sigma \notin bP_{\min}$, then Σ is a manifold with corners with maximal codimension no more than one less than that of P. Hence there is some boundary component Σ' of Σ which is minimal. Clearly Σ' is also a boundary component of P, and since it has no boundary, it must be minimal for P as well. □

Note that if $\Sigma \in bP^T_{\min}(L) = bP^T(L) \cap bP_{\min}(L)$, then either Σ is a point and all coefficients of L vanish at Σ, or else $\dim \Sigma > 0$ and L_Σ is a non-degenerate elliptic operator. It follows immediately that if $Lw \geq 0$ as above, and if $w \restriction_\Sigma$ attains a local maximum on $\Sigma \in bP^T_{\min}(L)$, then $w \restriction_\Sigma$ is constant.

Finally, the *terminal boundary* of P relative to L, denoted $bP_{\text{ter}}(L)$, consists of the union of boundary components $\Sigma \subset bP^T(L)$ such that L_Σ is transverse to all components of $b\Sigma$ (i.e., $b\Sigma \in b\Sigma^{\pitchfork}(L_\Sigma)$). In particular, if L is transverse to all components of bP itself, then $bP_{\text{ter}}(L) = P$. As elements of $bP^T_{\min}(L)$ have empty boundary, it follows that $bP^T_{\min}(L) \subset bP_{\text{ter}}(L)$.

There is a version of the Hopf boundary point lemma, adapted to this setting.

LEMMA 3.2.6. *Let P be a compact, connected manifold with corners, and L a generalized Kimura diffusion operator that meets bP cleanly. Suppose that $w \in \mathcal{D}^2_{WF}$ is a subsolution of L, $Lw \geq 0$, in a neighborhood, U, of a point $p_0 \in bP$ which lies in the interior of a boundary component $\Sigma \in bP^{\pitchfork}(L)$. If w attains a local maximum at p_0, then w is constant on U.*

This has an immediate and important consequence.

LEMMA 3.2.7. *Let P be a compact manifold with corners, and L a generalized Kimura diffusion on P, which meets bP cleanly. Suppose that L is transverse to every face of bP, and $w \in \mathcal{D}^2_{WF}(P)$ is a subsolution of L. Then w is constant.*

This follows directly from Lemma 3.2.6.

We defer the proof of Lemma 3.2.6 momentarily and derive its main consequence. The following result shows that, at least when L meets bP cleanly, the null-space of L on $\mathcal{D}^2_{WF}(P)$ is finite dimensional.

PROPOSITION 3.2.8. *Let P be a compact, connected manifold with corners and L a generalized Kimura diffusion which meets bP cleanly. Let $w \in \mathcal{D}^2_{WF}(P)$ be a* *solution to* $Lw = 0$*. Then* w *is determined by its (constant) values on the components of* $bP_{\text{ter}}(L)$*.*

PROOF. We prove this by induction on the dimension of P. If the dimension of P is 1, then P is an interval. The statement of the proposition in this case was established in [15].

Now suppose the result has been proved for all compact manifolds with corners P' of dimension $N - 1$, and all general Kimura diffusion operators L' on them. Assume that $\dim P = N$, and that $w \restriction_{\Sigma} = 0$ for all $\Sigma \in bP_{\text{ter}}(L)$. Obviously, if P has no boundary faces to which L is tangent, then P is itself a terminal boundary and we already have that $w \equiv 0$. We henceforth assume that $bP^T(L) \neq \emptyset$. If w does not vanish identically, then we can assume that it is positive somewhere, and therefore attains a positive maximum somewhere in P.

The induction hypothesis shows that for any boundary component $\Sigma \in bP^T(L)$, we have a solution $w \restriction_{\Sigma}$, which vanishes on $b\Sigma_{\text{ter}}(L_{\Sigma})$. This follows because the terminal components of bP relative to L, which are contained in the closure of Σ, are the same as the terminal components of $b\Sigma$ relative to L_{Σ}. Indeed, if L is tangent to some component Σ_0 of $b\Sigma$, then L_{Σ} is also tangent to Σ_0. Furthermore, the restriction of L_{Σ} to Σ_0 is the same as L_{Σ_0}, the restriction of L to Σ_0. Thus the condition that L_{Σ} be transverse to all components of $b\Sigma$ is the same, whether restricting from P or Σ. This means that $w \restriction_{\Sigma_0}$ vanishes on all $\Sigma_0 \in b\Sigma_{\text{ter}}(L_{\Sigma})$, and hence by induction, $w \restriction_{\Sigma} = 0$. Note that w vanishes on every hypersurface boundary component to which L is tangent.

Lemma 3.2.6 shows that w cannot attain a positive maximum on a boundary component to which L is transverse. Thus w must attain its maximum at a point p_0 lying in a boundary component, Σ_1 to which L is neither tangent, nor transverse. That is, Σ_1 lies in the intersection of boundary faces some of which belong to $bP^T(L)$ and some of which L is transverse to. This implies that there is a boundary hypersurface Σ_0 to which L is tangent, and such that $\Sigma_1 \subset b\Sigma_0$. As $w = 0$ on Σ_0 and $\Sigma_1 \subset \overline{\Sigma_0}$, this contradiction establishes that $w \equiv 0$ on P. □

Remark. This theorem is proved in [38] in the special case of the classical Kimura diffusion, without selection, acting on the N-simplex.

Lemma 3.2.7 is a special case of this proposition. Two other corollaries are:

COROLLARY 3.2.9. *If* L *is a generalized Kimura diffusion on the compact manifold with corners and* L *is everywhere tangent to* bP*, then any solution* $w \in \mathcal{D}^2_{WF}(P)$ *to* $Lw = 0$ *is determined by its constant values on* $bP_{\min} = bP^T_{\min}$*.*

COROLLARY 3.2.10. *If* P *is a compact manifold with corners, and* L *a generalized Kimura diffusion on* P *meeting* bP *cleanly, then the dimension of the null-space of* L *acting on* $\mathcal{D}^2_{WF}(P)$ *is bounded above by the cardinality of the set* $bP_{\text{ter}}(L)$*.*

Remark. These results give powerful support for our assertion that the regularity condition $w \in \mathcal{D}^2_{WF}(P)$ is a reasonable replacement for a local boundary condition, at least when L meets bP cleanly. In applications to probability, one often considers solutions of equations of the form $Lw = g$, where w satisfies a Dirichlet condition on bP. Frequently g is non-negative, and our uniqueness results easily imply that there cannot be a regular solution. The simplest example arises in the 1-dimensional case, with $L = x(1-x)\partial_x^2$. The solution, w to the equation

$$Lw = -1 \text{ with } w(0) = w(1) = 0, \tag{3.13}$$

gives the expected time to arrive at $\{0, 1\}$, for a path of the process starting at $0 < x < 1$. There cannot be a regular solution as $x(1-x)\partial_x^2 w(x) = -1$ cannot converge to 0 as $x \to 0^+, 1^-$. The solution, given by

$$w(x) = -[x \log x + (1-x)\log(1-x)],$$

is plainly not regular. In applications to probability this situation often pertains. The fact that $Lw \geq 0$ has no non-trivial regular solutions shows that the required solutions cannot be regular, and therefore involve the non-zero indicial root.

We now turn to the proof of the "Hopf lemma":

PROOF OF LEMMA 3.2.6. The proof of this lemma relies on the construction of barrier functions and a simple scaling argument. To motivate the argument we first give the proof for a model operator and a boundary point of codimension 1. Let $(x, \boldsymbol{y})$ denote normalized local coordinates so that

$$L = x\partial_x^2 + b\partial_x + \sum_{l=1}^{m} \partial_{y_l}^2, \text{ with } b > 0, \tag{3.14}$$

and w assumes a local max at the $(0, \mathbf{0})$. We assume that w is not constant. The strong maximum principle implies that there is a neighborhood U of 0 so that if $(x, y) \in U$ and $x > 0$, then

$$w(x, \boldsymbol{y}) < w(0, \mathbf{0}). \tag{3.15}$$

For R, r positive numbers and $\frac{1}{2} \leq \alpha < 1$, we define anisotropic balls in $\mathbb{R}_+^n \times \mathbb{R}^m$:

$$B_{R,r,\alpha}^+(n,m) = \{(\boldsymbol{x}, \boldsymbol{y}) \in \mathbb{R}_+^n \times \mathbb{R}^m : |\boldsymbol{x}^\alpha - \boldsymbol{r}^\alpha|^2 + |\boldsymbol{y}|^2 \leq nR^{2\alpha}\} \cap \{\boldsymbol{x} \geq \mathbf{0}\}. \tag{3.16}$$

Here $\boldsymbol{r} = (r, \ldots, r)$, and $\boldsymbol{x}^\alpha = (x_1^\alpha, \ldots, x_n^\alpha)$, etc. We now construct a non-negative local barrier, $v_{\lambda,\alpha,r}$ that satisfies

$$Lv_{\lambda,r} > 0 \text{ in } B_{r,r,\alpha}^+(1,m) \setminus B_{\frac{r}{2^\beta},r,\alpha}^+(1,m), \text{ and } v_{\lambda,r} \restriction_{bB_{r,r,\alpha}^+(1,m)} = 0,$$
$$\text{where } \beta = \frac{1}{2\alpha}. \tag{3.17}$$

Note that $B_{\frac{r}{2^\beta},r,\alpha}^+(1,m)$ is a compact subset of P lying a positive distance from bP. Figure 3.1 shows the set $B_{1,1,\frac{1}{2}}^+(2,0) \setminus B_{\frac{1}{4},1,\frac{1}{2}}^+(2,0) \subset \mathbb{R}_+^2$.

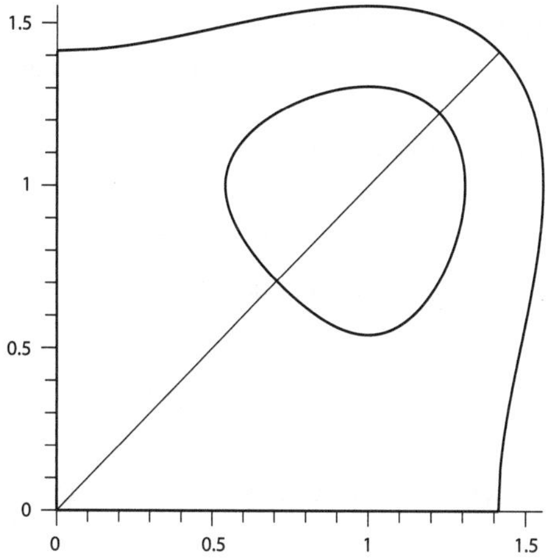

Figure 3.1: The set $B^+_{1,1,\frac{1}{2}}(2,0) \setminus B^+_{\frac{1}{4},1,\frac{1}{2}}(2,0) \subset \mathbb{R}^2_+$ lies in the positive quadrant between the black curves.

We first define barrier functions in the 1-codimensional case, letting

$$w_{\lambda,\alpha,r} = e^{-\lambda[(x^\alpha - r^\alpha)^2 + |y|^2]}, \tag{3.18}$$

then

$$Lw_{\lambda,r} = \Big[4\lambda^2[\alpha^2 x^{2\alpha-1}(r^\alpha - x^\alpha)^2 + |y|^2] + 2\lambda\Big(\alpha[b-(1-\alpha)]x^{\alpha-1}(r^\alpha - x^\alpha) - \alpha^2 x^{2\alpha-1} - m\Big)\Big]w_{\lambda,r}. \tag{3.19}$$

If $1-b < \alpha < 1$, then we see that $Lw_{\lambda,r}(x, \mathbf{y})$ tend to $+\infty$ as x tends to zero. As $[\alpha^2 x^{2\alpha-1}(r^\alpha - x^\alpha) + |y|^2]$ only vanishes at $(r, \mathbf{0})$ and $(0, \mathbf{0})$ (if $\alpha \neq \frac{1}{2}$) it is not difficult to see that for large enough λ we have that

$$Lw_{\lambda,\alpha,r} > 0 \text{ in } B^+_{r,r,\alpha}(1,m) \setminus B^+_{\frac{r}{2^\beta},r,\alpha}(1,m). \tag{3.20}$$

Let

$$v_{\lambda,\alpha,r} = w_{\lambda,\alpha,r} - w_{\lambda,\alpha,r}(0,\mathbf{0}), \tag{3.21}$$

so that the barrier vanishes on $bB^+_{r,r,\alpha}(1,m)$.

We fix an r so that $B^+_{r,r,\alpha}(1,m) \subset U$ and a λ and α as above. The hypothesis that w is non-constant shows that there is an $\epsilon > 0$ so that

$$(w + \epsilon v_{\lambda,\alpha,r}) \restriction_{bB^+_{\frac{r}{2^\beta},r,\alpha}(1,m)} < w(0,\mathbf{0}), \tag{3.22}$$

and therefore $w + \epsilon v_{\lambda,\alpha,r}$ assumes its maximum value at $(0, \mathbf{0})$. Thus for small positive x we have

$$(w + \epsilon v_{\lambda,r})(x,\mathbf{0}) - (w + \epsilon v_{\lambda,r})(0,\mathbf{0}) \leq 0. \tag{3.23}$$

A simple application of the mean value theorem shows that $\partial_x w(x, \mathbf{0})$ tends to $-\infty$ as $x \to 0^+$, contradicting the assumed regularity of w. This completes the proof that w must be constant in this case.

To treat a general (non-model operator) we might need to dilate the coordinates by setting:

$$x = \mu X \text{ and } \mathbf{y} = \sqrt{\mu}\mathbf{Y}, \text{ for } \mu > 0. \tag{3.24}$$

Under this change of variables, the model operator L becomes

$$\frac{1}{\mu}[X\partial_X^2 + b\partial_X + \sum_{l=1}^{m} \partial_{Y_l}^2]. \tag{3.25}$$

That is, up to a positive factor, this operator is invariant under these changes of coordinate.

If we let $W(X, \mathbf{Y}) = w(\mu X, \sqrt{\mu}\mathbf{Y})$, then evidently $LW \geq 0$, and W attains a local maximum at $(0, \mathbf{0})$. The ball

$$(X^\alpha - R^\alpha)^2 + |\mathbf{Y}|^2 \leq R^{2\alpha} \text{ where } R = \frac{r}{\mu} \tag{3.26}$$

is contained in this coordinate chart. In the original coordinates we have

$$\begin{aligned} L = x\partial_x^2 + b\partial_x + + \sum_{k=1}^{m} \partial_{y_k}^2 + \sum_{l=1}^{m} x a_l(x, \mathbf{y})\partial_x\partial_{y_l} + \\ \sum_{k,l=1}^{m} c_{kl}(\boldsymbol{x}, \mathbf{y})\partial_{y_k}\partial_{y_l} + \sum_{i=1}^{n} \widetilde{b}(x, \mathbf{y})\partial_x + \sum_{k=1}^{m} d_k(\boldsymbol{x}, \mathbf{y})\partial_{y_k}, \end{aligned} \tag{3.27}$$

where $\widetilde{b}(0, \mathbf{0}) = c_{kl}(0, \mathbf{0}) = 0$. Letting $x = \mu X$ and $\mathbf{y} = \sqrt{\mu}\mathbf{Y}$, we obtain:

$$\begin{aligned} L_\mu = \frac{1}{\mu}\Bigg\{ X\partial_X^2 + b\partial_X + \sum_{k=1}^{m} \partial_{Y_k}^2 + \sqrt{\mu}\sum_{l=1}^{m} X a_l(\mu X, \sqrt{\mu}\mathbf{Y})\partial_X\partial_{Y_l} + \\ \sum_{k,l=1}^{m} c_{kl}(\mu X, \sqrt{\mu}\mathbf{Y})\partial_{Y_k}\partial_{Y_l} + \widetilde{b}(\mu X, \sqrt{\mu}\mathbf{Y})\partial_X + \sqrt{\mu}\sum_{k=1}^{m} d_k(\mu \boldsymbol{X}, \sqrt{\mu}\mathbf{Y})\partial_{Y_k} \Bigg\}. \end{aligned} \tag{3.28}$$

Even though we may let μ get very small, we can fix a positive R. It is then not hard to see that, with a possibly larger $\alpha < 1$, by taking μ small enough we can arrange for

$$L_\mu w_{\lambda,\alpha,R} > 0 \text{ in } B^+_{R,R,\alpha}(1, m) \setminus B^+_{\frac{R}{2^\beta},R,\alpha}(1, m). \tag{3.29}$$

From this point the argument proceeds as before showing that if $Lw \geq 0$ and w attains a maximum at p_0, then w is constant.

Suppose that p_0 is a point on a stratum of codimension n, we can choose $(\boldsymbol{x}, \boldsymbol{y})$ adapted local coordinates, with p_0 corresponding to $(\mathbf{0}, \mathbf{0})$, so that the operator takes the form:

$$L = \sum_{i=1}^{n}[x_i\partial_{x_i}^2 + b_i\partial_{x_i}] + \sum_{k=1}^{m}\partial_{y_k}^2 +$$
$$\sum_{i\neq j=1}^{n} x_ix_ja_{ij}(\boldsymbol{x}, \boldsymbol{y})\partial_{x_i}\partial_{x_j} + \sum_{i=1}^{n}\sum_{l=1}^{m} x_ia_{il}(\boldsymbol{x}, \boldsymbol{y})\partial_{x_i}\partial_{y_l} + \sum_{k,l=1}^{m} c_{kl}(\boldsymbol{x}, \boldsymbol{y})\partial_{y_k}\partial_{y_l} +$$
$$\sum_{i=1}^{n}\widetilde{b}_i(\boldsymbol{x}, \boldsymbol{y})\partial_{x_i} + \sum_{k=1}^{m} d_k(\boldsymbol{x}, \boldsymbol{y})\partial_{y_k}, \quad (3.30)$$

where

$$c_{kl}(\mathbf{0}, \mathbf{0}) = \widetilde{b}_i(\mathbf{0}, \mathbf{0}) = 0. \quad (3.31)$$

Let $\boldsymbol{b} = (b_1, \ldots, b_n) > \mathbf{0}$. For the model operator $L_{\boldsymbol{b},m}$ we consider barrier functions of the form

$$w_{\lambda,\alpha,r} = \exp\left[-\lambda(|\boldsymbol{x}^{\alpha} - \boldsymbol{r}^{\boldsymbol{\alpha}}|^2 + |\boldsymbol{y}|^2)\right]. \quad (3.32)$$

Applying the model operator we see that

$$L_{\boldsymbol{b},m}w_{\lambda,\alpha,r} = 4\lambda^2\left[\alpha^2\sum_{j=1}^{n}x_j^{2\alpha-1}(x_j^{\alpha} - r^{\alpha})^2 + |\boldsymbol{y}|^2\right]w_{\lambda,\alpha,r} +$$
$$2\lambda\left[\alpha\sum_{j=1}^{n}\left([b_j - (1-\alpha)]x_j^{\alpha-1}(r^{\alpha} - x_j^{\alpha}) - \alpha x_j^{2\alpha-1}\right) - m\right]w_{\lambda,\alpha,r}. \quad (3.33)$$

In order for $w_{\lambda,\alpha,r}$ to be a subsolution, we need to choose $\frac{1}{2} \leq \alpha < 1$, so that

$$1 - \min\{b_1, \ldots, b_n\} < \alpha. \quad (3.34)$$

With such a choice of α, we see that $L_{\boldsymbol{b},m}w_{\lambda,\alpha,r}$ tends to $+\infty$ as any x_j tends to zero. We can therefore find λ_0 so that for $\lambda_0 < \lambda$ we have

$$L_{\boldsymbol{b},m}w_{\lambda,\alpha,r} > 0 \text{ in } B^+_{r,r,\alpha}(n, m) \setminus B^+_{\frac{r}{(2n)^{\beta}},r,\alpha}(n, m), \quad (3.35)$$

where, as before, $\beta = \frac{1}{2\alpha}$.

As before we can scale the variables $\boldsymbol{x} = \mu \boldsymbol{X}$ and $\boldsymbol{y} = \sqrt{\mu}\boldsymbol{Y}$ to obtain

$$L_\mu = \frac{1}{\mu}\Bigg\{\sum_{i=1}^{n}[X_i\partial_{X_i}^2 + b_i\partial_{X_i}] + \sum_{k=1}^{m}\partial_{Y_k}^2 + \mu \sum_{i\neq j=1}^{n} X_iX_ja_{ij}(\mu\boldsymbol{X}, \sqrt{\mu}\boldsymbol{Y})\partial_{X_i}\partial_{X_j} +$$
$$\sqrt{\mu}\sum_{i=1}^{n}\sum_{l=1}^{m} X_ia_{il}(\mu\boldsymbol{X}, \sqrt{\mu}\boldsymbol{Y})\partial_{X_i}\partial_{Y_l} + \sum_{k,l=1}^{m} c_{kl}(\mu\boldsymbol{X}, \sqrt{\mu}\boldsymbol{Y})\partial_{Y_k}\partial_{Y_l} +$$
$$\sum_{i=1}^{n}\widetilde{b}_i(\mu\boldsymbol{X}, \sqrt{\mu}\boldsymbol{Y})\partial_{X_i} + \sqrt{\mu}\sum_{k=1}^{m} d_k(\mu\boldsymbol{X}, \sqrt{\mu}\boldsymbol{Y})\partial_{Y_k}\Bigg\}. \tag{3.36}$$

A calculation shows that

$$L_\mu w_{\lambda,\alpha,r}(\boldsymbol{X}, \boldsymbol{Y}) = \frac{1}{\mu}\Bigg\{4\lambda^2\Bigg[\sum_{i=1}^{n}\alpha^2X_i^{2\alpha-1}(X_i^\alpha - r^\alpha)^2 + |\boldsymbol{Y}|^2 +$$
$$\sum_{i\neq j}\alpha^2a_{ij}(\mu\boldsymbol{X}, \sqrt{\mu}\boldsymbol{Y})X_iX_j(X_i^\alpha - r^\alpha)(X_j^\alpha - r^\alpha) +$$
$$\sqrt{\mu}\sum_{i=1}^{n}\sum_{l=1}^{m}\alpha a_{il}(\mu\boldsymbol{X}, \sqrt{\mu}\boldsymbol{Y})X_i^\alpha(X_i^\alpha - r^\alpha)Y_l + \sum_{k,l=1}^{m} c_{kl}(\mu\boldsymbol{X}, \sqrt{\mu}\boldsymbol{Y})Y_kY_l\Bigg] +$$
$$2\lambda\Bigg[\sum_{i=1}^{n}\Big[\big(b_i + \widetilde{b}_i(\mu\boldsymbol{X}, \sqrt{\mu}\boldsymbol{Y}) - (1-\alpha)\big)X_i^{\alpha-1}(r^\alpha - X_i^\alpha) - \alpha^2X_i^{2\alpha-1}\Big] -$$
$$\sum_{l=1}^{m}[1 + c_{ll}(\mu\boldsymbol{X}, \sqrt{\mu}\boldsymbol{Y}) + \sqrt{\mu}d_l(\mu\boldsymbol{X}, \sqrt{\mu}\boldsymbol{Y})Y_l]\Bigg]\Bigg\}. \tag{3.37}$$

If we take r and μ sufficiently small, then the $O(\lambda^2)$-term is bounded below by a positive multiple of

$$\sum_{i=1}^{n}\alpha^2X_i^{2\alpha-1}(X_i^\alpha - r^\alpha)^2 + |\boldsymbol{Y}|^2. \tag{3.38}$$

By taking $\alpha < 1$ a little larger and possibly reducing μ, we can assure that

$$\min\{b_i + \widetilde{b}_i(\mu\boldsymbol{X}, \sqrt{\mu}\boldsymbol{Y}) - (1-\alpha) : (\boldsymbol{X}, \boldsymbol{Y}) \in B^+_{r,r,\alpha}(n,m);\ i = 1, \ldots, n\} \tag{3.39}$$

is strictly positive. With these choices, there is a λ_0 so that if $\lambda > \lambda_0$, then

$$L_\mu w_{\lambda,\alpha,r} > 0 \text{ in } B^+_{r,r,\alpha}(n,m) \setminus B^+_{\frac{r}{(2n)^\beta},r,\alpha}(n,m). \tag{3.40}$$

Note also that $\partial_{X_i}w_{\lambda,\alpha,r}(\boldsymbol{X}, \boldsymbol{Y})$ tends to $+\infty$ as $X_i \to 0^+$. Finally we set

$$v_{\lambda,\alpha,r} = w_{\lambda,\alpha,r} - w_{\lambda,\alpha,r}(\boldsymbol{0}, \boldsymbol{0}). \tag{3.41}$$

The argument then proceeds as before. We are assuming that w is a non-constant solution to $Lw \geq 0$, which assumes a local maximum at $(\boldsymbol{0}, \boldsymbol{0})$. This implies that

$w(X, Y) < w(0, 0)$ for $(X, Y) \in \operatorname{int} P$. Thus we can choose an $\epsilon > 0$ so that for $(X, Y) \in bB^{+}_{\frac{r}{(2n)^{\beta}},r,\alpha}(n, m)$ we have the estimate

$$(w + \epsilon v_{\lambda,\alpha,r})(X, Y) < (w + \epsilon v_{\lambda,\alpha,r})(0, 0). \tag{3.42}$$

Since $v_{\lambda,\alpha,r}$ vanishes on $bB^{+}_{r,r,\alpha}(n, m)$, we see that $(w + \epsilon v_{\lambda,\alpha,r})$ must assume its maximum at a point p_0 on $bP \cap \overline{B^{+}_{r,r,\alpha}}(n, m)$. As before this implies that the derivatives $\partial_{X_j} w(p)$ tend to $-\infty$ as p approaches p_0, contradicting our assumptions about the smoothness of w. This completes the proof of the lemma. □

Remark. Because the derivatives of the barrier functions blowup at a rate dictated by the choice of α, the regularity hypotheses on w can be considerably weakened.

These results do not address the case when L fails to meet bP cleanly. While a different argument is needed, it seems likely that a result like that in Proposition 3.2.8 remains true. In particular, the dimension of the null-space of L acting on $\mathcal{D}^{2}_{WF}(P)$ should be finite dimensional.

3.3 MAXIMUM PRINCIPLES FOR THE HEAT EQUATION

We now turn to maximum principles for the heat equation:

PROPOSITION 3.3.1. *Let u be a subsolution of the Kimura diffusion equation $\partial_t u \leq Lu$ on $[0, T] \times P$, where P is a compact manifold with corners, such that*

$$u \in \mathscr{C}^0([0, T] \times P) \cap \mathscr{C}^1((0, T] \times P),$$

and $u(t, \cdot) \in \mathcal{D}^{2}_{WF}(P)$ for $t > 0$, then

$$\sup_{[0,T]\times P} u(p, t) = \sup_{P} u(p, 0).$$

PROOF. This is proved in almost exactly the same way as Proposition 3.1.1. Because P is compact and $u(\cdot, t)$ is continuous up to bP, there is no need to assume a growth condition on u. The hypotheses are such that we can verify as before that no local maximum occurs along $bP \times (0, T]$, and by the usual maximum principle, there is also no local maximum in the interior of P when $0 < t \leq T$. □

COROLLARY 3.3.2. *Let u_1 and u_2 be two solutions of $(\partial_t - L_{b,m})u = g$, $u(\boldsymbol{x}, \boldsymbol{y}, 0) = f(\boldsymbol{x}, \boldsymbol{y})$ in $\mathbb{R}_+ \times S_{n,m}$ such that $|u_j| \leq C_T e^{a(|\boldsymbol{x}|+|\boldsymbol{y}|^2)}$ for some $C > 0$ uniformly in any $[0, T] \times S_{n,m}$, satisfying the regularity hypotheses of Proposition 3.1.1. Then $u_1 \equiv u_2$.*

Similarly, if u_1 and u_2 are two solutions of $(\partial_t - L)u = f$, $u(\cdot, 0) = f(\cdot)$ in $\mathbb{R}_+ \times P$, satisfying the regularity hypotheses of Proposition 3.3.1, where P is a compact manifold with corners, then $u_1 \equiv u_2$.

Remark. The regularity assumption up to the boundaries where $x_j = 0$ are fundamental. For example, if $0 < b < 1$, then x^{1-b} is a stationary solution of $(\partial_t - L_b)u = 0$ on $\mathbb{R}_+ \times \mathbb{R}_+$ which is certainly subexponential as $x \to \infty$. However, from results of [15] it follows that there is another solution $w(x,t)$ to this equation with initial data $w(x,0) = x^{1-b}$. This solution is smooth up to $x = 0$ for $t > 0$, so that $w(x,t) \neq x^{1-b}$ for $t > 0$. Then $x^{1-b} - w(x,t)$ is a homogeneous solution with zero Cauchy data at $t = 0$ and which has subexponential growth. It is neither $\mathscr{C}^1$ up to $x = 0$, nor does it satisfy $\lim_{x\to 0^+} x\partial_x^2 u(x,t) = 0$.

We record one other easy extension of these results.

PROPOSITION 3.3.3. *Let L be a general Kimura operator on a compact manifold with corners P, and suppose that $c \in \mathscr{C}^0(P)$. Suppose that u is a subsolution of the diffusion equation associated to $L + c$, i.e., $\partial_t u \leq (L + c)u$, such that*

$$u \in \mathscr{C}^0([0,T] \times P) \cap \mathscr{C}^1((0,T] \times P),$$

and $u(\cdot,t) \in \mathscr{D}^2_{WF}(P)$ for $t > 0$, then

$$\sup_{[0,T]\times P} u(p,t) \leq e^{\alpha t} \sup_P u(p,0),$$

where $\alpha = ||c||_\infty$.

Consequently, if u_1 and u_2 are two solutions of $\partial_t u = (L + c)u$ which satisfy the regularity conditions above and which have the same initial condition at $t = 0$, then $u_1 \equiv u_2$.

Finally, we also state the corresponding maximum principle and uniqueness result for Kimura equations.

PROPOSITION 3.3.4. *Let L be a general Kimura operator on a compact manifold with corners P and that $c \in \mathscr{C}^0(P)$ is a non-positive function. Let f satisfy $0 \leq (L + c)f$ and $f \in \mathscr{D}^2_{WF}(P)$, then*

$$\sup_P f(p) = \sup_{bP} f(p).$$

If f_1 and f_2 are any two solutions of $(L + c)f = 0$ which satisfy all the regularity assumptions above and which agree on bP, then $f_1 \equiv f_2$.

There is an important special case, where a much sharper result is true.

PROPOSITION 3.3.5. *Let L be a general Kimura operator on a compact manifold with corners P and that $c \in \mathscr{C}^0(P)$ is a non-positive function, with $c \restriction_{bP}$ strictly negative. Let f satisfy $(L + c)f = 0$, and also suppose that*

$$f \in \mathscr{D}^2_{WF}(P),$$

then $f \equiv 0$.

PROOF. We show that f can neither attain a positive maximum nor a negative minimum, and hence is identically zero. Since we are considering the equation $(L + c)f = 0$, it suffices to show that f cannot attain a negative minimum. We suppose that f does attain a negative minimum. The regularity assumptions and the compactness of P show that f attains its minimum at some point $p_0 \in P$. The strict maximum principle implies that $p_0 \notin \text{int}\, P$. Suppose that p_0 belongs to a point of the boundary of codimension M, so that in local coordinates

$$Lf = \sum_{i=1}^{M} [x_i a_i(\boldsymbol{x}, \boldsymbol{y}) \partial_{x_i}^2 f + b_i(\boldsymbol{x}, \boldsymbol{y}) \partial_{x_i} f] + \sum_{i,j=1}^{M} x_i x_j a_{ij}(\boldsymbol{x}, \boldsymbol{y}) \partial_{x_i} \partial_{x_j} f +$$
$$\sum_{i=1}^{M} \sum_{l=1}^{n-M} x_i c_{il}(\boldsymbol{x}, \boldsymbol{y}) \partial_{x_i} \partial_{y_l} f + \sum_{l,m=1}^{n-M} c_{lm}(\boldsymbol{x}, \boldsymbol{y}) \partial_{y_m} \partial_{y_l} f + \sum_{l=1}^{n-M} d_m(\boldsymbol{x}, \boldsymbol{y}) \partial_{y_l} f. \quad (3.43)$$

At $p_0 = (\boldsymbol{0}, \boldsymbol{y}_0)$ we see that the regularity assumptions show that

$$Lf(p_0) = \sum_{i=1}^{M} b_i(\boldsymbol{0}, \boldsymbol{y}_0) \partial_{x_i} f + \sum_{l,m=1}^{n-M} c_{lm}(\boldsymbol{0}, \boldsymbol{y}_0) \partial_{y_m} \partial_{y_l} f. \quad (3.44)$$

As p_0 is a local minimum, the second order part is non-negative; since the vector field is inward pointing, so is the first order part. We therefore conclude that at p_0 we have $Lf(p_0) + c(p_0) f(p_0) > 0$, contradicting our assumption that f is in the null-space of $L + c$. □

Part II

Analysis of Model Problems

Chapter Four

The Model Solution Operators

In this chapter we introduce the model heat kernels $k^t_{b,m}$, i.e., the solution operators for the model problems $\partial_t - L_{b,m}$, and then prove a sequence of basic estimates for these operators which are direct generalizations of the estimates for the 1-dimensional version of this problem considered in [15]. We recall and slightly extend the $\mathscr{C}^0$ and $\mathscr{C}^k$ theory in the 1-dimensional case proved in [15] and then derive the straightforward extensions of these results to higher dimensions. This sets the stage for the more difficult Hölder estimates for solutions, which is carried out in the next several chapters, and which forms the technical heart of this monograph.

We also define the resolvent families $(L_{b,m} - \mu)^{-1}$, describe their holomorphic behavior as functions of μ and relate this to the analytic semi-group theory for the model parabolic problems. At the end of the chapter we describe why the estimates we prove here are not adequate for the perturbation theoretic arguments needed to construct the solution operator for general Kimura diffusions $\partial_t - L$.

4.1 THE MODEL PROBLEM IN 1-DIMENSION

First recall the 1-dimensional model operator,

$$L_b = x\partial_x^2 + b\partial_x \qquad \text{on } \mathbb{R}_+ \times \mathbb{R}_+, \tag{4.1}$$

where b is any non-negative constant, and the general inhomogeneous Cauchy problem

$$\begin{cases} \partial_t w - L_b w = g & \text{on } \mathbb{R}_+ \times (0, T] \\ w(x, 0) = f(x), & x \in \mathbb{R}_+. \end{cases} \tag{4.2}$$

So long as g and f have moderate growth, then Corollary 3.3.2 guarantees that there is a unique solution with moderate growth and satisfying certain regularity hypotheses at $x = 0$; it is given by the integral formula

$$w(x,t) = \int_0^t \int_0^\infty k^b_{t-s}(x,\tilde{x}) g(\tilde{x}, s)\, d\tilde{x} ds + \int_0^\infty k^b_t(x,\tilde{x}) f(\tilde{x})\, d\tilde{x}. \tag{4.3}$$

The precise form of the heat kernel for this problem was derived in [15]: for any $b > 0$,

$$k^b_t(x,\tilde{x}) = \frac{\tilde{x}^{b-1}}{t^b} e^{-\frac{x+\tilde{x}}{t}} \psi_b\left(\frac{x\tilde{x}}{t^2}\right), \tag{4.4}$$

where

$$\psi_b(z) = \sum_{j=0}^{\infty} \frac{z^j}{j!\Gamma(j+b)}. \tag{4.5}$$

When $b = 0$, the Schwartz kernel takes a somewhat different form:

$$k_t^0(x, \tilde{x}) = \left(\frac{x}{t^2}\right) e^{-\frac{x+\tilde{x}}{t}} \psi_2\left(\frac{x\tilde{x}}{t^2}\right) + e^{-\frac{x}{t}} \delta_0(y). \tag{4.6}$$

The defining equation for this kernel is that $(\partial_t - L_b)k_t^b = 0$ for $t > 0$ (along with the initial condition that $\lim_{t \to 0^+} k^b(x, \tilde{x}) = \delta(x - \tilde{x})$). However, it can be checked directly from the explicit expression that

$$(\partial_t - L_{b,\tilde{x}}^t)k_t^b(x, \tilde{x}) = 0, \quad \text{where } L_{b,\tilde{x}}^t = \partial_{\tilde{x}}(\partial_{\tilde{x}}\tilde{x} - b) \tag{4.7}$$

is the formal adjoint operator. Note too that we can verify directly from (4.4) that

$$\lim_{\tilde{x} \to 0^+} (\partial_{\tilde{x}}\tilde{x} - b)k_t^b(x, \tilde{x}) = 0 \text{ and } k_t^b(x, \tilde{x}) = \mathcal{O}(\tilde{x}^{b-1}) \tag{4.8}$$

when $t > 0$ and $b > 0$.

We write the solution as a sum $w = v + u$ where v is a solution to the problem with $g = 0$ and u is a solution to the problem with $f = 0$. We often call the first of these the homogeneous Cauchy problem and the second the inhomogeneous problem.

In the following, we describe the various estimates for solutions on integer order spaces. More specifically, we use the standard spaces of ℓ-times continuously differentiable functions $\mathscr{C}^\ell(\mathbb{R}_+)$, $\ell \in \mathbb{N}$, and their parabolic analogues, $\mathscr{C}^{\ell, \frac{\ell}{2}}(\mathbb{R}_+^2)$, which are the closures of $\mathscr{C}_c^\infty(\mathbb{R}_+^2)$ with respect to the norms

$$\|g\|_{\ell, \frac{\ell}{2}} = \sum_{0 \le p+2q \le \ell} \|\partial_t^q \partial_x^p g(x, t)\|_\infty. \tag{4.9}$$

Remark. To make the notation less cumbersome, in the context of these parabolic spaces, we always take $\frac{\ell}{2}$ to mean the greatest integer in $\ell/2$.

To keep track of behavior of solutions as $x \to \infty$, we also use $\dot{\mathscr{C}}^\ell(\mathbb{R}_+)$, which is the closure of $\mathscr{C}_c^\infty(\mathbb{R}_+)$ with respect to the norm

$$\|f\|_\ell = \sum_{p=0}^{\ell} \|\partial_x^p f(x)\|_\infty. \tag{4.10}$$

We first discuss solutions of the homogeneous problem and after that solutions of the inhomogeneous problem.

The first result is a slight improvement of a theorem from [15].

LEMMA 4.1.1. *For each $\ell \in \mathbb{N}$, if $f \in \dot{\mathscr{C}}^{\ell}(\mathbb{R}_+)$, then $v \in \mathscr{C}^{\ell,\frac{\ell}{2}}(\mathbb{R}_+ \times \mathbb{R}_+)$. Moreover if $p + 2q \leq \ell$, then*

$$\partial_t^q \partial_x^p v(x,t) = \int_0^\infty k_t^{b+p}(x,\tilde{x}) L_{b+p}^q \partial_{\tilde{x}}^p f(\tilde{x})\, d\tilde{x}. \tag{4.11}$$

PROOF. Let v be defined by (4.3) with $g = 0$. It is shown in [15] that

$$v \in \mathscr{C}^0([0,\infty)_t; \mathscr{C}_b^{\ell}([0,\infty)_x)) \cap \mathscr{C}^\infty((0,\infty)_t \times [0,\infty)_x),$$

and (4.11) is established for any $0 \leq p \leq \ell$, but only for $q = 0$. We establish the general case as follows. For $t > 0$, differentiate the representation formula for v (i.e., (4.3) with $g = 0$) and use (4.7) to obtain

$$\partial_t v(x,t) = \int_0^\infty \partial_t k_t^b(x,\tilde{x}) f(\tilde{x})\, d\tilde{x} = \lim_{\epsilon \to 0^+} \int_\epsilon^\infty L_{b,\tilde{x}}^t k_t^b(x,\tilde{x}) f(\tilde{x})\, d\tilde{x}. \tag{4.12}$$

If $\ell \geq 2$ and $b > 0$, we can integrate by parts twice in $\tilde{x}$ and let $\epsilon \to 0$ to obtain that

$$\partial_t v(x,t) = \int_0^\infty k_t^b(x,\tilde{x}) L_b f(\tilde{x})\, d\tilde{x}, \tag{4.13}$$

and hence $\partial_t v \in \mathscr{C}^0(\mathbb{R}_+ \times \mathbb{R}_+)$. In particular, if $f \in \mathscr{C}^2(\mathbb{R}_+)$, then $v \in \mathscr{C}^{2,1}(\mathbb{R}_+ \times \mathbb{R}_+)$. Using (4.11) and (4.13) inductively shows that v has the stated regularity and also gives (4.11) for all p, q with $p+2q \leq \ell$. The result for $b = 0$ follows from the formula (4.11) above when $b > 0$ and Proposition 7.8 in [15]. □

Now turn to the inhomogeneous equation, $(\partial_t - L_b)u = g$, $u(\cdot, 0) = 0$. The solution is given by the Duhamel formula

$$u(x,t) = \int_0^t \int_0^\infty k_{t-s}^b(x,\tilde{x}) g(\tilde{x},s)\, d\tilde{x} ds. \tag{4.14}$$

Since $k_t^b(x,\tilde{x}) \to \delta(x - \tilde{x})$ as $t \to 0^+$, we understand this to mean that

$$u(x,t) = \lim_{\epsilon \to 0^+} \int_0^{t-\epsilon} \int_0^\infty k_{t-s}^b(x,\tilde{x}) g(\tilde{x},s)\, d\tilde{x} ds. \tag{4.15}$$

Denote this Volterra operator by K_t^b. Lemma 4.1.1 implies the basic regularity result.

LEMMA 4.1.2. *If* $g \in \mathscr{C}^{\ell,\frac{\ell}{2}}(\mathbb{R}_+ \times [0,T])$, *then* $u = K_t^b g \in \mathscr{C}^{\ell,\frac{\ell}{2}}(\mathbb{R}_+ \times [0,T])$; *furthermore, for any* p, q *with* $p + 2q \leq \ell$,

$$\partial_t^q \partial_x^p u = K_t^{b+p} L_{b+p}^q \partial_{\tilde{x}}^p g + \sum_{l=0}^{q-1} L_{b+p}^l \partial_t^{q-l-1} \partial_x^p g(x,t), \tag{4.16}$$

where the sum is absent when $q = 0$.

PROOF. Let

$$G(\tau, s, x) = \int_0^\infty k_\tau^b(x, \tilde{x}) g(\tilde{x}, s)\, d\tilde{x}; \tag{4.17}$$

Lemma 4.1.1 implies that $\partial_x^i \partial_t^j \partial_\tau^k G \in \mathscr{C}^0(\mathbb{R}_+^3)$, provided $i + 2(j+k) \leq \ell$. From this it follows easily that

$$u(x,t) = \int_0^t G(t-s, s, x) g(\tilde{x}, s)\, ds \in \mathscr{C}^{\ell,\frac{\ell}{2}}(\mathbb{R}_+ \times \mathbb{R}_+). \tag{4.18}$$

Equation (4.16) with $q = 0$ follows directly from Lemma 4.1.1. Without loss of generality, we can therefore let $p = 0$, and prove the remainder of the formula by induction. The case $q = 1$ is simply the equation

$$\partial_t u = L_b u + g. \tag{4.19}$$

Assume that the formula holds for some q with $2q \leq \ell - 2$. The results in [15] show that we can differentiate the equation in (4.16) to obtain that

$$\partial_t^{q+1} u = \partial_t K_t^b L_b^q g + \sum_{l=0}^{q-1} L_b^l \partial_t^{q-l} g. \tag{4.20}$$

The first term on the right equals $L_b K_t^b L_b^q g + L_b^q g$. This completes the proof since $L_b K_t^b L_b^q g = K_t^b L_b^{q+1} g$. □

4.2 THE MODEL PROBLEM IN HIGHER DIMENSIONS

We now generalize these results and formulæ to the higher dimensional model operators $L_{\boldsymbol{b},m}$. Using the multiplicative nature of heat kernels, we can immediately write the model heat kernels in terms of the 1-dimensional ones,

$$k_t^{\boldsymbol{b},m}(\boldsymbol{x}, \boldsymbol{y}, \widetilde{\boldsymbol{x}}, \widetilde{\boldsymbol{y}}) = k_t^{\boldsymbol{b}}(\boldsymbol{x}, \widetilde{\boldsymbol{x}}) k_t^{\mathrm{euc},m}(\boldsymbol{y}, \widetilde{\boldsymbol{y}}), \tag{4.21}$$

where

$$k_t^{\boldsymbol{b}}(\boldsymbol{x}, \widetilde{\boldsymbol{x}}) = \prod_{i=1}^n k_t^{b_i}(x_i, \tilde{x}_i), \quad k_t^{\mathrm{euc},m}(\boldsymbol{y}, \widetilde{\boldsymbol{y}}) = \frac{1}{(4\pi t)^{\frac{m}{2}}} e^{-\frac{|\boldsymbol{y}-\widetilde{\boldsymbol{y}}|^2}{4t}}. \tag{4.22}$$

Note that this makes sense, even when $b_i = 0$ for some indices. This is because there is, at most, one δ-factor in each coordinate. The general problem is

$$\begin{cases} \partial_t w - L_{\boldsymbol{b},m} w = g & \text{on } S_{n,m} \times (0, T] \\ w(\boldsymbol{x}, \boldsymbol{y}, 0) = f(\boldsymbol{x}, \boldsymbol{y}), & (\boldsymbol{x}, \boldsymbol{y}) \in S_{n,m}. \end{cases} \tag{4.23}$$

Uniqueness of moderate growth solutions with appropriate regularity and moderate growth data is then given by Corollary 3.3.2, and this solution has the integral representation

$$\begin{aligned} w(\boldsymbol{x}, \boldsymbol{y}, t) = \int_{S_{n,m}} \int_0^t k_{t-s}^{\boldsymbol{b},m}(\boldsymbol{x}, \boldsymbol{y}, \widetilde{\boldsymbol{x}}, \widetilde{\boldsymbol{y}}) g(\widetilde{\boldsymbol{x}}, \widetilde{\boldsymbol{y}}, s)\, d\widetilde{\boldsymbol{x}} d\widetilde{\boldsymbol{y}} ds \\ + \int_{S_{n,m}} k_t^{\boldsymbol{b},m}(\boldsymbol{x}, \boldsymbol{y}, \widetilde{\boldsymbol{x}}, \widetilde{\boldsymbol{y}}) f(\widetilde{\boldsymbol{x}}, \widetilde{\boldsymbol{y}})\, d\widetilde{\boldsymbol{x}} d\widetilde{\boldsymbol{y}}. \end{aligned} \tag{4.24}$$

As before we discuss the homogeneous ($g = 0$) and inhomogeneous ($f = 0$) problems separately, analyzing regularity in the elementary spaces $\mathscr{C}^\ell(S_{n,m})$ and $\mathscr{C}^{\ell,\frac{\ell}{2}}(S_{n,m} \times [0, T])$, which are defined as the completions of the spaces $\mathscr{C}_c^\infty(S_{n,m})$ and $\mathscr{C}_c^\infty(S_{n,m} \times [0, T])$ with respect to the norms

$$||f||_\ell = \max_{|\boldsymbol{\alpha}|+|\boldsymbol{\beta}|\leq \ell} \|\partial_{\boldsymbol{x}}^{\boldsymbol{\alpha}} \partial_{\boldsymbol{y}}^{\boldsymbol{\beta}} f\|_{L^\infty(S_{n,m})} \quad \text{and}$$

$$||g||_{\ell,\frac{\ell}{2}} = \max_{2j+|\boldsymbol{\alpha}|+|\boldsymbol{\beta}|\leq \ell} \|\partial_t^j \partial_{\boldsymbol{x}}^{\boldsymbol{\alpha}} \partial_{\boldsymbol{y}}^{\boldsymbol{\beta}} g\|_{L^\infty(S_{n,m}\times[0,T])}, \tag{4.25}$$

respectively.

The following result will be helpful.

PROPOSITION 4.2.1. *Fix $\bar{\boldsymbol{b}} \in \mathbb{R}_+^n$ and $f \in \mathscr{C}^0(S_{n,m})$. For any $\boldsymbol{b} \in \mathbb{R}_+^n$, let $v^{\boldsymbol{b}}$ be the unique moderate growth solution of* (4.23) *with Cauchy data f (and with $g = 0$). Then $\lim_{\boldsymbol{b}\to\bar{\boldsymbol{b}}} v^{\boldsymbol{b}} = v^{\bar{\boldsymbol{b}}}$ in $\mathscr{C}^0(S_{n,m} \times [0, T])$, for every $T > 0$.*

PROOF. Note that the result is trivial if all entries of $\bar{\boldsymbol{b}}$ are strictly positive since $k_t^{\boldsymbol{b},m}$ varies smoothly with $\boldsymbol{b} \in (0, \infty)^n$ and is uniformly integrable, so we assume that some entries of $\bar{\boldsymbol{b}}$ vanish. This result is the multidimensional generalization of [15, Prop. 7.8], and we review the proof of this 1-dimensional case because we wish to use this same argument inductively. That proof both starts the induction and provides the inductive step.

So, first let $f \in \mathscr{C}^0(\mathbb{R}_+)$. If $f(0) = 0$, then we can approximate f uniformly by $f_\ell \in \mathscr{C}_c^0((0, \infty))$. Since k_t^b converges smoothly to k_t^0 away from $x = \tilde{x} = 0$, it is clear that $k_t^b f_\ell \to k_t^0 f_\ell$. Now estimate

$$|k_t^b f - k_t^0 f| \leq |k_t^b(f - f_\ell)| + |k_t^0(f - f_\ell)| + |(k_t^b - k_t^0) f_\ell|.$$

Given f, choose ℓ so that $\sup |f - f_\ell| < \epsilon$; by the maximum principle, the first and second terms are each less than ϵ. Then, for this ℓ, choose b sufficiently small so that the third term is less than ϵ too.

For arbitrary $f \in \mathscr{C}^0(\mathbb{R}_+)$, choose a smooth cutoff $\chi(x)$ which equals 1 for $x \leq 1$ and vanishes for $x \geq 2$, and write

$$f(x) = f(0) + \chi(x)(f(x) - f(0)) + (1 - \chi(x))(f(x) - f(0)).$$

Applying k_t^b to this sum, then $k_t^b f(0) = f(0)$ for all b, and the other two terms vanish at zero, so we may apply the previous reasoning to each of them. This proves the 1-dimensional case.

Now consider the higher dimensional case. For simplicity, assume that $b_n = 0$; write $\boldsymbol{b}' = (b_1, \ldots, b_{n-1})$ and $\bar{\boldsymbol{b}}' = (\bar{b}_1, \ldots, \bar{b}_{n-1})$, and also set $\boldsymbol{x}' = (x_1, \ldots, x_{n-1})$. Suppose that $f \in \mathscr{C}^0(S_{n,m})$. Decompose $f(\boldsymbol{x}, \boldsymbol{y})$ as in the 1-dimensional case, as

$$f(\boldsymbol{x}', 0, \boldsymbol{y}) + \chi(x_n)(f(\boldsymbol{x}, \boldsymbol{y}) - f(\boldsymbol{x}', 0, \boldsymbol{y})) + (1 - \chi(x_n))(f(\boldsymbol{x}, \boldsymbol{y}) - f(\boldsymbol{x}', 0, \boldsymbol{y})).$$

Since the first term is independent of x_n, we have

$$\int_{S_{n,m}} k_t^{\boldsymbol{b},m}(\boldsymbol{x}, \boldsymbol{y}, \widetilde{\boldsymbol{x}}, \widetilde{\boldsymbol{y}}) f(\widetilde{\boldsymbol{x}}', 0, \widetilde{\boldsymbol{y}})\, d\widetilde{\boldsymbol{x}} d\widetilde{\boldsymbol{y}} = \int_{S_{n-1,m}} k_t^{\boldsymbol{b}',m}(\boldsymbol{x}', \boldsymbol{y}', \widetilde{\boldsymbol{x}}', \widetilde{\boldsymbol{y}}') f(\widetilde{\boldsymbol{x}}', 0, \widetilde{\boldsymbol{y}})\, d\widetilde{\boldsymbol{x}}' d\widetilde{\boldsymbol{y}},$$

so we may apply the inductive hypothesis to see that this is continuous in $\boldsymbol{b}'$ up to $\bar{\boldsymbol{b}}'$. The third term is supported away from $x_n = 0$ already, so the result is clear for this term. Finally, for the second term, which we denote by f_2, we can argue exactly as in the 1-dimensional case, choosing a continuous function h_2 supported away from $x_n = 0$ and such that $\sup |f_2 - h_2| < \epsilon$. Then $|k_t^{\boldsymbol{b},m}(f_2 - h_2)| < \epsilon$ and $|k_t^{\bar{\boldsymbol{b}},m}(f_2 - h_2)| < \epsilon$, and we may choose $\boldsymbol{b}$ sufficiently close to $\bar{\boldsymbol{b}}$ so that $\sup |(k_t^{\boldsymbol{b},m} - k_t^{\bar{\boldsymbol{b}},m})h_2| < \epsilon$ too. □

Remark. In cases where some of the $\{b_j\}$ vanish, the solution kernel is quite a bit more complicated than when all the $b_j > 0$. Suppose $k < n$ and that $b_1 = \cdots = b_k = 0$, but $b_j > 0$ for $k = k + 1, \ldots, n$. The heat kernel for $L_{\boldsymbol{b},0}$ takes the form

$$k_t^{\boldsymbol{b}}(\boldsymbol{x}, \widetilde{\boldsymbol{x}}) = \prod_{j=1}^{k} [k_t^{0,D}(x_j, \widetilde{x}_j) + e^{-\frac{x_j}{t}} \delta_0(\widetilde{x}_j)] \prod_{j=k+1}^{n} k_t^{b_j}(x_j, \widetilde{x}_j). \tag{4.26}$$

If $H_j = \{\widetilde{x}_j = 0\}$, then this kernel has a δ-distribution on the incoming boundary strata

$$\{H_{i_1} \cap \cdots \cap H_{i_l} : \forall 1 \leq l \leq k, \text{ and sequences: } 1 \leq i_1 < \cdots < i_l \leq k\}.$$

A similar result is given in [37, 38] for the case of the Fleming-Viot operator defined on the simplex.

The analogue of Lemma 4.1.1 is

PROPOSITION 4.2.2. *For $\ell \in \mathbb{N}$, if $f \in \mathscr{C}^{\ell}(S_{n,m})$, then $v \in \mathscr{C}^{\ell,\frac{\ell}{2}}(S_{n,m} \times [0,\infty))$. For $j \in \mathbb{N}_0$ and any multi-index of non-negative integers $\boldsymbol{\alpha}$ and $\boldsymbol{\beta}$, if $2j+|\boldsymbol{\alpha}|+|\boldsymbol{\beta}| \leq \ell$, then*

$$\partial_t^j \partial_{\boldsymbol{x}}^{\boldsymbol{\alpha}} \partial_{\boldsymbol{y}}^{\boldsymbol{\beta}} v = \int_{S_{n,m}} k_t^{\boldsymbol{b}+\boldsymbol{\alpha},m}(\boldsymbol{x},\boldsymbol{y},\widetilde{\boldsymbol{x}},\widetilde{\boldsymbol{y}}) L_{\boldsymbol{b}+\boldsymbol{\alpha},m}^j \partial_{\widetilde{\boldsymbol{x}}}^{\boldsymbol{\alpha}} \partial_{\widetilde{\boldsymbol{y}}}^{\boldsymbol{\beta}} f(\widetilde{\boldsymbol{x}},\widetilde{\boldsymbol{y}})\, d\widetilde{\boldsymbol{x}} d\widetilde{\boldsymbol{y}}. \tag{4.27}$$

PROOF. We assume that all entries $b_i > 0$, since if some $b_i = 0$ then we can prove the result for an approximating sequence $\boldsymbol{b}^{(j)} \to \boldsymbol{b}$ with all $b_i^{(j)} > 0$ and then apply the previous proposition.

For $t > 0$ (and all $b_i > 0$) the kernel is smooth in $(t, \boldsymbol{x}, \boldsymbol{y})$ and we can differentiate under the integral sign. Using (4.8) and [15, Cor. 7.4], we obtain (4.27) with $j = 0$.

The argument needed to handle the ∂_t derivatives is slightly more delicate. We claim that

$$\partial_t k_t^b(x,\tilde{x}) = (x\partial_x^2 + b\partial_x)k_t^b(x,\tilde{x}) = (\partial_{\tilde{x}}^2 \tilde{x} - b\partial_{\tilde{x}})k_t^b(x,\tilde{x}), \tag{4.28}$$

but some care is needed because $\partial_{\tilde{x}}^2 \tilde{x} k_t^b$ and $\partial_{\tilde{x}} k_t^b$ are not integrable at 0 when $b \leq 1$. We overcome this issue with the following lemma.

LEMMA 4.2.3. *If $f \in \mathscr{C}^2(S_{n,m})$, then*

$$L_{b_i,x_i} \int_{S_{n,m}} k_t^{\boldsymbol{b},m}(\boldsymbol{x},\boldsymbol{y},\widetilde{\boldsymbol{x}},\widetilde{\boldsymbol{y}}) f(\widetilde{\boldsymbol{x}},\widetilde{\boldsymbol{y}})\, d\widetilde{\boldsymbol{x}} d\widetilde{\boldsymbol{y}} =$$

$$\int_{S_{n,m}} k_t^{\boldsymbol{b},m}(\boldsymbol{x},\boldsymbol{y},\widetilde{\boldsymbol{x}},\widetilde{\boldsymbol{y}}) L_{b_i,\tilde{x}_i} f(\widetilde{\boldsymbol{x}},\widetilde{\boldsymbol{y}})\, d\widetilde{\boldsymbol{x}} d\widetilde{\boldsymbol{y}}. \tag{4.29}$$

PROOF. To simplify notation, relabel so that $i = 1$, and write $S_{n-1,m} = \mathbb{R}_+^{n-1} \times \mathbb{R}^m$. First assume that $b_1 > 0$. If $f \in L^\infty(\mathbb{R}_+)$, then

$$\int_0^\infty L_{b_1,x_1} k_t^{b_1}(x_1,\tilde{x}_1) f(\tilde{x}_1)\, d\tilde{x}_1 \text{ and } \int_0^\infty \partial_t k_t^{b_1}(x_1,\tilde{x}_1) f(\tilde{x}_1)\, dy_1 \tag{4.30}$$

are both absolutely integrable when $t > 0$. Thus using (4.28), we re-express

$$L_{b_i,x_i} \int_{S_{n,m}} k_t^{\boldsymbol{b},m}(\boldsymbol{x},\boldsymbol{y},\widetilde{\boldsymbol{x}},\widetilde{\boldsymbol{y}}) f(\widetilde{\boldsymbol{x}},\widetilde{\boldsymbol{y}})\, d\widetilde{\boldsymbol{x}} d\widetilde{\boldsymbol{y}} =$$

$$\lim_{\epsilon \to 0^+} \int_\epsilon^\infty L_{b_1,\tilde{x}_1}^t k_t^{b_1}(x_1,\tilde{x}_1) \int_{S_{n-1,m}} \prod_{i=2}^n k_t^{b_i}(x_i,\tilde{x}_i) k_t^{\text{euc},m}(\boldsymbol{y},\widetilde{\boldsymbol{y}}) f(\widetilde{\boldsymbol{x}},\widetilde{\boldsymbol{y}})\, d\widetilde{\boldsymbol{x}} d\widetilde{\boldsymbol{y}}. \tag{4.31}$$

Since $f \in \mathscr{C}^2(S_{n,m})$ and $\lim_{\tilde{x}_1 \to 0^+} (\partial_{\tilde{x}_1} \tilde{x}_1 - b_1) k_t^{b_1}(x_1,\tilde{x}_1) = 0$, we can integrate by parts in the $\tilde{x}_1$-integral and let $\epsilon \to 0$ to obtain (4.29). The case $b_1 = 0$ follows by applying Proposition 4.2.1. $\square$

The assertion that

$$\partial_t \int\limits_{S_{n,m}} k_t^{b,m}(x, y, \widetilde{x}, \widetilde{y}) f(\widetilde{x}, \widetilde{y})\, d\widetilde{x} d\widetilde{y} = \int_{S_{n,m}} k_t^{b,m}(x, y, \widetilde{x}, \widetilde{y}) L_{b,m} f(\widetilde{x}, \widetilde{y})\, d\widetilde{x} d\widetilde{y} \quad (4.32)$$

follows directly from this lemma.

The fact that $v \in \mathscr{C}^{\ell,\frac{\ell}{2}}$ if $f \in \mathscr{C}^{\ell}$ then follows from these formulæ and the more elementary fact that if $f \in \mathscr{C}^0(S_{n,m})$, then $v \in \mathscr{C}^0(S_{n,m} \times [0, \infty))$. □

Finally consider the inhomogeneous problem

$$(\partial_t - L_{b,m})u = g, \qquad u(x, y, 0) = 0, \quad (4.33)$$

with solution given by (4.23). This is treated just as before. Let

$$G(\tau, s, x, y) = \int_{S_{n,m}} k_\tau^{b,m}(x, y, \widetilde{x}, \widetilde{y}) g(\widetilde{x}, \widetilde{y}, s)\, d\widetilde{x} d\widetilde{y}, \quad (4.34)$$

and observe that

$$u(x, y, t) = \lim_{\epsilon \to 0^+} \int\limits_0^{t-\epsilon} G(t - s, s, x, y)\, ds. \quad (4.35)$$

The following result is an immediate consequence of Proposition 4.2.2 and (4.35):

PROPOSITION 4.2.4. *If $g \in \mathscr{C}^{\ell,\frac{\ell}{2}}(S_{n,m} \times [0, T])$ for some $\ell \in \mathbb{N}$, then the unique bounded solution u to* (4.33) *is given by* (4.35) *and lies in $\mathscr{C}^{\ell,\frac{\ell}{2}}(S_{n,m} \times [0, T])$. If $2j + |\alpha| + |\beta| \leq l$, then*

$$\partial_t^j \partial_x^\alpha \partial_y^\beta u = \int_{S_{n,m}} k_t^{b+\alpha,m}(x, y, \widetilde{x}, \widetilde{y})(L_{b+\alpha,m})^j \partial_{\widetilde{x}}^\alpha \partial_{\widetilde{y}}^\beta g(\widetilde{x}, \widetilde{y}, s)\, d\widetilde{x} d\widetilde{y} + \sum_{r=0}^{j-1} \partial_t^{j-r-1} (L_{b+\alpha,m})^r \partial_y^\beta \partial_x^\alpha g. \quad (4.36)$$

We prove one final result, concerning the behavior at spatial infinity of solutions corresponding to compactly supported data f, g.

PROPOSITION 4.2.5. *Let w be given by* (4.23) *with data $f \in \mathscr{C}_c^0(S_{n,m})$ and $g \in \mathscr{C}_c^0(S_{n,m} \times [0, T])$. For any sets of non-negative integers α, β and N,*

$$\begin{aligned} \lim_{|(x,y)| \to \infty} (1 + |(x, y)|)^N |\partial_x^\alpha \partial_y^\beta v(x, y, t)| &= 0 \\ \lim_{|(x,y)| \to \infty} (1 + |(x, y)|)^N |\partial_x^\alpha \partial_y^\beta u(x, y, t)| &= 0 \end{aligned} \quad (4.37)$$

for each $t > 0$.

PROOF. This follows easily from the fact that the singularities of these kernels are located on the diagonal at $t = 0$, and from their exponential rates of decay at spatial infinity. If the incoming variables $\widetilde{x}, \widetilde{y}$ are confined to a compact set, this decay is uniform for $t \in [0, T]$ for any $T < \infty$. □

4.3 HOLOMORPHIC EXTENSION

The kernel functions $k_t^b(x, y)$ extend to be analytic for t lying in the right half plane $\operatorname{Re} t > 0$. By the permanence of functional relations, the functional equation

$$\int_0^\infty k_t^b(x, y) k_s^b(y, z) dy = k_{t+s}^b(x, y) \tag{4.38}$$

holds for s and t in this half plane. Therefore the solution to the homogeneous Cauchy problem,

$$v(x, t) = \int_0^\infty k_t^b(x, y) f(y) dy, \tag{4.39}$$

is analytic in $\operatorname{Re} t > 0$, and satisfies

$$\partial_t v - L_b v = 0 \text{ where } \operatorname{Re} t > 0, \tag{4.40}$$

where ∂_t is the complex derivative.

If we let $t = \tau e^{i\theta}$, where $\tau \in (0, \infty)$ and $|\theta| < \frac{\pi}{2}$, then

$$k_{\tau e^{i\theta}}^b(x, y) = \frac{y^{b-1} e^{-ib\theta}}{\tau^b} e^{-\frac{(x+y)e^{-i\theta}}{\tau}} \psi_b\left(\frac{xye^{-2i\theta}}{\tau^2}\right). \tag{4.41}$$

For any $\alpha > 0$, the asymptotic expansion

$$\psi_b(z) \sim \frac{z^{\frac{1}{4}-\frac{b}{2}} e^{2\sqrt{z}}}{\sqrt{4\pi}} \left(1 + \sum_{j=1}^\infty \frac{c_{b,j}}{z^{\frac{j}{2}}}\right) \tag{4.42}$$

holds uniformly for $|\arg z| \leq \pi - \alpha$. This shows that the kernel has an asymptotic expansion:

$$k_{\tau e^{i\theta}}^b(x, y) \sim \frac{e^{-\frac{i\theta}{2}}}{\tau^b \sqrt{y}} \left(\frac{y}{x}\right)^{\frac{b}{2}-\frac{1}{4}} e^{-\frac{(\sqrt{x}-\sqrt{y})^2 e^{-i\theta}}{\tau}} \left(1 + \sum_{j=1}^\infty c_{b,j} \frac{\tau^j e^{ij\theta}}{(xy)^{\frac{j}{2}}}\right). \tag{4.43}$$

This explicit expression shows that the qualitative behavior of this kernel, as $\tau \to 0$, is uniform in sectors

$$S_\phi = \{t : |\arg t| < \frac{\pi}{2} - \phi\}, \tag{4.44}$$

for any $\phi > 0$. Moreover, if $f \in \mathscr{C}_b^0([0,\infty))$, then

$$\lim_{\substack{t\to 0\\ t\in S_\phi}} v(x,t) = f(x) \tag{4.45}$$

uniformly in the $\mathscr{C}^0$-topology.

In the sequel we shall also be analyzing the resolvent $R(\mu) = (L_b - \mu)^{-1}$. If $\mu \in (0,\infty)$ and $f \in \mathscr{C}_b^0([0,\infty))$, then $R(\mu)f$ is defined by expression

$$R(\mu)f = \lim_{\epsilon\to 0+} \int_{\epsilon}^{\frac{1}{\epsilon}} \left(\int_0^\infty k_t^b(x,y) f(y)\, dy \right) e^{-\mu t}\, dt. \tag{4.46}$$

Using the asymptotics of k_t^b, we can apply Morera's theorem to show that for each fixed $x \in \mathbb{R}_+$, $R(\mu)f(x)$ is an analytic function of μ in the right half plane. Applying L_b to the integral on the right and integrating by parts, and using estimates proved below, we show that if f is Hölder continuous and bounded on $[0,\infty)$, then

$$(\mu - L_b)R(\mu)f = f. \tag{4.47}$$

Using Cauchy's theorem and the asymptotics of the heat kernel, the contour for the t-integral can be deformed to show that for $\mu \in (0,\infty)$ and $|\arg\theta| < \frac{\pi}{2}$ we also have

$$R(\mu)f = \lim_{\epsilon\to 0+} \int_{\epsilon}^{\frac{1}{\epsilon}} \left(\int_0^\infty k_{\tau e^{i\theta}}^b(x,y) f(y)\, dy \right) e^{-\mu\tau e^{i\theta}} e^{i\theta}\, d\tau. \tag{4.48}$$

This expression is analytic in μ in the region where $\operatorname{Re}(\mu e^{i\theta}) > 0$, and hence $R(\mu)f$ extends analytically to $\mathbb{C} \setminus (-\infty, 0]$, and this extension satisfies (4.47).

For $f \in \mathscr{C}^0$, we show that the following limits exist locally uniformly for $x \in (0,\infty)$:

$$\begin{gathered}\lim_{\epsilon\to 0+} \int_{\epsilon}^{\frac{1}{\epsilon}} \int_0^\infty \partial_x k_{\tau e^{i\theta}}^b(x,y) f(y) dy e^{-\mu\tau e^{i\theta}} e^{i\theta} d\tau\\ \lim_{\epsilon\to 0+} \int_{\epsilon}^{\frac{1}{\epsilon}} \int_0^\infty x\partial_x^2 k_{\tau e^{i\theta}}^b(x,y) f(y) dy e^{-\mu\tau e^{i\theta}} e^{i\theta} d\tau.\end{gathered} \tag{4.49}$$

This demonstrates the $R(\mu)f$ is twice differentiable in $(0,\infty)$. If f is only in $\mathscr{C}^0([0,\infty))$, then the limits $\lim_{x\to 0^+} \partial_x R(\mu)f(x)$, and $\lim_{x\to 0^+} x\partial_x^2 R(\mu)f(x) = 0$ may not exist.

For $|\theta| < \frac{\pi}{2}$ and $\epsilon > 0$, we let

$$\Gamma_{\theta,\epsilon} = \{te^{i\theta} : t \in [\epsilon, \epsilon^{-1}]\}. \tag{4.50}$$

For $t \in S_0$ the kernel function satisfies the equation

$$\partial_t k_t^b(x, y) = L_b k_t^b(x, y), \tag{4.51}$$

where ∂_t is the complex derivative. For $\epsilon > 0$ we see that

$$L_b \int\limits_{\Gamma_{\theta,\epsilon}} \int\limits_0^\infty k_t^b(x, y) f(y) dy e^{-\mu t} dy dt = \int\limits_{\Gamma_{\theta,\epsilon}} \int\limits_0^\infty \partial_t k_t^b(x, y) f(y) dy e^{-\mu t} dy dt. \tag{4.52}$$

For $\epsilon > 0$ we can interchange the order of the integrations and integrate by parts to obtain:

$$L_b \int\limits_{\Gamma_{\theta,\epsilon}} \int\limits_0^\infty k_t^b(x, y) f(y) dy e^{-\mu t} dy dt = \mu \int\limits_{\Gamma_{\theta,\epsilon}} \int\limits_0^\infty k_t^b(x, y) f(y) dy e^{-\mu t} dy dt +$$
$$\int\limits_0^\infty k^b_{\epsilon^{-1} e^{i\theta}}(x, y) f(y) dy e^{-\mu \epsilon^{-1} e^{i\theta}} dy - \int\limits_0^\infty k^b_{\epsilon e^{i\theta}}(x, y) f(y) dy e^{-\mu \epsilon e^{i\theta}} dy. \tag{4.53}$$

Provided that $\text{Re}(e^{i\theta}\mu) > 0$, and $f \in \mathscr{C}^0(\mathbb{R}_+)$, we can let $\epsilon \to 0^+$ to obtain that

$$(\mu - L_b) R(\mu) f = f. \tag{4.54}$$

We summarize these observations as a proposition:

PROPOSITION 4.3.1. *The solution to the homogeneous Cauchy problem* $(\partial_t - L_b)v = 0$ *with* $v(x, 0) = f(x) \in \mathscr{C}_b^0([0, \infty))$ *extends to an analytic function of* t *with* $\text{Re}\, t > 0$. *The resolvent operator* $R(\mu)$ *is analytic in the complement of* $(-\infty, 0]$, *and is given by the integral in* (4.48) *provided that* $\text{Re}(\mu e^{i\theta}) > 0$. *Moreover,* $(\mu - L_b) R(\mu) f = f$, *for* $\mu \in \mathbb{C} \setminus (-\infty, 0]$.

From the corresponding fact for its 1-dimensional factors, it is obvious that the kernel $k_t^{\boldsymbol{b},m}(\boldsymbol{x}, \boldsymbol{y}, \widetilde{\boldsymbol{x}}, \widetilde{\boldsymbol{y}})$ extends analytically to $\text{Re}\, t > 0$ and hence the solution $v^{\boldsymbol{b}}(\boldsymbol{x}, \boldsymbol{y}, t)$ does as well. Indeed, if $f \in \mathscr{C}_b^0(S_{n,m})$ then $t \mapsto v^{\boldsymbol{b}}(\cdot, \cdot, t)$ is an analytic function from the right half plane with values in $\mathscr{C}_b^\infty(\text{int}\, S_{n,m}) \cap \mathscr{C}_b^0(S_{n,m})$. The asymptotic formula (4.43) and the standard asymptotics for the Euclidean heat kernel then give that for any $\phi > 0$,

$$\lim_{t \to 0 \text{ in } S_\phi} v^{\boldsymbol{b}}(\cdot, \cdot, t) = f, \tag{4.55}$$

in the uniform topology.

For $f \in \mathscr{C}_b^0(S_{n,m})$ and $\text{Re}\, \mu > 0$, the Laplace transform is defined by the limit

$$R(\mu) f(\boldsymbol{x}, \boldsymbol{y}) = \lim_{\epsilon \to 0^+} \int\limits_\epsilon^{\frac{1}{\epsilon}} v^{\boldsymbol{b}}(\boldsymbol{x}, \boldsymbol{y}, t) e^{-\mu t}\, dt. \tag{4.56}$$

Assuming that $f \in \mathscr{C}^0_{WF}(S_{n,m})$ then again using the analyticity and asymptotic behavior of the kernel, we can use Cauchy's theorem to deform the contour of integration in (4.56). For $|\theta| < \frac{\pi}{2}$ and $\mu \in (0, \infty)$, we have that

$$R(\mu) f(\boldsymbol{x}, \boldsymbol{y}) = \lim_{\epsilon \to 0+} \int_{\epsilon}^{\frac{1}{\epsilon}} v^{\boldsymbol{b}}(\boldsymbol{x}, \boldsymbol{y}, \tau e^{i\theta}) e^{-\mu \tau e^{i\theta}} e^{i\theta} \, d\tau. \tag{4.57}$$

The expression in (4.57) defines an analytic function of μ where $\operatorname{Re} \mu e^{i\theta} > 0$. This in turn shows that $R(\mu) f$ has an analytic continuation to $\mathbb{C} \setminus (-\infty, 0]$.

In order to establish the identity

$$(\mu - L_{\boldsymbol{b},m}) R(\mu) f = f \tag{4.58}$$

in the higher dimensional case, it is simpler to assume that f is Hölder continuous. Specifically, in the next chapter we shall define Hölder spaces $\mathscr{C}^{0,\gamma}_{WF}(S_{n,m})$ which are specially adapted to this problem. For such data one can show that the individual terms in $L_{\boldsymbol{b},m} R(\mu) f$ are continuous on $S_{n,m}$, and satisfy (4.58). If f is only continuous, then arguing as before, one can show that the limit, as $\epsilon \to 0^+$ of

$$L_{\boldsymbol{b},m} \int_{\epsilon}^{\frac{1}{\epsilon}} v^{\boldsymbol{b}}(\boldsymbol{x}, \boldsymbol{y}, \tau e^{i\theta}) e^{-\mu \tau e^{i\theta}} e^{i\theta} \, d\tau, \tag{4.59}$$

exists in $\mathscr{C}^0(S_{n,m})$. It satisfies the identity $(\mu - L_{\boldsymbol{b},m}) R(\mu) f = f$ in the $\mathscr{C}^0$-graph closure sense. Generally the individual terms of $L_{\boldsymbol{b},m} R(\mu) f$ are not defined as $(\boldsymbol{x}, \boldsymbol{y})$ approaches the boundary of $S_{n,m}$.

We summarize these results in a proposition.

PROPOSITION 4.3.2. *The solution v to the homogeneous Cauchy problem, with initial data $v(\boldsymbol{x}, \boldsymbol{y}, 0) = f(\boldsymbol{x}, \boldsymbol{y}) \in \mathscr{C}^0_b(S_{n,m})$, extends to an analytic function of t with* $\operatorname{Re} t > 0$. *The resolvent operator $R(\mu)$ is analytic in the complement of $(-\infty, 0]$, and is given by the integral in* (4.57) *provided that* $\operatorname{Re}(\mu e^{i\theta}) > 0$. *If $f \in \mathscr{C}^0_{WF}(S_{n,m})$, then $(\mu - L_{\boldsymbol{b},m}) R(\mu) f = f$, for $\mu \in \mathbb{C} \setminus (-\infty, 0]$.*

4.4 FIRST STEPS TOWARD PERTURBATION THEORY

Our primary goal in this monograph is to construct the solution operator $\mathfrak{Q}^t$ for a general Kimura operator $\partial_t - L$ and to use it to study properties of the associated semi-group on various function spaces. This is done by perturbing an approximate solution obtained by patching together the solution operators $\mathfrak{Q}^t_{\boldsymbol{b},m}$ for the models $\partial_t - L_{\boldsymbol{b},m}$ associated to the normal forms of L in various coordinate charts. This strategy works very well in the 1-dimensional problem considered in [15], but turns out to be substantially more complicated in higher dimensions. We explain this now.

The relative simplicity of this method for operators in 1-dimension is not hard to explain. As already pointed out in Chapter 2.2, the normal form for the second order part of a 1-dimensional Kimura operator is *exactly* $x\partial_x^2$ in a full neighborhood of a boundary point. Thus we can choose coordinates and a constant $b \geq 0$ so that $L = L_b + W$, where W is a vector field which vanishes at the boundary point. Hence the error term incurred by using $\mathcal{Q}_b^t$ as an approximate solution operator is $W \circ \mathcal{Q}_b^t$. In an appropriate sense, this operator is smoothing of order 1, and restricted to data on suitably small time intervals $[0, T]$, it also has small norm acting on continuous functions. Hence it is easy to solve away this error term using a convergent Neumann series.

When carrying out the same procedure in higher dimensions, the difference between L and any one of the models $L_{\boldsymbol{b},m}$ is unavoidably second order. Hence the error term E incurred by applying $\partial_t - L$ to a parametrix formed by patching together these model heat kernels is no longer smoothing, since it is the result of applying a differential operator of order 2 to an operator which is smoothing of order 2. Even worse, this error is *not* bounded on $\mathcal{C}^0$. This is a well-known fact in classical potential theory, that $\mathcal{C}^0$ and higher $\mathcal{C}^k$ spaces are ill-suited for the study of regularity properties of elliptic and parabolic problems in higher dimensions, and that one should use Hölder spaces instead.

The applications of these Kimura diffusions in probability and biology demand that we study the semi-group for L on $\mathcal{C}^0$. This leaves us with a slightly unsatisfactory state of affairs. We are only able to construct the solution operator for $\partial_t - L$ on a suitable scale of Hölder spaces. We can still prove the existence and many properties of the semi-group on $\mathcal{C}^0$, but this must be done in an indirect fashion.

Chapter Five

Degenerate Hölder Spaces

The starting point to implement this perturbation theory is a description of the various function spaces we shall be using. As described above, we seek function spaces on the domain P for which the diffusion associated to a general Kimura diffusion operator L is well posed. More pragmatically, we wish to define spaces on which one can prove analogues of the standard parabolic Schauder estimates, so that we can pass from the model to more general operators. This chapter is devoted to a description of the various spaces on which this is possible, and to an explanation of the relationships between them.

Two familiar guiding principles when choosing the right function spaces for a problem are that one should choose spaces which respect the natural scaling properties of the operator, and in addition, that these spaces should be based on the geometry of an associated metric. In the classical setting, the operator $\partial_t - \sum \partial_{y_j}^2$ on $\mathbb{R}_+ \times \mathbb{R}^m$ is homogeneous with respect to the parabolic dilations $(t, y) \mapsto (\lambda^2 t, \lambda y)$, and is naturally associated to the Euclidean metric. The first of these principles indicates that t and y derivatives should be weighted differently; the second suggests that we formulate mapping properties in terms of Hölder spaces with semi-norms defined using the Euclidean distance function. This is indeed the case, and we review the definitions of the standard parabolic Hölder spaces below. Other examples where these principles are applied include [11, 12, 28, 23, 24] and [32].

To apply the same two principles in the present setting, we observe that $\partial_t - L_{\boldsymbol{b},m}$ is homogeneous with respect to the slightly different scaling,

$$(t, \boldsymbol{x}, \boldsymbol{y}) \mapsto (\lambda t, \lambda \boldsymbol{x}, \sqrt{\lambda}\, \boldsymbol{y}),$$

which indicates that derivatives with respect to t and the x_i should be twice as strong as derivatives with respect to the y_j. On the other hand, when all x_j are strictly positive, we must revert to the standard scaling corresponding to the interior problem. In other words, whatever function spaces we use must incorporate both types of homogeneity. The metric naturally associated to $L_{\boldsymbol{b},m}$ is

$$ds^2 = \sum_{i=1}^{n} \frac{dx_i^2}{x_i} + \sum_{j=1}^{m} dy_j^2; \tag{5.1}$$

note that this metric is homogeneous with respect to $(\boldsymbol{x}, \boldsymbol{y}) \mapsto (\lambda \boldsymbol{x}, \sqrt{\lambda}\, \boldsymbol{y})$, and that the associated Laplacian is simply $L_{\boldsymbol{b},m}$ with $\boldsymbol{b} = (\frac{1}{2}, \ldots, \frac{1}{2})$.

Before embarking on the many definitions below, we make two remarks. First, the basic definition of a Hölder semi-norm with respect to a given metric g is

$$[u]_{g;0,\gamma} := \sup_{z\neq z'} \frac{|u(z)-u(z')|}{d_g(z,z')^\gamma}, \tag{5.2}$$

where d_g is the Riemannian distance between the two points. It is very useful to observe that instead of taking the supremum over all distinct pairs z, z', it suffices to take the supremum only over pairs with $z \neq z'$ and $d_g(z, z') \leq 1$. This is simply because if $d_g(z, z') > 1$, then the quotient on the right, evaluated at z, z', is bounded by $2 \sup |u|$. For this reason, we introduce the notation

$$\sup_{z\neq z'}^{1} \equiv \sup_{\{z\neq z' : d(z,z')<1\}}, \tag{5.3}$$

which will be used throughout the rest of this book. This makes the semi-norm monotonely increasing, as a function of γ. For functions depending on both z and t, we use this same notation to denote the supremum over pairs $(z, t) \neq (z', t')$ with $d_g(z, z') + \sqrt{|t-t'|} \leq 1$.

Second, although our main focus is on generalized Kimura diffusion operators L on compact regions P, it is convenient from certain technical points of view to study the model operators $L_{\boldsymbol{b},m}$ on the unbounded region $\mathbb{R}_+^n \times \mathbb{R}^m$. In addition, there are some practical motivations for this since certain problems arising in biological applications actually occur on such unbounded orthants. We handle spatial infinity by defining appropriate Hölder norms and then taking spaces which are the completions of the subspaces of smooth compactly supported functions with respect to these norms. The functions obtained in this way must tend to zero at infinity along with an appropriate number of scaled derivatives. This requires us to check that the solution operators for these heat equations preserve this property. We denote the spaces obtained by this closure procedure with a superscript dot. Thus, for example, $\dot{\mathscr{C}}^0(\mathbb{R}^m)$ (resp., $\dot{\mathscr{C}}^0(\mathbb{R}^m \times [0, T])$) denotes the closure in $\mathscr{C}^0(\mathbb{R}^m)$ (resp., $\mathscr{C}^0(\mathbb{R}^m \times [0, T])$) of the subspace of compactly supported smooth functions. The space $\dot{\mathscr{C}}^0(\mathbb{R}^m)$ consists of continuous functions which tend to zero at infinity.

5.1 STANDARD HÖLDER SPACES

To be clear about notation and definitions, we briefly recall the classical interior Hölder spaces and their parabolically scaled "heat" analogues. All spaces here are subspaces of $\dot{\mathscr{C}}^0(\mathbb{R}^m)$ (resp., $\dot{\mathscr{C}}^0(\mathbb{R}^m \times [0, T])$). Here and in the remainder of the book γ denotes a number in the interval $(0, 1)$.

The space $\mathscr{C}^{0,\gamma}(\mathbb{R}^m)$ is the subspace of $\dot{\mathscr{C}}^0(\mathbb{R}^m)$ consisting of functions f for which the norm

$$\|f\|_{0,\gamma} := \|f\|_{L^\infty(\mathbb{R}^m)} + [f]_{0,\gamma} \tag{5.4}$$

is finite. Here

$$[f]_{0,\gamma} = \sup_{y\neq y'}^{1} \frac{|f(y)-f(y')|}{|y-y'|^\gamma} \tag{5.5}$$

is the Hölder semi-norm of order γ. Note that this is different from the so-called little Hölder space $\dot{\mathcal{C}}^{0,\gamma}(\mathbb{R}^m)$, which is the closure of the space of smooth functions with bounded supported in this Hölder norm and which consists of $\mathcal{C}^{0,\gamma}$ functions such that

$$\lim_{\delta\to 0} \sup_{|y-y'|<\delta} \frac{|f(y)-f(y')|}{|y-y'|^{\gamma}} = 0. \tag{5.6}$$

Similarly, a function $f \in \dot{\mathcal{C}}^k(\mathbb{R}^m)$ belongs to $\mathcal{C}^{k,\gamma}(\mathbb{R}^m)$ if the norm

$$\|f\|_{k,\gamma} = \|f\|_{\mathcal{C}^k(\mathbb{R}^m)} + \sup_{|\boldsymbol{\alpha}|=k} \left[\partial^{\alpha} f\right]_{0,\gamma} \tag{5.7}$$

is finite. (This sup is over multi-indices $\boldsymbol{\alpha} \in \mathbb{N}^m$, where $|\boldsymbol{\alpha}| = \alpha_1 + \ldots + \alpha_m$.)

Now consider functions which depend on both $\boldsymbol{y}$ and t. The heat Hölder spaces $\mathcal{C}^{0,\gamma}(\mathbb{R}^m \times [0,T])$ are defined as the set of functions $g \in \dot{\mathcal{C}}^0(\mathbb{R}^m \times [0,T])$ such that

$$\|g\|_{0,\gamma} := \|g\|_\infty + [g]_{0,\gamma} < \infty, \tag{5.8}$$

where now

$$[g]_{0,\gamma} = \sup^{1}_{(y,t)\neq(y',t')} \frac{|g(\boldsymbol{y},t)-g(\boldsymbol{y}',t')|}{[|\boldsymbol{y}-\boldsymbol{y}'|+\sqrt{|t-t'|}]^{\gamma}}. \tag{5.9}$$

Finally, letting $\dot{\mathcal{C}}^{k,\frac{k}{2}}(\mathbb{R}^m \times [0,T])$ denote the closure of $\mathcal{C}^\infty_c(\mathbb{R}^m \times [0,T])$ with respect to the norm

$$\|g\|_{k,\frac{k}{2}} := \sup_{|\boldsymbol{\alpha}|+2j\leq k} \|\partial_t^j \partial_{\boldsymbol{y}}^{\boldsymbol{\alpha}} g\|_\infty, \tag{5.10}$$

then $\mathcal{C}^{k,\gamma}(\mathbb{R}^m \times [0,T])$ consists of functions $g \in \dot{\mathcal{C}}^{k,\frac{k}{2}}(\mathbb{R}^m \times [0,T])$ such that

$$\|g\|_{k,\gamma} := \|g\|_{k,\frac{k}{2}} + \sup_{|\boldsymbol{\alpha}|+2j=k} \left[\partial_t^j \partial_{\boldsymbol{y}}^{\boldsymbol{\alpha}} g\right]_{0,\gamma} < \infty. \tag{5.11}$$

Note that in all these cases, the Euclidean metric appears through the quantity $|\boldsymbol{y}-\boldsymbol{y}'|$ (which is comparable to $d_{\text{euc}}(\boldsymbol{y},\boldsymbol{y}')$) and that the parabolic scaling is reflected not only in the definition of $\mathcal{C}^{k,\frac{k}{2}}$, but also by the quantity $|\boldsymbol{y}-\boldsymbol{y}'|+\sqrt{|t-t'|}$.

5.2 WF-HÖLDER SPACES IN 1-DIMENSION

We now turn to the definitions of the degenerate Hölder spaces associated to the 1-dimensional model operator L_b. As indicated above, one guide is the geometry on $\mathbb{R}_+$ with the incomplete metric

$$ds^2_{WF} = \frac{dx^2}{x}. \tag{5.12}$$

Note that the change coordinates $\xi = 2\sqrt{x}$ transforms this to the standard Euclidean metric $d\xi^2$, and that the model operator $L_{1/2}$ is simply the Laplacian ∂_ξ^2. This allows us to transform all the standard Hölder theory for functions of ξ (or ξ and t) to obtain the corresponding spaces and estimates for this particular operator $L_{1/2}$. As we eventually show, these spaces and estimates also adapt to the other operators L_b, although this requires more than a simple coordinate transformation to verify.

This identification makes certain basic geometric formulæ trivial to verify. We record these here, although they will not be used until a later chapter. First, the distance function has the explicit expression

$$\rho = d_{WF}(x_1, x_2) = 2|\sqrt{x_2} - \sqrt{x_1}|. \tag{5.13}$$

Next, the midpoint of the interval $[x_1, x_2]$ with respect to ds^2_{WF} is

$$\bar{x} = \frac{(\sqrt{x_1} + \sqrt{x_2})^2}{4}.$$

Finally, the WF-ball $B_\rho(\bar{x})$ centered at the point $\bar{x}$ and with radius ρ is the interval $[\alpha, \beta]$, where

$$\sqrt{\alpha} = \max\left\{0, \frac{3\sqrt{x_1} - \sqrt{x_2}}{2}\right\}, \quad \sqrt{\beta} = \frac{3\sqrt{x_2} - \sqrt{x_1}}{2}. \tag{5.14}$$

5.2.1 WF-Hölder Spaces on $\mathbb{R}_+$

We now proceed to the definitions of the associated function spaces. Following the dictum in the beginning of this chapter, the WF-Hölder semi-norm is given by

$$[f]_{WF,0,\gamma} = \sup_{x \neq x'} \frac{|f(x) - f(x')|}{|\sqrt{x} - \sqrt{x'}|^\gamma} = 2 \sup_{x \neq x'} \frac{|f(x) - f(x')|}{d_{WF}(x, x')^\gamma}. \tag{5.15}$$

Then $\mathscr{C}^{0,\gamma}_{WF}(\mathbb{R}_+)$ is the subspace of $\dot{\mathscr{C}}^0(\mathbb{R}_+)$ on which the norm

$$\|f\|_{WF,\gamma} = ||f||_\infty + [f]_{WF,0,\gamma} \tag{5.16}$$

is finite. This is clearly a Banach space. We also define $\mathscr{C}^{0,1}_{WF}(\mathbb{R}_+)$ to be the closure of $\mathscr{C}^1(\mathbb{R}_+)$ with respect to the norm:

$$\|f\|_{WF,0,1} = \|f\|_{L^\infty} + \|\sqrt{x}\partial_x f\|_{L^\infty}. \tag{5.17}$$

Note that if $f \in \mathscr{C}^{0,1}_{WF}(\mathbb{R}_+)$, then

$$\lim_{x \to 0^+} \sqrt{x}\partial_x f(x) = 0, \tag{5.18}$$

since this is true for every $f \in \mathscr{C}^1$. Moreover, integration gives

$$|f(x_2) - f(x_1)| \leq 2\|f\|_{WF,0,1}|\sqrt{x_2} - \sqrt{x_1}|, \tag{5.19}$$

and hence, for any $0 < \gamma < 1$, the inclusion

$$\mathscr{C}^{0,1}_{WF}(\mathbb{R}_+) \subset \mathscr{C}^{0,\gamma}_{WF}(\mathbb{R}_+) \tag{5.20}$$

is compact.

Two simple facts will be used repeatedly below. First, if $f \in \mathscr{C}^{0,\gamma}(\mathbb{R}_+)$, then directly from the definition,

$$|f(x) - f(x')| \leq 2\|f\|_{WF,\gamma} |\sqrt{x} - \sqrt{x'}|^{\gamma}. \tag{5.21}$$

Second, the basic inequality

$$|f(x)g(x) - f(x')g(x')| \leq |f(x)(g(x) - g(x'))| + |g(x')(f(x) - f(x'))|$$

implies that these Hölder semi-norms satisfy a standard Leibniz rule: $f, g \in \mathscr{C}^{0,\gamma}_{WF}(\mathbb{R}_+)$, then

$$[fg]_{WF,0,\gamma} \leq \|f\|_\infty [g]_{WF,0,\gamma} + \|g\|_\infty [f]_{WF,0,\gamma}, \tag{5.22}$$

(where $||\cdot||_\infty$ is the L^∞ norm).

There are in fact a couple of slightly different ways to define WF spaces which capture higher regularity. The ultimate goal is to capture the precise gain in regularity for elliptic and parabolic problems, which leads us to the various definitions below.

The first set of spaces is meant to capture the fact that if $L_b u = f$, then we wish to be able to estimate u, $\partial_x u$ and $x\partial_x^2 u$ separately in terms of f. Define $\dot{\mathscr{C}}^2_{WF}(\mathbb{R}_+)$ as the closure of $\mathscr{C}^2_c(\mathbb{R}_+)$ with respect to the norm:

$$\|f\|_{WF,2} = \|f\|_\infty + \|\partial_x f\|_\infty + \|x\partial_x^2 f\|_\infty, \tag{5.23}$$

and then let $\mathscr{C}^{0,2+\gamma}_{WF}(\mathbb{R}_+)$ be the subspace of $\dot{\mathscr{C}}^2_{WF}(\mathbb{R}_+)$ on which the norm

$$\|f\|_{WF,0,2+\gamma} = \|f\|_{WF,0,\gamma} + \|\partial_x f\|_{WF,0,\gamma} + \|x\partial_x^2 f\|_{WF,0,\gamma} \tag{5.24}$$

is finite.

As a matter of convention, we write $\mathbb{R}_+$ for the closed half-line $[0, \infty)$ and denote the open half-line by $(0, \infty)$. Clearly $\mathscr{C}^2_{WF}(\mathbb{R}_+) \subset \mathscr{C}^2((0, \infty)) \cap \mathscr{C}^1(\mathbb{R}_+)$. Furthermore, analogous to (5.6), since $\dot{\mathscr{C}}^2_{WF}(\mathbb{R}_+)$ is the closure of $\mathscr{C}^2_c(\mathbb{R}_+)$ with respect to (5.23), then for any $f \in \mathscr{C}^2_{WF}$,

$$\lim_{x \to 0^+} x\partial_x^2 f(x) = 0, \quad \lim_{x \to \infty} \Big(|f(x)| + |\partial_x f(x)| + |x\partial_x^2 f(x)|\Big) = 0. \tag{5.25}$$

The first assertion is an important part of the characterization of the domain of L_b on $\mathscr{C}^0$.

There is an elementary characterization of $\mathscr{C}^{0,2+\gamma}_{WF}$, which also gives a simple proof that it is a Banach space.

LEMMA 5.2.1. *Suppose that* $f \in \mathscr{C}^2((0, \infty)) \cap \mathscr{C}^1(\mathbb{R}_+)$ *satisfies* (5.25), *and that* $\|f\|_{WF,0,2+\gamma} < \infty$. *Then* $f \in \mathscr{C}^2_{WF}(\mathbb{R}_+)$.

PROOF. We must find a sequence $\{f_n\}$ in $\dot{\mathscr{C}}^2(\mathbb{R}_+)$ such that $||f_n - f||_{WF,0,2+\gamma} \to 0$. However, we know that

$$|x\partial_x^2 f(x)| \leq Mx^{\frac{\gamma}{2}} \tag{5.26}$$

for $x < 1$ and some $M < \infty$. Letting $f_n(x) = f(x + 1/n)$, then clearly $\|f_n - f\|_\infty + \|\partial_x(f_n - f)\|_\infty \to 0$.

It remains to show that

$$\lim_{n\to\infty} \|x\partial_x^2(f - f_n)\|_\infty = 0. \tag{5.27}$$

For any $0 < \delta < C$, the uniform convergence of these second derivatives on $[\delta, C]$ is obvious; in fact, by the second part of (5.25), there is also no difficulty as $x \to \infty$. Now observe that

$$|x\partial_x^2 f(x) - x\partial_x^2 f_n(x)| = \\ \left| x\partial_x^2 f(x) - \left(x + \frac{1}{n}\right)\partial_x^2 f\left(x + \frac{1}{n}\right) + \left(\frac{1}{n}\right)\partial_x^2 f\left(x + \frac{1}{n}\right)\right|.$$

Using (5.26) and the fact that $\|x\partial_x^2 f\|_{WF,0,\gamma} < \infty$, we see that

$$|x\partial_x^2(f(x) - f_n(x))| \leq \frac{1}{n^{\frac{\gamma}{2}}}\|x\partial_x^2 f\|_{WF,0,\gamma} + M\left(x + \frac{1}{n}\right)^{\frac{\gamma}{2}}, \tag{5.28}$$

which implies that $\|x\partial_x^2(f(x) - f_n(x))\|_\infty \to 0$ as $n \to \infty$. □

COROLLARY 5.2.2. *If $0 < \gamma < 1$, then the topological vector space $\mathscr{C}_{WF}^{0,2+\gamma}(\mathbb{R}_+)$ is a Banach space.*

PROOF. If f_n is a sequence in $\mathscr{C}_{WF}^{0,2+\gamma}(\mathbb{R}_+)$ which converges to some f in the $\mathscr{C}_{WF}^{0,2+\gamma}$-norm, then f satisfies the hypotheses of the previous lemma. This shows that f is the $\mathscr{C}_{WF}^2$-limit of a sequence of functions in $\dot{\mathscr{C}}^2([0,\infty))$ and hence $f \in \mathscr{C}_{WF}^{0,2+\gamma}(\mathbb{R}_+)$ as well. □

5.2.2 Parabolic WF-Hölder Spaces on $\mathbb{R}_+ \times [0, T]$

We now introduce the parabolic (or heat) WF-Hölder spaces $\mathscr{C}_{WF}^{0,\gamma}(\mathbb{R}_+ \times [0,T])$ and $\mathscr{C}_{WF}^{0,2+\gamma}(\mathbb{R}_+ \times [0,T])$. To define these, first let $\dot{\mathscr{C}}_{WF}^{2,1}(\mathbb{R}_+ \times [0,T])$ be the closure of $\mathscr{C}_c^{2,1}(\mathbb{R}_+ \times [0,T])$ with respect to the norm

$$\|g\|_{WF,0,2,1} = \|g\|_\infty + \|\partial_x g\|_\infty + \|\partial_t g\|_\infty + \|x\partial_x^2 g\|_\infty. \tag{5.29}$$

As before, if $g \in \dot{\mathscr{C}}_{WF}^{2,1}(\mathbb{R}_+ \times [0,T])$, then

$$\begin{gathered} g \in \mathscr{C}^1(\mathbb{R}_+ \times [0,T]) \cap \mathscr{C}^{2,1}((0,\infty) \times [0,T]), \\ \lim_{x\to 0^+} x\partial_x^2 g(x,t) = 0, \text{ and} \\ \lim_{x\to\infty} [|g(x,t)| + |\partial_x g(x,t)| + |\partial_t g(x,t)| + |x\partial_x^2 g(x,t)|] = 0,\ 0 \leq t \leq T. \end{gathered} \tag{5.30}$$

Next, define the WF semi-norm of order γ

$$[g]_{WF,0,\gamma} = \overset{1}{\sup_{(x,t)\neq(x',t')}} \frac{|g(x,t)-g(x',t')|}{(|\sqrt{x}-\sqrt{x'}|+\sqrt{|t-t'|})^{\gamma}};$$

this has a Leibniz formula,

$$[gh]_{WF,0,\gamma} \leq \|g\|_{\infty}\,[h]_{WF,0,\gamma} + \|h\|_{\infty}\,[g]_{WF,0,\gamma}, \tag{5.31}$$

and provides the constant in the estimate

$$|g(x,t)-g(x',t')| \leq 2\|g\|_{WF,0,\gamma}(|\sqrt{x}-\sqrt{x'}|+\sqrt{|t-t'|})^{\gamma}. \tag{5.32}$$

Finally,

$$\mathscr{C}^{0,\gamma}_{WF}(\mathbb{R}_+\times[0,T]) \subset \dot{\mathscr{C}}^{0}(\mathbb{R}_+\times[0,T]) \text{ and } \mathscr{C}^{0,2+\gamma}_{WF}(\mathbb{R}_+\times[0,T]) \subset \dot{\mathscr{C}}^{2,1}_{WF}(\mathbb{R}_+\times[0,T])$$

are the respective subspaces on which the norms

$$\begin{aligned} \|g\|_{WF,0,\gamma} &= \|g\|_{\infty} + [g]_{WF,0,\gamma}, \quad \text{and} \\ \|g\|_{WF,0,2+\gamma} &= \|g\|_{WF,0,\gamma} + \|\partial_x g\|_{WF,0,\gamma} + \|\partial_t g\|_{WF,0,\gamma} + \|x\partial_x^2 g\|_{WF,0,\gamma} \end{aligned} \tag{5.33}$$

are finite. As before, there is a characterization of elements in $\mathscr{C}^{0,2+\gamma}_{WF}(\mathbb{R}_+\times[0,T])$.

LEMMA 5.2.3. *Suppose that* $g \in \mathscr{C}^{2,1}((0,\infty)\times[0,T]) \cap \mathscr{C}^{1}(\mathbb{R}_+\times[0,T])$ *satisfies* (5.30) *and* $\|g\|_{WF,0,2+\gamma} < \infty$. *Then* $g \in \mathscr{C}^{2,1}_{WF}(\mathbb{R}_+\times[0,T])$.

PROOF. The proof is essentially identical to that of Lemma 5.2.1. The hypotheses imply that there is a constant M so that

$$|x\partial_x^2 g(x,t)| \leq M x^{\frac{\gamma}{2}}, \qquad \text{when } x \leq 1. \tag{5.34}$$

This implies that g is the $\mathscr{C}^{2,1}_{WF}$-limit of

$$g_n(x,t) = g\left(x+\frac{1}{n},t\right). \tag{5.35}$$

□

COROLLARY 5.2.4. *For* $0 < \gamma < 1$, *the spaces* $\mathscr{C}^{0,2+\gamma}_{WF}(\mathbb{R}_+\times[0,T])$ *are Banach spaces.*

5.2.3 Hybrid Spaces

For $k \in \mathbb{N}$, we define analogues of all the spaces above which have k full derivatives in the x direction. We call these hybrid since they mix ordinary with WF regularity.

First let $\mathscr{C}^{k,\gamma}_{WF}(\mathbb{R}_+)$ be the subspace of $\dot{\mathscr{C}}^k(\mathbb{R}_+)$ on which

$$\|f\|_{WF,k,\gamma} = \|f\|_{\mathscr{C}^k(\mathbb{R}_+)} + \|\partial_x^k f\|_{WF,0,\gamma} < \infty; \tag{5.36}$$

next, $\dot{\mathscr{C}}^{k,2}_{WF}(\mathbb{R}_+)$ is the closure of $\mathscr{C}^\infty_c(\mathbb{R}_+)$ with respect to

$$\|f\|_{WF,k,2} = \|f\|_{\mathscr{C}^{k-1}(\mathbb{R}_+)} + \|\partial_x^k f\|_{WF,2}; \tag{5.37}$$

in terms of this, $\mathscr{C}^{k,2+\gamma}_{WF}(\mathbb{R}_+)$ is the subspace of this space on which

$$\|f\|_{WF,k,2+\gamma} = \|f\|_{\mathscr{C}^k(\mathbb{R}_+)} + \|\partial_x^k f\|_{WF,0,2+\gamma} < \infty. \tag{5.38}$$

We could equally well substitute other spaces in place of $\mathscr{C}^{0,\gamma}_{WF}$ or $\mathscr{C}^{0,2+\gamma}_{WF}$ here. In particular, we can define $\mathscr{C}^{k,\ell+\gamma}_b(\mathbb{R}_+)$ to consist of all functions f such that

$$\|f\|_{b,k,\ell+\gamma} = \sup_{\substack{j\le k\\ i\le \ell}} \|\partial_x^j (x\partial_x)^i f\|_\infty + \left[\!\left[\partial_x^k (x\partial_x)^\ell\right]\!\right]_{b,0,\gamma}. \tag{5.39}$$

Similarly, we can define analogous parabolic versions of these hybrid spaces. For example, $\mathscr{C}^{k,\gamma}_{WF}(\mathbb{R}_+\times[0,T])$ and $\mathscr{C}^{k,2+\gamma}_{WF}(\mathbb{R}_+\times[0,T])$ are the spaces on which

$$\begin{aligned}\|g\|_{WF,k,\gamma} &= \|g\|_{\mathscr{C}^{k,\frac{k}{2}}(\mathbb{R}_+\times[0,T])} + \sum_{2i+j=k} \|\partial_t^i \partial_x^j g\|_{WF,0,\gamma} < \infty,\\ \|g\|_{WF,k,2+\gamma} &= \|g\|_{\mathscr{C}^{k,\frac{k}{2}}(\mathbb{R}_+\times[0,T])} + \sum_{2i+j=k} \|\partial_t^i \partial_x^j g\|_{WF,0,2+\gamma} < \infty,\end{aligned} \tag{5.40}$$

respectively. These can be proved to be Banach spaces exactly as for the case $k = 0$.

Remark. In the 1-dimensional case, we use the formulæ in (4.11), and (4.16) to deduce the higher order regularity of the solutions to the Cauchy and inhomogeneous problems, respectively, when the data has more regularity. A little thought shows that these formulæ involve expressions of the form $x^l \partial_x^{l+p+q} f$. This suggests that the higher order norms should include terms involving these weighted derivatives, i.e., terms like $\|x^l \partial_x^{l+p+q} f\|_{WF,0,\gamma}$. The estimates in Lemmas 7.1.2 and A.1.10 strongly suggest that the desired weighted estimates are also correct. To avoid further proliferation of an already very large number of cases, we have decided to omit these terms from our norms.

For our applications to the analysis of generalized Kimura diffusions on compact manifolds with corners it suffices to assume that the data has support in a fixed compact set. With this assumption, the Leibniz formula leads to a bound on a term like $\|x^l \partial_x^{l+p+q} f\|_{WF,0,\gamma}$ by a multiple of $\|\partial_x^{l+p+q} f\|_{WF,0,\gamma}$. To generalize the results in this monograph to the case of P non-compact, it would be natural to modify the definitions of the higher norms spaces to include terms of this type.

5.2.4 Multidimensional WF-Hölder Spaces

Following this detailed presentation of these various function spaces in one and $1+1$ dimensions, we can follow much the same path in defining the WF-Hölder spaces in higher dimensions. As before, we work on the model space, either

$$S_{n,m} = \mathbb{R}_+^n \times \mathbb{R}^m, \text{ or } S_{n,m} \times [0, T].$$

We denote points in these spaces by $(\boldsymbol{x}, \boldsymbol{y}, t)$, where $\boldsymbol{x} = (x_1, \dots, x_n)$ and $\boldsymbol{y} = (y_1, \dots, y_m)$, with all $x_j \geq 0$.

The metric on which the WF-Hölder spaces are based is

$$ds_{WF}^2 = \sum_{j=1}^{n} \frac{dx_j^2}{x_j} + \sum_{k=1}^{m} dy_k^2. \tag{5.41}$$

Note that this is incomplete as any $x_j \to 0$. The Riemannian distance function is equivalent to

$$d_{WF}((\boldsymbol{x}, \boldsymbol{y}), (\boldsymbol{x}', \boldsymbol{y}')) = \sum_{j=1}^{n} |\sqrt{x_j} - \sqrt{x_j'}| + \sum_{k=1}^{m} |y_k - y_k'|; \tag{5.42}$$

we sometimes write the right-hand side as $\rho_s(\boldsymbol{x}, \boldsymbol{x}') + \rho_e(\boldsymbol{y}, \boldsymbol{y}')$. We also set

$$d_{WF}((\boldsymbol{x}, \boldsymbol{y}, t), (\boldsymbol{x}', \boldsymbol{y}', t')) = d_{WF}((\boldsymbol{x}, \boldsymbol{y}), (\boldsymbol{x}', \boldsymbol{y}')) + \sqrt{|t - t'|}.$$

The function $f \in \dot{\mathscr{C}}^0(S_{n,m})$ belongs to $\mathscr{C}_{WF}^{0,\gamma}(S_{n,m})$ if

$$\|f\|_{WF,0,\gamma} = \|f\|_\infty + \sup_{(\boldsymbol{x},\boldsymbol{y}) \neq (\boldsymbol{x}',\boldsymbol{y}')} \frac{|f(\boldsymbol{x}, \boldsymbol{y}) - f(\boldsymbol{x}', \boldsymbol{y}')|}{[\rho_s(\boldsymbol{x}, \boldsymbol{x}') + \rho_e(\boldsymbol{y}, \boldsymbol{y}')]^\gamma} < \infty. \tag{5.43}$$

The semi-norm $[f]_{WF,0,\gamma}$ is the second term on the right.

The space $\dot{\mathscr{C}}_{WF}^2(S_{n,m})$ is the closure of $\mathscr{C}_c^2(S_{n,m})$ with respect to the norm:

$$\|f\|_{WF,2} = \|f\|_\infty + \|\nabla f\|_\infty + \sup_{|\alpha|+|\beta|=2} \|(\sqrt{\boldsymbol{x}}\partial_{\boldsymbol{x}})^\alpha \partial_{\boldsymbol{y}}^\beta f\|_\infty.$$

We are introducing here the notation

$$(\sqrt{\boldsymbol{x}}\partial_{\boldsymbol{x}})^\alpha = (\sqrt{x_1}\partial_{x_1})^{\alpha_1} \dots (\sqrt{x_n}\partial_{x_n})^{\alpha_n},$$

and $\partial_{\boldsymbol{y}}^\beta = \partial_{y_1}^{\beta_1} \dots \partial_{y_m}^{\beta_m}$ as usual. To be even more specific, we are measuring the L^∞ norms of all second derivatives

$$\sqrt{x_i x_j}\partial_{x_i x_j}^2 f, \ \sqrt{x_j}\partial_{x_j y_p}^2 f, \ \partial_{y_p y_q}^2 f, \quad i, j \leq n, \ p, q \leq m.$$

We are also implicitly extending any of these norms to vector-valued functions (e.g., ∇f) in the obvious way. A function $f \in \dot{\mathscr{C}}_{WF}^2(S_{n,m})$ belongs to $\mathscr{C}_{WF}^{0,2+\gamma}(S_{n,m})$ provided

$$\|f\|_{WF,0,2+\gamma} = \|f\|_{WF,2} + \|\nabla f\|_{WF,0,\gamma} + \sum_{|\alpha|+|\beta| \leq 2} \|(\sqrt{\boldsymbol{x}}\partial_{\boldsymbol{x}})^\alpha \partial_{\boldsymbol{y}}^\beta f\|_{WF,0,\gamma} < \infty.$$

We prove once again the basic characterization lemma.

LEMMA 5.2.5. *If* $f \in \mathscr{C}^1(S_{n,m}) \cap \dot{\mathscr{C}}^2((0,\infty)^n \times \mathbb{R}^m)$ *has* $||f||_{WF,0,2+\gamma} < \infty$ *and satisfies*

$$\lim_{x_j \text{ or } x_k \to 0^+} \sqrt{x_j x_k}\, \partial^2_{x_j x_k} f(\boldsymbol{x}, \boldsymbol{y}) = 0$$
$$\text{and} \lim_{x_j \to 0^+} \sqrt{x_j}\, \partial^2_{x_j y_p} f(\boldsymbol{x}, \boldsymbol{y}) = 0, \tag{5.44}$$

for all $j, k \leq n$, $p \leq m$, *and in addition,*

$$\lim_{(\boldsymbol{x},\boldsymbol{y})\to\infty} \left(|f(\boldsymbol{x}, \boldsymbol{y})| + |\nabla f(\boldsymbol{x}, \boldsymbol{y})| + \sum_{|\alpha|+|\beta|\leq 2} |(\sqrt{\boldsymbol{x}}\partial_{\boldsymbol{x}})^\alpha \partial_{\boldsymbol{y}}^\beta f| \right) = 0. \tag{5.45}$$

Then $f \in \mathscr{C}^2_{WF}(S_{n,m})$.

PROOF. The hypotheses imply that

$$\sqrt{x_j x_k}\, |\partial^2_{x_j x_k} f(\boldsymbol{x}, \boldsymbol{y})| \leq \|f\|_{WF,0,2+\gamma} \min\{x_j^{\frac{\gamma}{2}}, x_k^{\frac{\gamma}{2}}\}$$
$$\sqrt{x_j}\, |\partial^2_{x_j y_p} f(\boldsymbol{x}, \boldsymbol{y})| \leq \|f\|_{WF,0,2+\gamma}\, x_j^{\frac{\gamma}{2}}, \tag{5.46}$$

and in addition that each scaled second derivative has a continuous extension to a certain part of the boundary. For example, $\sqrt{x_j}\, \partial^2_{x_j y_p} f$ extends continuously to that subset of the boundary of $S_{n,m}$ where $\{x_j > 0\}$.

Let $\mathbf{1} = (1, \ldots, 1)$ and choose any sequence of positive numbers $\eta_i \to 0$. Then define

$$f_i(\boldsymbol{x}, \boldsymbol{y}) = f(\boldsymbol{x} + \eta_i \mathbf{1}, \boldsymbol{y}). \tag{5.47}$$

The definition $||\cdot||_{WF,0,2+\gamma}$ and (5.45) imply that

$$\lim_{i\to\infty} \left(\|f - f_i\|_\infty + \|\nabla_{\boldsymbol{x},\boldsymbol{y}}(f - f_i)\|_\infty + \sup_{|\beta|=2} \|\partial_{\boldsymbol{y}}^\beta (f - f_i)\|_\infty \right) = 0. \tag{5.48}$$

Hence it remains to study the terms $||(\sqrt{\boldsymbol{x}}\partial_{\boldsymbol{x}})^\alpha \partial_{\boldsymbol{y}}^\beta (f - f_i)||_\infty$ for $|\alpha| + |\beta| = 2$ and $\alpha \neq (0, \ldots, 0)$.

We begin with $\sqrt{x_j}\partial^2_{x_j y_p}(f - f_i)$. For $\delta > 0$, define

$$W_{j,\delta} = \{(\boldsymbol{x}, \boldsymbol{y}) : \delta \leq x_j\} \subset S_{n,m}.$$

From the hypotheses again, it is clear that if $\delta > 0$, then

$$\lim_{i\to\infty} \|\sqrt{x_j}\, \partial^2_{x_j y_p}(f - f_i)\|_{\infty, W_{j,\delta}} = 0, \tag{5.49}$$

so we must only show that $|\sqrt{x_j}\, \partial^2_{x_j y_p}(f - f_i)(\boldsymbol{x}, \boldsymbol{y})|$ is uniformly small when i is large and x_j is small. We have

$$\begin{aligned}
&|\sqrt{x_j}\, \partial^2_{x_j y_p}(f - f_i)(\boldsymbol{x}, \boldsymbol{y})| \\
&\qquad \leq |\sqrt{x_j}\, \partial^2_{x_j y_p} f(\boldsymbol{x}, \boldsymbol{y}) - (\sqrt{x_j + \eta_i})\partial^2_{x_j y_p} f(\boldsymbol{x} + \eta_i, \boldsymbol{y})| \\
&\qquad\qquad + \frac{|\sqrt{x_j} - \sqrt{x_j + \eta_i}|}{\sqrt{x_j + \eta_i}} |(\sqrt{x_j + \eta_i})\partial^2_{x_j y_p} f(\boldsymbol{x} + \eta_i \mathbf{1}, \boldsymbol{y})|.
\end{aligned}$$

By definition of the $\mathscr{C}^{0,2+\gamma}_{WF}$-norm again, and using (5.46), this gives:

$$|\sqrt{x_j}\,\partial^2_{x_j y_p}(f - f_i)(\boldsymbol{x}, \boldsymbol{y})| \leq \|f\|_{WF,0,2+\gamma}\left[n^\gamma\,\eta_i^{\gamma/2} + (x_j + \eta_i)^{\frac{\gamma}{2}}\right]. \tag{5.50}$$

Together with (5.49), this implies that

$$\lim_{i\to\infty}\|\sqrt{x_j}\,\partial_{x_j y_p}(f - f_i)\|_\infty = 0. \tag{5.51}$$

Finally, we must consider terms of the form $\sqrt{x_j x_k}\,|\partial^2_{x_j x_k}(f - f_i)|$. Once again, for any $\delta > 0$,

$$\lim_{i\to\infty}\|\sqrt{x_j x_k}\,\partial^2_{x_j x_k}(f - f_i)\|_{\infty, W_{j,\delta}\cap W_{k,\delta}} = 0. \tag{5.52}$$

Near the boundary, we have

$$\begin{aligned}|\sqrt{x_j x_k}\,\partial^2_{x_j x_k}(f - f_i)(\boldsymbol{x}, \boldsymbol{y})| \leq \\ |\sqrt{x_j x_k}\,\partial^2_{x_j x_k} f(\boldsymbol{x}, \boldsymbol{y}) - \sqrt{(x_j + \eta_i)(x_k + \eta_i)}\,\partial^2_{x_j x_k} f_i(\boldsymbol{x} + \eta_i \mathbf{1}, \boldsymbol{y})| + \\ \frac{|\sqrt{(x_j + \eta_i)(x_k + \eta_i)} - \sqrt{x_j x_k}|}{\sqrt{(x_j + \eta_i)(x_k + \eta_i)}}\sqrt{(x_j + \eta_i)(x_k + \eta_i)}|\partial^2_{x_j x_k} f_i(\boldsymbol{x} + \eta_i \mathbf{1}, \boldsymbol{y})|,\end{aligned} \tag{5.53}$$

whence, by (5.46),

$$\begin{aligned}|\sqrt{x_j x_k}\,\partial^2_{x_j x_k}(f - f_i)(\boldsymbol{x}, \boldsymbol{y})| \leq \\ \|f\|_{WF,0,2+\gamma}\left[n^\gamma\,\eta_i^{\gamma/2}\min\left\{\left|x_j + \eta_i\right|^{\frac{\gamma}{2}}, |x_k + \eta_i|^{\frac{\gamma}{2}}\right\}\right].\end{aligned} \tag{5.54}$$

This implies that

$$\lim_{i\to\infty}\|\sqrt{x_j x_k}\,\partial^2_{x_j x_k}(f - f_i)\|_\infty = 0, \tag{5.55}$$

and proves the lemma. □

A function $f \in \dot{\mathscr{C}}^k(S_{n,m})$ belongs to $\mathscr{C}^{k,\gamma}_{WF}(S_{n,m})$ if

$$\|f\|_{WF,k,\gamma} = \|f\|_{\mathscr{C}^k} + \sup_{\{|\boldsymbol{\alpha}|+|\boldsymbol{\beta}|=k\}}\ \sup_{(\boldsymbol{x},\boldsymbol{y})\neq(\boldsymbol{x}',\boldsymbol{y}')}\frac{|\partial^{\boldsymbol{\alpha}}_{\boldsymbol{x}}\partial^{\boldsymbol{\beta}}_{\boldsymbol{y}} f(\boldsymbol{x}, \boldsymbol{y}) - \partial^{\boldsymbol{\alpha}}_{\boldsymbol{x}}\partial^{\boldsymbol{\beta}}_{\boldsymbol{y}} f(\boldsymbol{x}', \boldsymbol{y}')|}{[\rho_s(\boldsymbol{x}, \boldsymbol{x}') + \rho_e(\boldsymbol{y}, \boldsymbol{y}')]^\gamma} \tag{5.56}$$

is finite. Similarly, $\dot{\mathscr{C}}^{k,2}_{WF}(S_{n,m})$ is the closure of $\mathscr{C}^{k+2}_c(S_{n,m})$ with respect to the norm

$$\|f\|_{WF,k,2} = \|f\|_{\mathscr{C}^{k-1}} + \sup_{\{|\boldsymbol{\alpha}|+|\boldsymbol{\beta}|=k\}}\|\partial^{\boldsymbol{\alpha}}_{\boldsymbol{x}}\partial^{\boldsymbol{\beta}}_{\boldsymbol{y}} f\|_{WF,2}, \tag{5.57}$$

and a function $f \in \dot{\mathscr{C}}^{k,2}_{WF}(S_{n,m})$ belongs to $\mathscr{C}^{k,2+\gamma}_{WF}(S_{n,m})$ if

$$\|f\|_{WF,k,\gamma} = \|f\|_{\mathscr{C}^k} + \sup_{\{|\boldsymbol{\alpha}|+|\boldsymbol{\beta}|=k\}}\|\partial^{\boldsymbol{\alpha}}_{\boldsymbol{x}}\partial^{\boldsymbol{\beta}}_{\boldsymbol{y}} f\|_{WF,0,2+\gamma} < \infty. \tag{5.58}$$

The analogue of Lemma 5.2.5 is straightforward and shows that these are Banach spaces.

The parabolic Hölder spaces are defined similarly. A function $g \in \dot{\mathscr{C}}^0(S_{n,m} \times [0,T])$ belongs to $\mathscr{C}^{0,\gamma}_{WF}(S_{n,m} \times [0,T])$ provided

$$\|g\|_{WF,0,\gamma} = \|g\|_\infty + \sup_{(x,y,t)\neq(x',y',t')} \frac{|g(x,y,t)-g(x',y',t')|}{[\rho_s(x,x')+\rho_e(y,y')+\sqrt{|t-t'|}]^\gamma} < \infty. \tag{5.59}$$

The semi-norm $[g]_{WF,0,\gamma}$ is the second term on the right. When it is important to emphasize the maximum time T, we use the notation $[g]_{WF,0,\gamma,T}$ for this semi-norm. A function $g \in \dot{\mathscr{C}}^k(S_{n,m} \times [0,T])$ belongs to $\mathscr{C}^{k,\gamma}_{WF}(S_{n,m} \times [0,T])$ if

$$\|g\|_{WF,k,\gamma} = \|g\|_{\mathscr{C}^k} + \sup_{\{|\boldsymbol{\alpha}|+|\boldsymbol{\beta}|+2j=k\}} \sup_{\substack{(x,y,t)\\ \neq(x',y',t')}} \frac{|\partial_t^j\partial_x^{\boldsymbol{\alpha}}\partial_y^{\boldsymbol{\beta}} g(x,y,t)-\partial_t^j\partial_x^{\boldsymbol{\alpha}}\partial_y^{\boldsymbol{\beta}} g(x',y',t')|}{[\rho_s(x,x')+\rho_e(y,y')+\sqrt{|t-t'|}]^\gamma} < \infty. \tag{5.60}$$

We now list several basic estimates and facts. First, for functions $f \in \mathscr{C}^{0,\gamma}_{WF}(S_{n,m})$ and $g \in \mathscr{C}^{0,\gamma}_{WF}(S_{n,m} \times [0,T])$, we have

$$\begin{aligned} |f(x,y)-f(x',y')| &\leq 2\|f\|_{WF,0,\gamma} d_{WF}((x,y),(x',y'))^\gamma \\ |g(x,y,t)-g(x',y',t')| &\leq 2\|g\|_{WF,0,\gamma} d_{WF}((x,y,t),(x',y',t'))^\gamma. \end{aligned} \tag{5.61}$$

Furthermore, there are Leibniz formulæ for these semi-norms: if $f, g \in \mathscr{C}^{0,\gamma}_{WF}(S_{n,m})$, or $f, g \in \mathscr{C}^{0,\gamma}_{WF}(S_{n,m} \times [0,T])$, then

$$[fg]_{WF,0,\gamma} \leq [f]_{WF,0,\gamma}\, \|g\|_{L^\infty} + [g]_{WF,0,\gamma}\, \|f\|_{L^\infty}. \tag{5.62}$$

LEMMA 5.2.6. *Let $0 < \gamma' < \gamma < 1$ and suppose that $f \in \mathscr{C}^{0,\gamma}_{WF}(S_{n,m})$ and $g \in \mathscr{C}^{0,\gamma}_{WF}(S_{n,m} \times [0,T])$. Then*

$$[f]_{WF,0,\gamma'} \leq 2\,[f]_{WF,0,\gamma}^{\frac{\gamma'}{\gamma}}\, \|f\|_\infty^{1-\frac{\gamma'}{\gamma}}, \tag{5.63}$$

$$[g]_{WF,0,\gamma'} \leq 2\,[g]_{WF,0,\gamma}^{\frac{\gamma'}{\gamma}}\, \|g\|_\infty^{1-\frac{\gamma'}{\gamma}}. \tag{5.64}$$

PROOF. These follow directly from the identity

$$\frac{|h(x,y,t)-h(x',y',t')|}{d_{WF}((x,y,t),(x',y',t'))^{\gamma'}} = \left[\frac{|h(x,y,t)-h(x',y',t')|}{d_{WF}((x,y,t),(x',y',t'))^\gamma}\right]^{\frac{\gamma}{\gamma}\gamma} \left(|h(x,y,t)-h(x',y',t')|\right)^{1-\frac{\gamma'}{\gamma}} \tag{5.65}$$

where h is defined on $S_{n,m} \times [0,T]$, or the analogous identity for functions defined on $S_{n,m}$. □

The space $\dot{\mathscr{C}}^{2,1}_{WF}(S_{n,m} \times [0,T])$ is the closure of $\mathscr{C}^{2,1}_c(S_{n,m} \times [0,T])$ with respect to the norm

$$\|g\|_{WF,2,1} = \|g\|_\infty + \|\nabla_{x,y,t} g\|_\infty + \sup_{|\alpha|+|\beta|=2} \|(\sqrt{x}\partial_x)^\alpha \partial_y^\beta g\|_\infty, \tag{5.66}$$

and $\mathscr{C}^{0,2+\gamma}_{WF}(S_{n,m} \times [0,T])$ is the subspace on which

$$\begin{aligned}\|g\|_{WF,0,2+\gamma} &= \|g\|_{WF,2,1} \\ &+ \|\nabla_{x,y,t} g\|_{WF,0,\gamma} + \sup_{|\alpha|+|\beta|=2} \|(\sqrt{x}\partial_x)^\alpha \partial_y^\beta g\|_{WF,0,\gamma} < \infty.\end{aligned} \tag{5.67}$$

The basic lemma now reads:

LEMMA 5.2.7. *Let* $g \in \mathscr{C}^1(S_{n,m} \times [0,T]) \cap \dot{\mathscr{C}}^{2,1}_{WF}((0,\infty)^n \times \mathbb{R}^m \times [0,T])$ *satisfy*

$$\lim_{x_j \text{ or } x_k \to 0^+} \sqrt{x_j x_k}\, \partial^2_{x_j x_k} g(x,y,t) = 0 \text{ and } \lim_{x_j \to 0^+} \sqrt{x_j}\, \partial^2_{x_j y_p} g(x,y,t) = 0$$

for $j,k \le n$, $p \le m$, *and*

$$\lim_{(x,y)\to\infty} \left[|g(x,y,t)| + |\nabla_{x,y,t} g(x,y,t)| + \sup_{|\alpha|+|\beta|=2} |(\sqrt{x}\partial_x)^\alpha \partial_y^\beta g(x,y,t)| \right] = 0.$$

If $\|g\|_{WF,0,2+\gamma} < \infty$, *then* $g \in \mathscr{C}^{0,2+\gamma}_{WF}(S_{n,m} \times [0,T])$.

The proof is nearly identical to the one for Lemma 5.2.5, and this implies as before that $\mathscr{C}^{0,2+\gamma}_{WF}(S_{n,m} \times [0,T])$ is a Banach space.

We finally define the higher parabolic Hölder spaces in the expected way. Namely, $\dot{\mathscr{C}}^{k+2,\frac{k}{2}+1}_{WF}(S_{n,m} \times [0,T])$ is the closure of $\mathscr{C}^\infty_c(S_{n,m} \times [0,T])$ with respect to the norm

$$||g||_{WF,k+2,k/2+1} = ||g||_{\mathscr{C}^{k,\frac{k}{2}}} + \sup_{|\alpha|+|\beta|+2j=k} ||\partial_x^\alpha \partial_y^\beta \partial_t^j g||_{WF,2,1}. \tag{5.68}$$

We define $\mathscr{C}^{k,2+\gamma}_{WF}(S_{n,m} \times [0,T])$ to be the subspace of $\dot{\mathscr{C}}^{k+2,\frac{k}{2}+1}_{WF}(S_{n,m} \times [0,T])$ on which

$$||g||_{WF,k,2+\gamma} = ||g||_{\mathscr{C}^{k,\frac{k}{2}}} + \sup_{|\alpha|+|\beta|+2j=k} \left[\!\left[\partial_x^\alpha \partial_y^\beta \partial_t^j g \right]\!\right]_{WF,0,2+\gamma} < \infty.$$

As before, if the upper limit T, for the time variable, is important we sometimes denote these norms by $||g||_{WF,k,\gamma,T}$, and $||g||_{WF,k,2+\gamma,T}$, respectively.

These various spaces satisfy some obvious inclusions: if $k' > k$, or $k' = k$ and $1 \ge \gamma' > \gamma > 0$, then

$$\mathscr{C}^{k',\gamma'}_{WF}(S_{n,m}) \subset \mathscr{C}^{k,\gamma}_{WF}(S_{n,m}), \quad \mathscr{C}^{k',2+\gamma'}_{WF}(S_{n,m}) \subset \mathscr{C}^{k,2+\gamma}_{WF}(S_{n,m}) \tag{5.69}$$

and

$$\begin{aligned}&\mathscr{C}^{k',\gamma'}_{WF}(S_{n,m} \times [0,T]) \subset \mathscr{C}^{k,\gamma}_{WF}(S_{n,m} \times [0,T]) \\ &\mathscr{C}^{k',2+\gamma'}_{WF}(S_{n,m} \times [0,T]) \subset \mathscr{C}^{k,2+\gamma}_{WF}(S_{n,m} \times [0,T]).\end{aligned} \tag{5.70}$$

PROPOSITION 5.2.8. *If $k < k'$ or $k' = k$ and $\gamma < \gamma'$, then the restrictions of the inclusions in* (5.69) *and* (5.70) *to subspaces of functions which are supported in a ball of finite radius in $S_{n,m}$ or $S_{n,m} \times [0, T]$ are compact.*

PROOF. These facts can all be deduced in a fairly straightforward manner from the Arzela-Ascoli theorem. We illustrate this by considering the inclusion

$$\{u \in \mathscr{C}^{0,\gamma'}_{WF}(S_{n,m}) : u = 0 \text{ for } |(\boldsymbol{x}, \boldsymbol{y})| > R\} \hookrightarrow \mathscr{C}^{0,\gamma}_{WF}(S_{n,m}).$$

If $\{u_j\}$ is a sequence in the space on the left with uniformly bounded norm, then by (5.61), this sequence is uniformly bounded and equicontinuous, hence some subsequence converges in $\mathscr{C}^0$ to a limit function u. Now apply (5.63) to see that this subsequence is Cauchy in $\mathscr{C}^{0,\gamma'}_{WF}$. □

As described in remark 5.2.3 in the 1-dimensional case, the higher order estimates in the general case are deduced by using formulæ (4.27) and (4.36). Again this suggests that the higher order norms should include weighted derivatives. As noted above, for our applications to Kimura operators on compact manifolds with corners we only need these results for data with fixed bounded support. To somewhat shorten this already long text, we have omitted these terms from the definitions of the higher order norms. Using the Leibniz formula we easily deduce the following estimates:

PROPOSITION 5.2.9. *Fix an $R > 0$, a $k \in \mathbb{N}$, a non-negative vector $\mathbf{0} \leq \boldsymbol{b}$, and a $0 < \gamma < 1$. There is a constant C_R so that*

1. *If $f \in \mathscr{C}^{k,\gamma}_{WF}(S_{n,m})$ has support in the set $\{(\boldsymbol{x}; \boldsymbol{y}) : \|\boldsymbol{x}\| \leq R\}$, then if $2q + |\alpha| + |\beta| \leq k$, we have the estimate*

$$\|L^q_{\boldsymbol{b},m}\partial^{\boldsymbol{\alpha}}_{\boldsymbol{x}}\partial^{\boldsymbol{\beta}}_{\boldsymbol{y}} f\|_{WF,0,\gamma} \leq C_R\|f\|_{WF,k,\gamma}. \tag{5.71}$$

2. *If $f \in \mathscr{C}^{k,2+\gamma}_{WF}(S_{n,m})$ has support in the set $\{(\boldsymbol{x}; \boldsymbol{y}) : \|\boldsymbol{x}\| \leq R\}$, then if $2q + |\alpha| + |\beta| \leq k$, we have the estimate*

$$\|L^q_{\boldsymbol{b},m}\partial^{\boldsymbol{\alpha}}_{\boldsymbol{x}}\partial^{\boldsymbol{\beta}}_{\boldsymbol{y}} f\|_{WF,0,2+\gamma} \leq C_R\|f\|_{WF,k,2+\gamma}. \tag{5.72}$$

3. *If $g \in \mathscr{C}^{k,\gamma}_{WF}(S_{n,m} \times [0, T])$ has support in the set $\{(\boldsymbol{x}; \boldsymbol{y}, t) : \|\boldsymbol{x}\| \leq R\}$, then if $2p + 2q + |\alpha| + |\beta| \leq k$, we have the estimate*

$$\|\partial^p_t L^q_{\boldsymbol{b},m}\partial^{\boldsymbol{\alpha}}_{\boldsymbol{x}}\partial^{\boldsymbol{\beta}}_{\boldsymbol{y}} g\|_{WF,0,\gamma,T} \leq C_R\|g\|_{WF,k,\gamma,T}. \tag{5.73}$$

4. *If $g \in \mathscr{C}^{k,2+\gamma}_{WF}(S_{n,m} \times [0, T])$ has support in the set $\{(\boldsymbol{x}; \boldsymbol{y}, t) : \|\boldsymbol{x}\| \leq R\}$, then if $2p + 2q + |\alpha| + |\beta| \leq k$, we have the estimate*

$$\|\partial^p_t L^q_{\boldsymbol{b},m}\partial^{\boldsymbol{\alpha}}_{\boldsymbol{x}}\partial^{\boldsymbol{\beta}}_{\boldsymbol{y}} g\|_{WF,0,2+\gamma,T} \leq C_R\|g\|_{WF,k,2+\gamma,T}. \tag{5.74}$$

Chapter Six

Hölder Estimates for the 1-dimensional Model Problems

In this and the following three chapters we establish Hölder estimates for the solutions of the model problems, i.e., w such that

$$(\partial_t - L_{\boldsymbol{b},m})w = g \text{ with } w(p,0) = f(p), \tag{6.1}$$

where f and g belong to the anisotropic Hölder spaces introduced in Chapter 5. It may appear that we are taking a circuitous path, by first considering the 1-dimensional case, then pure corner models, $\mathbb{R}_+^n$, followed by Euclidean models ($\mathbb{R}^m$) before finally treating the general case, $\mathbb{R}_+^n \times \mathbb{R}^m$. In fact, all cases need to be treated, and in the end nothing is really wasted. We give a detailed treatment of the 1-dimensional case, both because it establishes a pattern that will be followed in the subsequent cases, and because all of the higher dimensional estimates are reduced to estimates on heat kernels for the 1-dimensional model problems.

The derivation of parabolic Schauder estimates is now an old subject, and there are many possible approaches to follow. Our proof of these estimates for the model operator $\partial_t - L_b$ is elementary. It uses the explicit formula for the heat kernel, (1.29), along with standard tools of analysis, like Taylor's formula and Laplace's method. The paper [11] considers a similar degenerate diffusion operator in $2 + 1$-dimensions, and contains proofs of parabolic Schauder estimates for that problem. We present different arguments to derive the analogous estimates here. This allows us to handle the case $b = 0$, which is somewhat different than the situation in [11].

It is straightforward from the definitions that for any $k \in \mathbb{N}_0$ and $0 < \gamma < 1$,

$$\partial_t - L_b : \{u \in \mathscr{C}_{WF}^{k,2+\gamma}(\mathbb{R}_+ \times [0,T]) : u(x,0) = 0\} \longrightarrow \mathscr{C}_{WF}^{k,\gamma}(\mathbb{R}_+ \times [0,T]).$$

Our goal is to prove the converse, and of course also to study the regularity effects of nontrivial initial data. We shall prove the following two results:

PROPOSITION 6.0.10. *Fix $k \in \mathbb{N}_0$, $\gamma \in (0,1)$, $0 < R$ and $b \geq 0$. Suppose that $f \in \mathscr{C}_{WF}^{k,\gamma}(\mathbb{R}_+)$, and let v be the unique solution to* (4.2), *with $g = 0$. If $k > 0$, then also assume that f has support in* $[0, R]$. *Then $v \in \mathscr{C}_{WF}^{k,\gamma}(\mathbb{R}_+ \times [0,T])$ for any $T > 0$ and there is a constant $C_{k,\gamma,b,R}$ so that*

$$\|v\|_{WF,k,\gamma} \leq C_{k,\gamma,b,R}\|f\|_{WF,k,\gamma}. \tag{6.2}$$

If $0 < \gamma' < \gamma$, then

$$\lim_{t\to 0^+} \|v(\cdot,t) - f\|_{WF,k,\gamma'} = 0. \tag{6.3}$$

If $f \in \mathscr{C}^{k,2+\gamma}_{WF}(\mathbb{R}_+)$, *then*

$$\|v\|_{WF,k,2+\gamma} \leq C_{k,\gamma,b,R}\|f\|_{WF,k,2+\gamma}. \tag{6.4}$$

The constants $C_{k,\gamma,b,R}$ *are uniformly bounded on any finite interval* $0 \leq b \leq B$. *If* $0 < \gamma' < \gamma$, *then*

$$\lim_{t\to 0^+} \|v(\cdot,t) - f\|_{WF,k,2+\gamma'} = 0. \tag{6.5}$$

If $k = 0$, *then the constants in these estimates do not depend on* R.

PROPOSITION 6.0.11. *Fix* $k \in \mathbb{N}_0$, $\gamma \in (0,1)$, $0 < R$, *and* $b \geq 0$. *Let* u *be the unique solution to* (4.2), *with* $f = 0$ *and* $g \in \mathscr{C}^{k,\gamma}_{WF}(\mathbb{R}_+ \times [0,T])$. *If* $k > 0$ *we assume that* $g(x,t)$ *is supported in* $\{(x,t) : x \leq R\}$. *Then* $u \in \mathscr{C}^{k,2+\gamma}_{WF}(\mathbb{R}_+ \times [0,T])$ *and there is a constant* $C_{k,\gamma,b,R}$ *so that*

$$\|u\|_{WF,k,2+\gamma,T} \leq C_{k,\gamma,b,R}(1+T)\|g\|_{WF,k,\gamma,T}. \tag{6.6}$$

The constants $C_{k,\gamma,b,R}$ *are uniformly bounded for* $0 \leq b \leq B$. *For any* $0 < \gamma' < \gamma$, *the solution* $u(\cdot,t)$ *tends to zero in* $\mathscr{C}^{k,2+\gamma'}_{WF}(\mathbb{R}_+)$. *If* $k = 0$, *then the constant is independent of* R.

The assertions about the behavior of solutions as $t \to 0^+$ follow easily from Proposition 5.2.8, the following lemma, and the obvious facts that $v(\cdot,t)$ tends to f and $u(\cdot,t)$ tends to zero in $\mathscr{C}^0(\mathbb{R}_+)$.

LEMMA 6.0.12. *Let* $X_2 \subset X_1 \subset X_0$ *be Banach spaces with the first inclusion precompact, and the second bounded. If for some* M, *the family* $v(t) \in X_2$ *satisfies:*

$$\sup_{t\in[0,1]} \|v(t)\|_{X_2} \leq M \text{ and } \lim_{t\to 0^+} \|v(t)\|_{X_0} = 0, \tag{6.7}$$

then

$$\lim_{t\to 0^+} \|v(t)\|_{X_1} = 0. \tag{6.8}$$

PROOF. If $\lim_{t\to 0^+} \|v(t)\|_{X_1} \neq 0$, then, by compactness, we can choose a sequence $< t_n >$, tending to zero so that $< v(t_n) >$ converges, in X_1, to $v^* \neq 0$. The boundedness of the inclusion $X_1 \subset X_0$, implies that $< v(t_n) >$ must also converge, in X_0, to v^*, but then v^* must equal 0. □

Our final results concern the resolvent operator defined, for $\mu \in (0,\infty)$, by

$$R(\mu)f = \lim_{\epsilon\to 0^+} \int\limits_{\epsilon}^{\frac{1}{\epsilon}} e^{-\mu t} v(x,t)dt. \tag{6.9}$$

As noted in Proposition 4.3.2, $R(\mu)f$ extends to define an analytic function for $\mu \in \mathbb{C} \setminus (-\infty, 0]$. Our final proposition gives a more refined statement of the mapping properties of $R(\mu)$ for the 1-dimensional model problem.

PROPOSITION 6.0.13. *The resolvent operator* $R(\mu)$ *is analytic in the complement of* $(-\infty, 0]$, *and is given by the integral in* (4.48) *provided that* $\mathrm{Re}(\mu e^{i\theta}) > 0$. *For* $\alpha \in (0, \pi]$, *there are constants* $C_{b,\alpha}$ *so that if*

$$\alpha - \pi \leq \arg \mu \leq \pi - \alpha, \tag{6.10}$$

then for $f \in \mathscr{C}_b^0(\mathbb{R}_+)$ *we have:*

$$\|R(\mu) f\|_{L^\infty} \leq \frac{C_{b,\alpha}}{|\mu|} \|f\|_{L^\infty}, \tag{6.11}$$

with $C_{b,\pi} = 1$. *Moreover, for* $0 < \gamma < 1$, *there is a constant* $C_{b,\alpha,\gamma}$ *so that if* $f \in \mathscr{C}_{WF}^{0,\gamma}(\mathbb{R}_+)$, *then*

$$\|R(\mu) f\|_{WF,0,\gamma} \leq \frac{C_{b,\alpha,\gamma}}{|\mu|} \|f\|_{WF,0,\gamma}. \tag{6.12}$$

If for a $k \in \mathbb{N}_0$, *and* $0 < \gamma < 1$, $f \in \mathscr{C}_{WF}^{k,\gamma}(\mathbb{R}_+)$, *then* $R(\mu) f \in \mathscr{C}_{WF}^{k,2+\gamma}(\mathbb{R}_+)$, *and, we have*

$$(\mu - L_b) R(\mu) f = f. \tag{6.13}$$

If $f \in \mathscr{C}_{WF}^{0,2+\gamma}(\mathbb{R}_+)$, *then*

$$R(\mu)(\mu - L_b) f = f. \tag{6.14}$$

There are constants $C_{b,k,\alpha}$ *so that, for* μ *satisfying* (6.10), *we have*

$$\|R(\mu) f\|_{WF,k,2+\gamma} \leq C_{b,k,\alpha} \left[1 + \frac{1}{|\mu|}\right] \|f\|_{WF,k,\gamma}. \tag{6.15}$$

For any $B > 0$, *these constants are uniformly bounded for* $0 \leq b \leq B$.

Remark. Unlike the results for the heat equations, the higher order estimates for the resolvent do not require an assumption about the support of the data. This is because the estimates for this operator only involve spatial derivatives; it is the time derivatives that lead to the x^j-weights.

6.1 KERNEL ESTIMATES FOR DEGENERATE MODEL PROBLEMS

The proofs of the estimates in one and higher dimensions rely upon estimates for the kernel functions $k_t^b(x, y)$ and their derivatives. These kernels are analytic in the right half plane $\mathrm{Re}\, t > 0$, and many of these estimates are stated and proved for this analytic continuation. Since we often need to refer to these results, we first list these estimates as a series of lemmas. Most of the proofs are given in Appendix A.

Throughout this book we let C, C_b or $C_{b,*}$ (where $*$ are other parameters) denote positive constants that are uniformly bounded for $0 \leq b \leq B$, and a fixed value of γ. We often make use of the following elementary inequalities.

LEMMA 6.1.1. *For each $k \in \mathbb{N}$, and $0 < \gamma < 1$, there is a constant $C_{k,\gamma}$ such that for non-negative numbers $\{a_1, \dots, a_k\}$ we have*

$$C_{k,\gamma}^{-1} \sum_{j=1}^{k} a_j^{\gamma} < \left[\sum_{j=1}^{k} a_j \right]^{\gamma} < C_{k,\gamma} \sum_{j=1}^{k} a_j^{\gamma}. \tag{6.16}$$

PROOF. As everything is homogeneous of degree γ it suffices to consider non-negative k-tuples, $(a_1, \dots, a_k)$, with

$$\sum_{j=1}^{k} a_j = 1, \tag{6.17}$$

for which the statement is obvious. □

LEMMA 6.1.2. *For $0 < \gamma < 1$, there is a constant m_γ so that, if x and y are non-negative, then*

$$|x^{\gamma} - y^{\gamma}| \leq m_{\gamma} |x - y|^{\gamma}. \tag{6.18}$$

PROOF. We can assume that $x > y$, and therefore the inequality is equivalent to the assertion that, for $1 < x$, we have

$$|x^{\gamma} - 1| \leq m_{\gamma} |x - 1|^{\gamma}. \tag{6.19}$$

The existence of m_γ follows easily from the observation that

$$x^{\gamma} - 1 = \gamma (x - 1) + O((x - 1)^2). \tag{6.20}$$

□

The remaining lemmas are divided according to the order of the derivative being estimated. Proofs are given in Appendix A. The reader can skip the rest of this subsection and refer to it later, as needed. Recall that, for $0 < b$,

$$k_t^b(x, y) = \frac{y^{b-1}}{t^b} e^{-\frac{(x+y)}{t}} \psi_b \left(\frac{xy}{t^2} \right), \tag{6.21}$$

where

$$\psi_b(z) = \sum_{j=0}^{\infty} \frac{z^j}{j! \Gamma(j + b)}. \tag{6.22}$$

This heat kernel is a smooth function in $[0, \infty)_x \times (0, \infty)_y \times (0, \infty)_t$, which has an analytic extension in the t variable to the right half plane S_0, where the sectors S_ϕ are defined in (4.44). The asymptotic expansion (4.43) is valid in any sector S_ϕ, with $\phi > 0$.

6.1.1 Basic Kernel Estimates

Recall that as $b \to 0^+$, the kernels k_t^b converge, in the sense of distributions, to

$$k_t^0(x,y) = k_t^{0,D}(x,y) + e^{-\frac{x}{t}}\delta_0(y), \tag{6.23}$$

where

$$k_t^{0,D}(x,y) = \left(\frac{x}{t^2}\right) e^{-\frac{x+y}{t}} \psi_2\left(\frac{xy}{t^2}\right) \tag{6.24}$$

is the solution operator for the equation $\partial_t v = x\partial_x^2 v$ with $v(0,t) = 0$. As we will see, the solutions to the equations $\partial_t u - L_b u = g$, and their higher dimensional analogues satisfy Hölder estimates with constants uniformly bounded as $b \to 0^+$. The kernel estimates are therefore proved for $0 < b$, and the properties of solutions to the PDE with $b = 0$ are obtained by taking limits of solutions.

A trivial but crucial fact is the following:

LEMMA 6.1.3. *For $t \in S_0$ and $b > 0$ we have:*

$$\int_0^\infty k_t^b(x,y)dy = 1. \tag{6.25}$$

There is a constant C_ϕ so that, for $t \in S_\phi$,

$$\int_0^\infty |k_t^b(x,y)|dy \leq C_\phi. \tag{6.26}$$

PROOF. The integral is absolutely convergent for any $t \in S_0$, and clearly defines an analytic function of t. For $t \in (0,\infty)$, the integral equals 1, which proves the first assertion of the lemma. For the second, suppose that $t = \tau e^{i\theta}$, and change variables, setting $w = y/\tau$ and $\lambda = x/\tau$, to obtain:

$$\int_0^\infty |k_t^b(x,y)|dy = \int_0^\infty w^b e^{-\cos\theta(w+\lambda)} |\psi_b(w\lambda e^{-2i\theta})| \frac{dw}{w}. \tag{6.27}$$

We split the integral into the part from 0 to $1/\lambda$ and the rest. In the compact part we use the estimate

$$|\psi_b(w\lambda e^{-2i\theta})| \leq \left(\frac{1}{\Gamma(b)} + C_b|w\lambda|\right). \tag{6.28}$$

Inserting this into the integral from 0 to $1/\lambda$, it is clear that this term is uniformly bounded. In the non-compact part we use the asymptotic expansion for ψ_b to see that this term is bounded by

$$I_+ = C_b \int_{\frac{1}{\lambda}}^\infty \left(\frac{w}{\lambda}\right)^{\frac{b}{2}-\frac{1}{4}} e^{-\cos\theta(\sqrt{w}-\sqrt{\lambda})^2} \frac{dw}{\sqrt{w}}. \tag{6.29}$$

This integral is $O(e^{-\cos\theta/(2\lambda)})$ as $\lambda \to 0$. As $\lambda \to \infty$, we let $z = \sqrt{w} - \sqrt{\lambda}$, to obtain that

$$I_+ = C_b \int\limits_{\frac{1}{\sqrt{\lambda}}-\sqrt{\lambda}}^{\infty} \left(\frac{z}{\sqrt{\lambda}} + 1\right)^{b-\frac{1}{2}} e^{-\cos\theta z^2} dz. \tag{6.30}$$

It is elementary to see that this integral is bounded by a constant depending only on θ. □

Remark. The proofs of the remaining estimates are in Appendix A.

LEMMA 6.1.4. *For $b > 0$, $0 < \gamma < 1$, and $0 < \phi < \frac{\pi}{2}$, there are constants $C_{b,\phi}$ uniformly bounded with b, so that for $t \in S_\phi$*

$$\int\limits_0^\infty |k_t^b(x, y) - k_t^b(0, y)| y^{\frac{\gamma}{2}} dy \leq C_{b,\phi} x^{\frac{\gamma}{2}}. \tag{6.31}$$

LEMMA 6.1.5. *For $b > 0$ there is a constant $C_{b,\phi}$ so that for $t \in S_\phi$*

$$\int\limits_0^\infty |k_t^b(x, y) - k_t^b(0, y)| dy \leq C_{b,\phi} \frac{x/|t|}{1 + x/|t|}. \tag{6.32}$$

For $0 < c < 1$ there is a constant $C_{b,c,\phi}$ so that, if $cx_2 < x_1 < x_2$, and $t \in S_\phi$, then

$$\int\limits_0^\infty |k_t^b(x_2, y) - k_t^b(x_1, y)| dy \leq C_{b,c,\phi} \left(\frac{\frac{\sqrt{x_2}-\sqrt{x_1}}{\sqrt{|t|}}}{1 + \frac{\sqrt{x_2}-\sqrt{x_1}}{\sqrt{|t|}}} \right). \tag{6.33}$$

LEMMA 6.1.6. *For $b > 0$, $0 < \gamma < 1$ and $t \in S_\phi$, $0 < \phi < \frac{\pi}{2}$, there is a $C_{b,\phi}$ so that*

$$\int\limits_0^\infty |k_t^b(x, y)| |\sqrt{x} - \sqrt{y}|^\gamma dy \leq C_{b,\phi} |t|^{\frac{\gamma}{2}}. \tag{6.34}$$

For fixed $0 < \phi$, and B, these constants are uniformly bounded for $0 < b < B$.

For several estimates we need to split $\mathbb{R}_+$ into a collection of subintervals. We let $J = [\alpha, \beta]$, where

$$\sqrt{\alpha} = \max\left\{\frac{3\sqrt{x_1} - \sqrt{x_2}}{2}, 0\right\} \text{ and } \sqrt{\beta} = \frac{3\sqrt{x_2} - \sqrt{x_1}}{2}. \tag{6.35}$$

LEMMA 6.1.7. *We assume that $x_1/x_2 > 1/9$ and $J = [\alpha, \beta]$, as defined in* (6.35). *For $b > 0$, $0 < \gamma < 1$ and $0 < \phi < \frac{\pi}{2}$, there is a $C_{b,\phi}$ so that if $t \in S_\phi$, then*

$$\int\limits_{J^c} |k_t^b(x_2, y) - k_t^b(x_1, y)| |\sqrt{y} - \sqrt{x_1}|^\gamma dy \leq C_{b,\phi} |\sqrt{x_2} - \sqrt{x_1}|^\gamma. \tag{6.36}$$

LEMMA 6.1.8. *For $b > 0$, $0 < \gamma < 1$ and $c < 1$ there is a C_b such that if $c < s/t < 1$, then*

$$\int_0^\infty \left| k_t^b(x,y) - k_s^b(x,y) \right| |\sqrt{x} - \sqrt{y}|^\gamma \, dy \le C_b |t-s|^{\frac{\gamma}{2}}. \tag{6.37}$$

We also have the simpler result, which holds without restriction on $s < t$, and when $\gamma = 0$.

LEMMA 6.1.9. *For $b > 0$ there is a C_b such that if $s < t$, then*

$$\int_0^\infty \left| k_t^b(x,y) - k_s^b(x,y) \right| dy \le C_b \left(\frac{t/s - 1}{1 + [t/s - 1]} \right). \tag{6.38}$$

6.1.2 First Derivative Estimates

The following lemma is central to many of the results in this paper.

LEMMA 6.1.10. *For $b > 0$, $0 \le \gamma < 1$, and $0 < \phi < \frac{\pi}{2}$, there is a $C_{b,\phi}$ so that for $t \in S_\phi$ we have*

$$\int_0^\infty |\partial_x k_t^b(x,y)| |\sqrt{y} - \sqrt{x}|^\gamma \, dy \le C_{b,\phi} \frac{|t|^{\frac{\gamma}{2} - 1}}{1 + \lambda^{\frac{1}{2}}}, \tag{6.39}$$

where $\lambda = x/|t|$.

The case $\gamma = 0$ is Lemma 8.1 in [15].

LEMMA 6.1.11. *For $b > 0$, $0 < \gamma < 1$, $0 < \phi < \frac{\pi}{2}$, and $0 < c < 1$, there is a constant $C_{b,c,\phi}$ so that for $cx_2 < x_1 < x_2$, $t \in S_\phi$,*

$$\int_0^\infty |\sqrt{x_1}\partial_x k_t^b(x_1,y) - \sqrt{x_2}\partial_x k_t^b(x_2,y)| |\sqrt{x_1} - \sqrt{y}|^\gamma \, dy \le$$

$$C_{b,c,\phi} |t|^{\frac{\gamma-1}{2}} \frac{\left(\frac{|\sqrt{x_2} - \sqrt{x_1}|}{\sqrt{|t|}} \right)}{1 + \left(\frac{|\sqrt{x_2} - \sqrt{x_1}|}{\sqrt{|t|}} \right)}. \tag{6.40}$$

LEMMA 6.1.12. *For $b > 0$, $0 < \gamma < 1$, there is a constant C_b so that for $t_1 < t_2 < 2t_1$, we have:*

$$\int_{t_2 - t_1}^{t_1} \int_0^\infty |\partial_x k_{t_2 - t_1 + s}^b(x,y) - \partial_x k_s^b(x,y)| |\sqrt{x} - \sqrt{y}|^{\frac{\gamma}{2}} dy ds < C_b |t_2 - t_1|^{\frac{\gamma}{2}}. \tag{6.41}$$

This result follows from the more basic:

LEMMA 6.1.13. *For* $b > 0$, $0 \leq \gamma < 1$, *and* $0 < t_1 < t_2 < 2t_1$, *we have for* $s \in [t_2 - t_1, t_1]$ *that there is a constant* C *so that*

$$\int_0^\infty |\partial_x k^b_{t_2-t_1+s}(x,y) - \partial_x k^b_s(x,y)||\sqrt{x} - \sqrt{y}|^\gamma \, dy < C \frac{(t_2 - t_1)s^{\frac{\gamma}{2}-1}}{(t_2 - t_1 + s)(1 + \sqrt{x/s})}. \tag{6.42}$$

6.1.3 Second Derivative Estimates

LEMMA 6.1.14. *For* $b > 0$, $0 < \gamma < 1$, *and* $0 < \phi < \frac{\pi}{2}$, *there is a* $C_{b,\phi}$ *so that for* $t = |t|e^{i\theta}$ *with* $|\theta| < \frac{\pi}{2} - \phi$,

$$\begin{aligned} &\int_0^{|t|}\int_0^\infty |x\partial_x^2 k^b_{se^{i\theta}}(x,y)||\sqrt{y} - \sqrt{x}|^\gamma \, dy ds \leq C_{b,\phi} x^{\frac{\gamma}{2}} \text{ and} \\ &\int_0^{|t|}\int_0^\infty |x\partial_x^2 k^b_{se^{i\theta}}(x,y)||\sqrt{y} - \sqrt{x}|^\gamma \, dy ds \leq C_{b,\phi} |t|^{\frac{\gamma}{2}}. \end{aligned} \tag{6.43}$$

This follows from the more basic result:

LEMMA 6.1.15. *For* $b > 0$, $0 \leq \gamma < 1$, $0 < \phi < \frac{\pi}{2}$, *there is a* $C_{b,\phi}$ *so that if* $t \in S_\phi$, *then*

$$\int_0^\infty |x\partial_x^2 k^b_t(x,y)||\sqrt{x} - \sqrt{y}|^\gamma \, dy \leq C_{b,\phi} \frac{\lambda |t|^{\frac{\gamma}{2}-1}}{1 + \lambda}, \tag{6.44}$$

where $\lambda = x/|t|$.

LEMMA 6.1.16. *For* $b > 0$, $0 < \gamma < 1$, $0 < \phi < \frac{\pi}{2}$, *and* $0 < x_2/3 < x_1 < x_2$, *there is a constant* $C_{b,\phi}$ *so that, for* $t \in S_\phi$, *we have*

$$\int_0^{|t|} \left| (\partial_y y - b) k^b_{se^{i\theta}}(x_2, \alpha) - (\partial_y y - b) k^b_{se^{i\theta}}(x_2, \beta) \right| ds \leq C_{b,\phi}, \tag{6.45}$$

where α *and* β *are defined in* (6.35).

LEMMA 6.1.17. *For* $b > 0$, $0 < \gamma < 1$, $\phi < \frac{\pi}{2}$, *and* $0 < x_2/3 < x_1 < x_2$, *if* $J = [\alpha, \beta]$, *with the endpoints given by* (6.35), *there is a constant* $C_{b,\phi}$ *so that if*

$|\theta| < \frac{\pi}{2} - \phi$, *then*

$$\begin{aligned} I_1 &= \int_0^{|t|}\int_\alpha^\beta |L_b k^b_{se^{i\theta}}(x_2, y)||\sqrt{y} - \sqrt{x_2}|^\gamma \, dyds \le C_{b,\phi}|\sqrt{x_2} - \sqrt{x_1}|^\gamma \\ I_2 &= \int_0^{|t|}\int_\alpha^\beta |L_b k^b_{se^{i\theta}}(x_1, y)||\sqrt{y} - \sqrt{x_1}|^\gamma \, dyds \le C_{b,\phi}|\sqrt{x_2} - \sqrt{x_1}|^\gamma . \end{aligned} \tag{6.46}$$

LEMMA 6.1.18. *For* $b > 0$, $0 < \gamma < 1$, $0 < \phi < \frac{\pi}{2}$, *and* $0 < x_2/3 < x_1 < x_2$, *if* $J = [\alpha, \beta]$, *with the endpoints given by* (6.35), *there is a constant* $C_{b,\phi}$ *so that if* $|\theta| < \frac{\pi}{2} - \phi$, *then*

$$\int_0^{|t|}\int_{J^c} |L_b k^b_{se^{i\theta}}(x_2, y) - L_b k^b_{se^{i\theta}}(x_1, y)||\sqrt{y} - \sqrt{x_1}|^\gamma \, dyds \le C_{b,\phi}|\sqrt{x_2} - \sqrt{x_1}|^\gamma . \tag{6.47}$$

LEMMA 6.1.19. *For* $b > 0$, $0 < \gamma < 1$, *and* $t_1 < t_2 < 2t_1$ *there is a constant* C_b *so that*

$$\int_{t_2 - t_1}^{t_1}\int_0^\infty |L_b k^b_{t_2 - t_1 + s}(x, y) - L_b k^b_s(x, y)||\sqrt{x} - \sqrt{y}|^\gamma \, dyds \le C_b |t_2 - t_1|^{\frac{\gamma}{2}} . \tag{6.48}$$

This lemma follows from the more basic result:

LEMMA 6.1.20. *For* $b > 0$, $0 < \gamma < 1$, *and* $t_1 < t_2 < 2t_1$ *and* $s > t_2 - t_1$, *there is a constant* C_b *so that*

$$\int_0^\infty |L_b k^b_{t_2 - t_1 + s}(x, y) - L_b k^b_s(x, y)||\sqrt{x} - \sqrt{y}|^\gamma \, dy \le C_b (t_2 - t_1) s^{\frac{\gamma}{2} - 2}. \tag{6.49}$$

6.1.4 Large t Behavior

To study the resolvent kernel of L_b, which is formally given by

$$(\mu - L_b)^{-1} = \int_0^\infty e^{-\mu t} e^{t L_b} dt, \tag{6.50}$$

and the off-diagonal behavior of the heat kernel in many variables, it is useful to have estimates for

$$\int_0^\infty |\partial_x^j k^b_t(x, y)| dx, \text{ and } \int_0^\infty |x^{\frac{j}{2}} \partial_x^j k^b_t(x, y)| dy \tag{6.51}$$

valid for $0 < t$. In the previous section we gave such results, but these were intended to study the behavior of these kernels as $t \to 0^+$, and assumed the Hölder continuity of the data. To study the resolvent we also need estimates as $t \to \infty$, valid for bounded, continuous data.

LEMMA 6.1.21. *For $0 < b < B$, $0 < \phi < \frac{\pi}{2}$, and $j \in \mathbb{N}$ there is a constant $C_{j,B,\phi}$ so that if $t \in S_\phi$, then*

$$\int_0^\infty |\partial_x^j k_t^b(x,y)|dy \leq \frac{C_{j,B,\phi}}{|t|^j}, \tag{6.52}$$

and

$$\int_0^\infty |x^{\frac{j}{2}} \partial_x^j k_t^b(x,y)|dy \leq \frac{C_{j,B}}{|t|^{\frac{j}{2}}}. \tag{6.53}$$

6.1.5 The Structure of the Proofs of the Lemmas

We close this subsection by considering the structure of the proofs of these estimates. Recall that

$$k_t^b(x,y) = \frac{1}{y}\left(\frac{y}{t}\right)^b e^{-\frac{x+y}{t}} \psi_b\left(\frac{xy}{t^2}\right). \tag{6.54}$$

In most of the estimates that follow we set $w = y/|t|$, $\lambda = x/|t|$, and $e_\theta = e^{-i\theta}$; in these variables

$$k_t^b(x,y)dy = (e_\theta w)^{b-1} e^{-(w+\lambda)e_\theta} \psi_b(w\lambda e_{2\theta})dw. \tag{6.55}$$

Using Taylor's theorem when $w\lambda < 1$, and the asymptotic expansions for the functions, ψ_b, ψ_b' when $w\lambda \geq 1$, we repeatedly reduce our considerations to the estimation of a small collection of types of integrals. Most of these are integrals that extend from 0 to $1/\lambda$, or from $1/\lambda$ to ∞. We need to consider what happens as λ itself varies from 0 to ∞. The following results are used repeatedly in the proofs of the foregoing lemmas.

LEMMA 6.1.22. *For $\gamma > 0$, $0 < \phi < \frac{\pi}{2}$, there are constants $C_{b,\phi}$, $C'_{b,\phi}$ uniformly bounded for $0 < b < B$, so that for $0 < \lambda < \infty$, $|\theta| \leq \frac{\pi}{2} - \phi$, we have*

$$\int_0^{\frac{1}{\lambda}} w^{b-1} e^{-\cos\theta w} |\sqrt{w} - \sqrt{\lambda}|^\gamma dw \leq \begin{cases} \frac{C_{b,\phi}}{b} \lambda^{\frac{\gamma}{2}-b} \text{ as } \lambda \to \infty \\ \frac{C_{b,\phi}}{b} \lambda^{\frac{\gamma}{2}+b} + C'_{b,\phi} \text{ as } \lambda \to 0. \end{cases} \tag{6.56}$$

PROOF. The proof of this estimate follows easily from the change of variables $w = \lambda\sigma$. □

LEMMA 6.1.23. *For $\gamma \geq 0$, $0 < \phi < \frac{\pi}{2}$, and $\nu \in \mathbb{R}$, there are constants $C_{\nu,\gamma,\phi}$ and $a_{\nu,\gamma}$, uniformly bounded for $|\nu| < B$, so that for $0 < \lambda < \infty$, $|\theta| < \frac{\pi}{2} - \phi$, we* *have*

$$\int_{\frac{1}{\lambda}}^{\infty} w^{\frac{\nu}{2}} e^{-\cos\theta(\sqrt{w}-\sqrt{\lambda})^2} |\sqrt{w}-\sqrt{\lambda}|^{\gamma} \frac{dw}{\sqrt{w}} \leq$$

$$\begin{cases} C_{\nu,\gamma,\phi}\lambda^{a_{\nu,\gamma}} e^{-\frac{\cos\theta}{\lambda}} & \text{as } \lambda \to 0^+ \\ C_{\nu,\gamma,\phi}\lambda^{\frac{\nu}{2}} & \text{as } \lambda \to \infty. \end{cases} \tag{6.57}$$

PROOF. Setting $z = \sqrt{\frac{w}{\lambda}} - 1$, the integral in (6.57) becomes:

$$I = \frac{\lambda^{\frac{\nu+\gamma+1}{2}}}{2} \int_{\frac{1}{\lambda}-1}^{\infty} (1+z)^{\nu} e^{-\cos\theta\lambda z^2} |z|^{\gamma}\, dz. \tag{6.58}$$

The estimate as $\lambda \to 0$ follows easily from this and Lemma 6.1.24, proved below. To prove the result as $\lambda \to \infty$, we need to split the integral into the part from $\frac{1}{\lambda} - 1$ to $-\frac{1}{2}$, and the rest. A simple application of Laplace's method shows that the unbounded part is estimated by $C_{\nu,\gamma,\phi}\lambda^{\frac{\nu}{2}}$. We can estimate the compact part by

$$\frac{e^{-\cos\theta\frac{\lambda}{4}}\lambda^{\frac{\nu+\gamma+1}{2}}}{2} \int_{\frac{1}{\lambda}-1}^{-\frac{1}{2}} (1+z)^{\nu} =$$

$$\frac{e^{-\cos\theta\frac{\lambda}{4}}\lambda^{\frac{\nu+\gamma+1}{2}}}{2} \begin{cases} \frac{1}{\nu+1}\left(\left(\frac{1}{2}\right)^{\nu+1} - \left(\frac{1}{\lambda}\right)^{\nu+1}\right) & \text{if } \nu \neq -1 \\ \log\left(\frac{\lambda}{2}\right) & \text{if } \nu = -1. \end{cases} \tag{6.59}$$

In all cases this quantity is bounded by $C_{\nu,\gamma,\phi}e^{-\cos\theta\frac{\lambda}{8}}$, completing the proof of the lemma. □

The following lemma is used to prove these estimates:

LEMMA 6.1.24. *Letting $\mu \in \mathbb{R}$ and $a > 0$, we define*

$$G_{\mu}(\lambda, a) = \int_{a}^{\infty} e^{-\lambda z^2} z^{\mu}\, dz. \tag{6.60}$$

There are constants C_{μ} so that

$$G_{\mu}(\lambda, a) \leq C_{\mu}\frac{e^{-\lambda a^2}}{\lambda a^{1-\mu}} \quad \textit{for } a\sqrt{\lambda} > \frac{1}{2}. \tag{6.61}$$

For $\mu > -1$,

$$G_{\mu}(\lambda, a) \leq C_{\mu}\frac{1}{\lambda^{\frac{1+\mu}{2}}} \quad \textit{for } a\sqrt{\lambda} \leq \frac{1}{2}, \tag{6.62}$$

if $\mu = -1$, *then*

$$G_\mu(\lambda, a) \leq C_{-1}|\log(a\sqrt{\lambda})| \quad \textit{for } a\sqrt{\lambda} \leq \frac{1}{2}, \tag{6.63}$$

if $\mu < -1$, *then*

$$G_\mu(\lambda, a) \leq C_\mu a^{1+\mu} \quad \textit{for } a\sqrt{\lambda} \leq \frac{1}{2}. \tag{6.64}$$

PROOF. The proofs are elementary. A simple change of variables shows that

$$G_\mu(\lambda, a) = \frac{1}{\lambda^{\frac{1+\mu}{2}}} G_\mu(1, a\sqrt{\lambda}). \tag{6.65}$$

The second estimate is immediate from this formula and the fact that $G_\mu(1, 0)$ is finite, for $\mu > -1$. To prove the first relation we integrate by parts to obtain that

$$\int_w^\infty e^{-z^2} dz = -\frac{e^{-z^2}}{2z}\Big|_w^\infty + \int_w^\infty \frac{e^{-z^2}}{2z^2}. \tag{6.66}$$

This easily implies that

$$\int_w^\infty e^{-z^2} dz \leq \frac{e^{-w^2}}{w}, \tag{6.67}$$

which implies the first estimate. The final two estimates follow from the fact that $G_\mu(1, a\sqrt{\lambda})$ diverges as $a\sqrt{\lambda} \to 0$ at a rate determined by $\mu \leq -1$. □

6.2 HÖLDER ESTIMATES FOR THE 1-DIMENSIONAL MODEL PROBLEMS

With these rather extensive preliminaries out of the way, we now give the proofs for the Hölder estimates on solutions stated above.

PROOF OF PROPOSITION 6.0.10. We first assume that $b > 0$, and begin with (6.2) for the case $k = 0$. Using Proposition 5.2.8, Lemma 4.1.1, the $b = 0$ case follows from the $b > 0$ case. To prove the higher order estimates we need to assume that the data has support in $[0, R]$, then these results follow easily from the $k = 0$ case by using Propositions 5.2.8 and 5.2.9. From the maximum principle it is immediate that the sup-norm of v is bounded by $\|f\|_{WF,0,\gamma}$. In light of Lemma 6.1.1 it suffices to separately prove that

$$\begin{aligned} |v(x,t) - v(y,t)| &\leq C\|f\|_{WF,0,\gamma}|\sqrt{x} - \sqrt{y}|^\gamma \text{ and} \\ |v(x,t) - v(x,s)| &\leq C\|f\|_{WF,0,\gamma}|t-s|^{\frac{\gamma}{2}}. \end{aligned} \tag{6.68}$$

We start the spatial estimate, by estimating $|v(x,t)-v(0,t)|$. Because for every x, and $t>0$, (6.25) holds, we use the formula for k_t^b, to deduce that

$$v(x,t)-v(0,t)=\int_0^\infty \left[k_t^b(x,y)-k_t^b(0,y)\right](f(y)-f(0))dy. \tag{6.69}$$

The basic estimate (5.21) shows that

$$|v(x,t)-v(0,t)|=2\|f\|_{WF,0,\gamma}\int_0^\infty \left|k_t^b(x,y)-k_t^b(0,y)\right| y^{\frac{\gamma}{2}}dy; \tag{6.70}$$

we apply Lemma 6.1.4 to see that

$$|v(x,t)-v(0,t)|\leq C_b x^{\frac{\gamma}{2}}\|f\|_{WF,0,\gamma}, \tag{6.71}$$

for all $t>0$, and that, for any B, the $\{C_b\}$ are uniformly bounded for $0<b<B$.

This is a very useful estimate, for observe that if $M>1$, then

$$y^{\frac{\gamma}{2}}\leq\frac{M-1}{M}x^{\frac{\gamma}{2}}\Longrightarrow x^{\frac{\gamma}{2}}\leq M(x^{\frac{\gamma}{2}}-y^{\frac{\gamma}{2}}). \tag{6.72}$$

Thus (6.71) implies that if $y^{\frac{\gamma}{2}}\leq\frac{M-1}{M}x^{\frac{\gamma}{2}}$, then there is a constant $C_{\gamma,b}$, depending on M, so that v satisfies

$$\begin{aligned}|v(x,t)-v(y,t)|&\leq|v(x,t)-v(0,t)|+|v(0,t)-v(y,t)|\\&\leq 2C_{\gamma,b}\|f\|_{WF,0,\gamma}x^{\frac{\gamma}{2}}\\&\leq C_{\gamma,b}\|f\|_{WF,0,\gamma}(x^{\frac{\gamma}{2}}-y^{\frac{\gamma}{2}}).\end{aligned} \tag{6.73}$$

Applying Lemma 6.1.2 we see that (6.73) implies that

$$|v(x,t)-v(y,t)|\leq C_{\gamma,b}\|f\|_{WF,0,\gamma}|\sqrt{x}-\sqrt{y}|^{\gamma}. \tag{6.74}$$

To complete the spatial part of the estimate we just need to show that (6.74) holds, as $\frac{x}{t}\to\infty$, for pairs (x,y) so that

$$c\leq\frac{y}{x}<1, \tag{6.75}$$

with c a positive number less than 1. To that end we introduce a device, familiar from the Euclidean case, that will allow us to obtain the needed estimate. For pairs of points $0\leq x_1<x_2$ we define $J=[\alpha,\beta]$ where α and β are defined by

$$\sqrt{\alpha}=\max\left\{\frac{3\sqrt{x_1}-\sqrt{x_2}}{2},0\right\}\text{ and }\sqrt{\beta}=\frac{3\sqrt{x_2}-\sqrt{x_1}}{2}. \tag{6.76}$$

As noted above, this is the WF ball centered on the WF midpoint of $[x_1,x_2]$, with radius equal to $d_{WF}(x_1,x_2)$.

Using the fact that $k_t^b(x,y)$ has y-integral 1, we easily deduce that

$$\begin{gathered}
v(x_2,t) - v(x_1,t) = (f(x_2) - f(x_1)) + \\
\int\limits_J k_t^b(x_2,y)(f(y) - f(x_2))dy - \int\limits_J k_t^b(x_1,y)(f(y) - f(x_1))dy + \\
\int\limits_{J^c} k_t^b(x_2,y)(f(x_1) - f(x_2))dy + \int\limits_{J^c} (k_t^b(x_2,y) - k_t^b(x_1,y))(f(y) - f(x_1))dy.
\end{gathered} \tag{6.77}$$

It is a simple matter to see that the first four terms are estimated by

$$C\|f\|_{WF,0\gamma}|\sqrt{x_2} - \sqrt{x_1}|^\gamma, \tag{6.78}$$

leaving just the second integral over J^c. Terms of this type are estimated, for $c > 1/9$, in Lemma 6.1.7. Thus for $f \in \mathscr{C}^{0,\gamma}_{WF}(\mathbb{R}_+)$ Lemma 6.1.7 shows that there is a constant C independent of $b \leq B$ so that v satisfies (6.74).

We now turn to the time estimate. We begin by estimating $|v(x,t) - v(x,0)|$. Arguing as above we see that we have the estimate:

$$\begin{aligned}
|v(x,t) - v(x,0)| &= \left| \int\limits_0^\infty k_t^b(x,y)(f(y) - f(x))dy \right| \\
&\leq 2\|f\|_{WF,0,\gamma} \int\limits_0^\infty k_t^b(x,y)|\sqrt{x} - \sqrt{y}|^\gamma dy.
\end{aligned} \tag{6.79}$$

Integrals of this type are estimated in Lemma 6.1.6, which shows that

$$|v(x,t) - v(x,0)| \leq Ct^{\frac{\gamma}{2}}\|f\|_{WF,0,\gamma}. \tag{6.80}$$

Using the estimate in (6.72), we see that for $M > 1$, if $Ms^{\frac{\gamma}{2}} \leq (M-1)t^{\frac{\gamma}{2}}$, then (6.80) implies that

$$|v(x,t) - v(x,s)| \leq 2MC\|f\|_{WF,0,\gamma}|t^{\frac{\gamma}{2}} - s^{\frac{\gamma}{2}}|. \tag{6.81}$$

Using Lemma 6.1.2 this estimate gives

$$|v(x,t) - v(x,s)| \leq C_{\gamma,b}\|f\|_{WF,0,\gamma}|t - s|^{\frac{\gamma}{2}}, \tag{6.82}$$

for a constant $C_{\gamma,b}$ uniformly bounded for $b \leq B$. This leaves only the case of of pairs s, t with $c < s/t < 1$, for a $c < 1$.

To complete the last case, we write

$$\begin{aligned}
|v(x,t) - v(x,s)| &\leq \int\limits_0^\infty \left|k_t^b(x,y) - k_s^b(x,y)\right| |f(y) - f(x)|dy \\
&\leq 2\|f\|_{WF,0,\gamma} \int\limits_0^\infty \left|k_t^b(x,y) - k_s^b(x,y)\right| |\sqrt{x} - \sqrt{y}|^\gamma dy.
\end{aligned} \tag{6.83}$$

This case follows from Lemma 6.1.8. Using this lemma we easily complete the proof of the Proposition 6.0.10 for the case $k = 0$. The assertion that $v \in \mathscr{C}^{0,\gamma}_{WF}(\mathbb{R}_+ \times \mathbb{R}_+)$ follows easily from these estimates. Notice that Lemma 6.1.15 applies to show that even if f is only in $\mathscr{C}^{0,\gamma}_{WF}(\mathbb{R}_+)$ then

$$\lim_{x\to 0} x\partial_x^2 v(x,t) = 0 \text{ for any } t > 0. \tag{6.84}$$

To show that

$$\lim_{x\to\infty} v(x,t) = 0 \text{ for any } t > 0, \tag{6.85}$$

we fix an $R >> 0$ and write

$$v(x,t) = \int_0^R k_t^b(x,y)f(y)dy + \int_R^\infty k_t^b(x,y)f(y)dy. \tag{6.86}$$

Proposition 4.2.5 shows that for any fixed R the first term tends uniformly to zero as $x \to \infty$. As $f \in \mathscr{C}^{0,\gamma}_{WF}(\mathbb{R}_+)$ it follows that $\lim_{x\to\infty} f(x) = 0$. Hence given $\epsilon > 0$ we can choose R_0 so that $|f(x)| < \epsilon$ for $x > R_0$. For this choice of R_0, the second integral is at most ϵ, for all x, and the first tends to zero as $x \to \infty$. Thus

$$\limsup_{x\to\infty} |v(x,t)| \leq \epsilon, \tag{6.87}$$

which proves (6.85).

The estimates for the $(2+\gamma)$-spaces follow easily from what we have just proved and Lemma 4.1.1. This shows that if $f \in \dot{\mathscr{C}}^m_b([0,\infty))$, and $2j + k \leq m$, then

$$\partial_t^j \partial_x^k v(x,t) = \int_0^\infty k_t^{b+k}(x,y) L^j_{b+k} \partial_y^k f(y)dy. \tag{6.88}$$

In particular, the relations

$$\begin{aligned}
\partial_x v(x,t) &= \int_0^\infty k_t^{b+1}(x,y)\partial_y f(y)dy, \\
\partial_t v(x,t) &= \int_0^\infty k_t^b(x,y)L_b f(y)dy \text{ and} \\
x\partial_x^2 v(x,t) &= (\partial_t - b\partial_x)v,
\end{aligned} \tag{6.89}$$

and the γ-case, show that $\|v\|_{WF,0,2+\gamma}$ is bounded by $\|f\|_{WF,0,2+\gamma}$. Using these identities along with (6.84) and (6.85) allows us to conclude that

$$\lim_{x\to\infty} [|v(x,t)| + |\partial_x v(x,t)| + |x\partial_x^2 v(x,t)|] = 0. \tag{6.90}$$

We can therefore apply Lemma 5.2.3 to see that $v \in \mathscr{C}^{0,2+\gamma}_{WF}(\mathbb{R}_+ \times \mathbb{R}_+)$.

For the $k > 0$ cases, we need to assume that f is supported in $[0, R]$. Now, using (6.88) and Proposition 5.2.9 we easily derive (6.2), and (6.4) for any $k \in \mathbb{N}$, and can again conclude that $v \in \mathscr{C}^{k,2+\gamma}_{WF}(\mathbb{R}_+ \times \mathbb{R}_+)$, provided that $f \in \mathscr{C}^{k,2+\gamma}_{WF}(\mathbb{R}_+)$, and $\operatorname{supp} f \subset [0, R]$.

Finally we consider what happens as $b \to 0^+$. We begin with the $k = 0$ case; Proposition 7.8 in [15] shows that the solutions to the Cauchy problem for $b > 0$ converge uniformly to the solution with $b = 0$ in sets of the form $\mathbb{R}_+ \times [0, T]$. If $v^b(x, t)$ denotes these solutions, then we have established the existence of constants C_γ so that for $0 < b < 1$, $x \neq y$ and $t \neq s$, the following estimates hold:

$$|v^b(x,t) - v^b(y,s)| \leq C_\gamma \|f\|_{WF,0,\gamma} [|\sqrt{x} - \sqrt{y}|^\gamma + |t-s|^{\frac{\gamma}{2}}]. \tag{6.91}$$

As the constants C_γ are independent of b, we can let b tend to zero, and apply Proposition 7.8 of [15] to conclude that this estimate continues to hold for $b = 0$. Using (6.88) as above we can extend all the remaining estimates for the $\mathscr{C}^{k,\gamma}_{WF}$-spaces to the $b = 0$ case as well.

To treat the $\mathscr{C}^{k,2+\gamma}_{WF}$-spaces, we use Proposition 5.2.8. If $f \in \mathscr{C}^{k,2+\gamma}_{WF}(\mathbb{R}_+)$, then the solutions v^b to (4.2), with $g = 0$, $v^b(x, 0) = f(x)$ and $0 < b < 1$, are a bounded family in $\mathscr{C}^{k,2+\gamma}_{WF}(\mathbb{R}_+ \times \mathbb{R}_+)$. Thus for any $0 < \gamma' < \gamma$ there is a subsequence $< v^{b_n} >$ with $b_n \to 0$, which converges to $v^* \in \mathscr{C}^{k,2+\gamma'}_{WF}(\mathbb{R}_+ \times \mathbb{R}_+)$. Evidently v^* satisfies

$$(\partial_t - L_0)v^* = 0 \text{ with } v^*(x, 0) = f(x). \tag{6.92}$$

The uniqueness theorem implies that $v^* = v^0$, and therefore the family $< v^b >$ converges in $\mathscr{C}^{k,2+\gamma'}_{WF}(\mathbb{R}_+ \times \mathbb{R}_+)$ to v^0. Since each element of $\{v^b : 0 < b < 1\}$ satisfies the estimates in (6.4), with uniformly bounded constants, we conclude that $v^0 \in \mathscr{C}^{k,2+\gamma}_{WF}(\mathbb{R}_+ \times \mathbb{R}_+)$, and v^0 also satisfies the estimate in (6.4). □

We now turn to the proof of Proposition 6.0.11. Many parts of the foregoing argument can be recycled.

PROOF OF PROPOSITION 6.0.11. We begin by studying the operator:

$$K^b_t g(x) = \int_0^t \int_0^\infty k^b_{t-s}(x, y) g(y, s)\, dy\, ds, \tag{6.93}$$

assuming that $g \in \mathscr{C}^{0,\gamma}_{WF}(\mathbb{R}_+ \times [0, T])$. We want to show that

$$K^b_t : \mathscr{C}^{0,\gamma}_{WF}(\mathbb{R}_+ \times [0,T]) \to \mathscr{C}^{0,2+\gamma}_{WF}(\mathbb{R}_+ \times [0,T])$$

is bounded. This entails differentiating under the integral defining $K^b_t g$, which is somewhat subtle near to $s = t$. If $g \in \mathscr{C}^\infty$, then we can apply Corollary 7.6 of [15] to conclude that

$$\partial_t^j \partial_x^k L_b^l u(x,t) = \lim_{\epsilon\to 0+} \partial_t^j \partial_x^k L_b^l \int_0^{t-\epsilon}\int_0^{\infty} k_{t-s}^b(x,y)g(y,s)dyds. \tag{6.94}$$

In the arguments that follow we show that if g is sufficiently smooth, then the derivatives, $\partial_t^j \partial_x^k L_b^l u$, exist and can be defined by this limit. Once cancellations are taken into account, the limits are, in fact, absolutely convergent. Provided that g is sufficiently smooth, we may use Lemma 4.1.2 to bring derivatives past the kernel onto g.

Of special import is the case $g \in \mathscr{C}_{WF}^{0,\gamma}(\mathbb{R}_+ \times [0,T])$. For $0 < \epsilon < t$, we let:

$$u_\epsilon(x,t) = \int_0^{t-\epsilon}\int_0^{\infty} k_{t-s}^b(x,y)g(y,s)dyds. \tag{6.95}$$

It follows easily that u_ϵ converges uniformly to u in $\mathbb{R}_+ \times [0,T]$. Using the standard estimate on the difference

$$|g(x,t) - g(y,t)| \le \|g\|_{WF,0,\gamma,T} |\sqrt{x} - \sqrt{y}|^\gamma \tag{6.96}$$

and the facts that

$$\begin{aligned} \partial_x u_\epsilon(x,t) &= \int_0^{t-\epsilon}\int_0^{\infty} \partial_x k_{t-s}^b(x,y)[g(y,s) - g(x,s)]dyds \\ x\partial_x^2 u_\epsilon(x,t) &= \int_0^{t-\epsilon}\int_0^{\infty} x\partial_x^2 k_{t-s}^b(x,y)[g(y,s) - g(x,s)]dyds, \end{aligned} \tag{6.97}$$

we can apply Lemmas 6.1.10 and 6.1.15 to establish the uniform convergence of $\partial_x u_\epsilon$ and $x\partial_x^2 u_\epsilon$ on $\mathbb{R}_+ \times [0,T]$. This establishes the continuous differentiability of u in x on $[0,\infty) \times [0,T]$, and the twice continuous differentiability of u in x on $(0,\infty)\times[0,T]$. We can differentiate u_ϵ in t to obtain that

$$\partial_t u_\epsilon(x,t) = \int_0^{\infty} k_\epsilon^b(x,y)g(y,t-\epsilon) + [x\partial_x^2 + b\partial_x]u_\epsilon(x,t). \tag{6.98}$$

The right-hand side converges uniformly to $g(x,t) + (x\partial_x^2 + b\partial_x)u(x,t)$, thereby establishing the continuous differentiability of u in t and the fact that

$$[\partial_t - (x\partial_x^2 + b\partial_x)]u = g \text{ for } (x,t) \in \mathbb{R}_+ \times [0,T]. \tag{6.99}$$

This argument, or a variant thereof, is used repeatedly to establish the differentiability of u, the formulæ for its derivatives:

$$\begin{aligned}\partial_x u(x,t) &= \int_0^t\int_0^\infty \partial_x k^b_{t-s}(x,y)[g(y,s)-g(x,s)]dyds\\ x\partial_x^2 u(x,t) &= \int_0^t\int_0^\infty x\partial_x^2 k^b_{t-s}(x,y)[g(y,s)-g(x,s)]dyds,\end{aligned} \tag{6.100}$$

along with the fact that, for $g \in \mathscr{C}^{0,\gamma}_{WF}(\mathbb{R}_+ \times [0,T])$, these are absolutely convergent integrals.

We let $u(x,t) = K^b_t g(x)$. From the maximum principle it is evident that

$$|u(x,t)| \leq t\|g\|_{L^\infty}. \tag{6.101}$$

The estimate in (6.74) can be integrated to prove that

$$|u(x,t)-u(y,t)| \leq C_b t\|g\|_{WF,0,\gamma}|\sqrt{x}-\sqrt{y}|^\gamma. \tag{6.102}$$

Using (6.105) and (6.110), proved below, and the equation $\partial_t u = L_b u + g$, we see that, for $s < t$,

$$\begin{aligned}|u(x,t)-u(x,s)| \leq C(1+t^{\frac{\gamma}{2}})\|g\|_{WF,0,\gamma}|t-s|+&\\ Ct^{1-\frac{\gamma}{2}}(1+t^{\frac{\gamma}{2}})\|g\|_{WF,0,\gamma}|t-s|^{\frac{\gamma}{2}}&.\end{aligned} \tag{6.103}$$

Note that (6.101), (6.102) and (6.103) show that there is a constant C so that

$$\|u\|_{WF,0,\gamma,T} \leq CT^{1-\frac{\gamma}{2}}(1+T^{\frac{\gamma}{2}})\|g\|_{WF,0,\gamma,T}. \tag{6.104}$$

Below we show that there is a constant C_b so that

$$|x\partial_x^2 u(x,t)| \leq C_b\min\{x^{\frac{\gamma}{2}},t^{\frac{\gamma}{2}}\}\|g\|_{WF,0,\gamma}; \tag{6.105}$$

dividing by x and integrating gives the Hölder estimate for the first spatial derivative:

$$|\partial_x u(x,t)-\partial_x u(y,t)| \leq C_b\|g\|_{WF,0,\gamma}|x^{\frac{\gamma}{2}}-y^{\frac{\gamma}{2}}|. \tag{6.106}$$

Lemma 6.1.2 then implies that

$$|\partial_x u(x,t)-\partial_x u(y,t)| \leq C_b\|g\|_{WF,0,\gamma}|\sqrt{x}-\sqrt{y}|^\gamma. \tag{6.107}$$

To complete the analysis of $\partial_x u$ we need to show that there is a constant C_b so that

$$|\partial_x u(x,t)-\partial_x u(x,s)| \leq C_b\|g\|_{WF,0,\gamma}|t-s|^{\frac{\gamma}{2}}. \tag{6.108}$$

In Lemma 6.1.10 it is shown that there are constants C_b, uniformly bounded for $b < B$, so that, with $\lambda = x/t$, we have

$$|\partial_x v(x,t)| \leq C\|f\|_{WF,0,\gamma}\frac{t^{\frac{\gamma}{2}-1}}{1+\lambda^{\frac{1}{2}}} = Cx^{\frac{\gamma}{2}-1}\|f\|_{WF,0,\gamma}\frac{\lambda^{1-\frac{\gamma}{2}}}{1+\lambda^{\frac{1}{2}}}. \tag{6.109}$$

It follows by integrating that

$$|\partial_x u(x,t)| \leq C_b\|g\|_{WF,0,\gamma}t^{\frac{\gamma}{2}}, \tag{6.110}$$

and therefore, for any $c < 1$, there is a C so that if $s < ct$, then (6.108) holds with $C_b = C$. We are left to consider $ct_2 < t_1 < t_2$, for any $c < 1$. For $\frac{1}{2}t_2 < t_1 < t_2$ we have:

$$\partial_x u(x,t_2) - \partial_x u(x,t_1) = \int\limits_0^{t_2-t_1}\int\limits_0^{\infty}\partial_x k_s^b(x,y)[g(y,t_2-s)-g(y,t_1-s)]dyds+$$

$$\int\limits_0^{2t_1-t_2}\int\limits_0^{\infty}[\partial_x k_{t_2-s}^b(x,y)-\partial_x k_{t_1-s}^b(x,y)](g(y,s)-g(x,s))dyds+$$

$$\int\limits_{2t_1-t_2}^{t_1}\int\limits_0^{\infty}\partial_x k_{t_2-s}^b(x,y)(g(y,s)-g(x,s))dyds. \tag{6.111}$$

To handle the first term, we observe that, for $j = 1,2$ we have

$$\int\limits_0^{t_2-t_1}\int\limits_0^{\infty}\partial_x k_s^b(x,y)g(y,t_j-s)dyds =$$

$$\int\limits_0^{t_2-t_1}\int\limits_0^{\infty}\partial_x k_s^b(x,y)[g(y,t_j-s)-g(x,t_j-s)]dyds, \tag{6.112}$$

which can be estimated by

$$\|g\|_{WF,0,\gamma}\int\limits_0^{t_2-t_1}\int\limits_0^{\infty}|\partial_x k_s^b(x,y)||\sqrt{x}-\sqrt{y}|^{\frac{\gamma}{2}}dyds. \tag{6.113}$$

Using Lemma 6.1.10, we see that these terms are bounded by the right-hand side of (6.108). In the third integral in (6.111) we use (5.32) to estimate $(g(y,s)-g(x,s))$, and again apply Lemma 6.1.10 to see that this term is also bounded by the right-hand side of (6.108). This leaves only the second integral in (6.111). To estimate this term we use Lemma 6.1.12.

We now establish (6.105), and then the Hölder continuity of $x\partial_x^2 u(x,t)$. Because $k_t^b(x,y)$ integrates to 1 w.r.t. y, for any x, and $t > 1$, it follows from (6.94) that

$$x\partial_x^2 u(x,t) = \int_0^t \int_0^\infty x\partial_x^2 k_s^b(x,y)[g(y,t-s) - g(x,t-s)]dyds. \tag{6.114}$$

Using the estimate

$$|g(y,t-s) - g(x,t-s)| \leq 2\|g\|_{WF,0,\gamma} |\sqrt{x} - \sqrt{y}|^\gamma \tag{6.115}$$

and Lemma 6.1.14 gives:

$$\begin{aligned} |x\partial_x^2 u(x,t)| &\leq \|g\|_{WF,0,\gamma} \int_0^t \int_0^\infty |x\partial_x^2 k_s^b(x,y)||\sqrt{y} - \sqrt{x}|^\gamma dyds \\ &\leq C_b \|g\|_{WF,0,\gamma} \min\{t^{\frac{\gamma}{2}}, x^{\frac{\gamma}{2}}\}. \end{aligned} \tag{6.116}$$

This completes the proof of (6.105), and therefore the proof of the spatial Hölder continuity of $\partial_x u$. This argument also establishes that

$$\lim_{x \to 0^+} x\partial_x^2 u(x,t) = 0. \tag{6.117}$$

To verify the hypotheses of Lemma 5.2.3 we also need to show that

$$\lim_{x \to \infty} [|u(x,t)| + |\partial_x u(x,t)| + |x\partial_x^2 u(x,t)|] = 0. \tag{6.118}$$

The claim for $|u(x,t)|$ follows as in the proof of (6.85). To estimate the derivatives we need to split the integral defining u into a compact and non-compact part, though more carefully than before.

Let $\varphi \in \mathscr{C}^\infty$ satisfy

$$\varphi(x) = 1 \text{ for } x \leq 0, \quad \varphi(x) = 0 \text{ for } x > 1. \tag{6.119}$$

For $R, m \in (0, \infty)$ we let

$$\varphi_{R,m}(x) = \varphi\left(\frac{x-R}{m}\right). \tag{6.120}$$

Using the mean value theorem we can easily show that there is a constant, C, independent of γ, R, m so that

$$\left[\varphi_{R,m}\right]_{WF,0,\gamma} \leq \frac{C\sqrt{R+m}}{m}. \tag{6.121}$$

We let $m = R$, so that $\lim_{R\to\infty} \left[\varphi_{R,m}\right]_{WF,0,\gamma} = 0$.

Define

$$
\begin{aligned}
u_R^0(x,t) &= \int_0^T \int_0^\infty k_s^b(x,y)\varphi_{R,R}(y)g(y,t-s)dyds \\
u_R^\infty(x,t) &= \int_0^T \int_0^\infty k_s^b(x,y)(1-\varphi_{R,R}(y))g(y,t-s)dyds.
\end{aligned} \tag{6.122}
$$

For any fixed R, it follows from Proposition 4.2.5 that

$$\lim_{x\to\infty}[|\partial_x u_R^0(x,t)| + |x\partial_x^2 u_R^0(x,t)|] = 0. \tag{6.123}$$

Given $\epsilon > 0$ we can choose R so that

$$\|(1-\varphi_{R,R}(y))g\|_{L^\infty} < \epsilon. \tag{6.124}$$

Fix a $0 < \gamma' < \gamma$. Applying (5.31) with (6.121) and (6.124) along with Lemma 5.2.6 we see that

$$\lim_{R\to\infty}\|(1-\varphi_{R,R}(y))g\|_{WF,0,\gamma'} = 0. \tag{6.125}$$

It now follows from (6.105) and (6.110) that, for a possibly larger R, we have the estimate:

$$|\partial_x u_R^\infty(x,t)| + |x\partial_x^2 u_R^\infty(x,t)| \leq \epsilon. \tag{6.126}$$

Combining this with (6.123) we easily complete the proof of (6.118).

To finish the spatial estimate, we need only show that $x\partial_x^2 u$ is Hölder continuous. As before, the estimate (6.105) implies that for any $c < 1$, there is a constant C so that

$$x_1 < cx_2 \Longrightarrow |x_2\partial_x^2 u(x_2,t) - x_1\partial_x^2 u(x_1,t)| \leq C\|g\|_{WF,0,\gamma}|\sqrt{x_2}-\sqrt{x_1}|^\gamma. \tag{6.127}$$

We are left to consider pairs x_1, x_2 with

$$cx_2 < x_1 < x_2. \tag{6.128}$$

Since we have already established the Hölder continuity of the first derivative, it suffices to show that $L_b u(x,t)$ is Hölder continuous, which technically, is a little easier.

This is a rather delicate estimate; we need to decompose the integral expression for $L_b u(x_1,t) - L_b u(x_2,t)$ as in (6.77). We use the notation introduced there, with

$J = [\alpha, \beta]$, etc.

$$
\begin{aligned}
L_b u(x_2, t) - L_b u(x_1, t) = \\
&\int_0^t \Bigg[\int_J L_b k_s^b(x_2, y)(g(y, t-s) - g(x_2, t-s))dy - \\
&\int_J L_b k_s^b(x_1, y)(g(y, t-s) - g(x_1, t-s))dy - \\
&\int_{J^c} L_b k_s^b(x_2, y)(g(x_2, t-s) - g(x_1, t-s))dy + \\
&\int_{J^c} (L_b k_s^b(x_2, y) - L_b k_s^b(x_1, y))(g(y, t-s) - g(x_1, t-s))dy \Bigg] ds \\
&= I_1 + I_2 + I_3 + I_4.
\end{aligned}
\tag{6.129}
$$

In this formula the operator L_b acts in the x-variable. The justification for this formula is essentially identical to that given for (6.114).

We begin by estimating I_3. For this purpose we observe that, for $t > 0$, we have:

$$
L_{b,x} k_t^b(x, y) = \partial_t k_t^b(x, y) = L_{b,y}^t k_t^b(x, y). \tag{6.130}
$$

The operator $L_{b,y}^t = \partial_y(\partial_y y - b)$, so we can perform the y-integral to obtain that

$$
I_3 = \int_0^t \left[(\partial_y y - b)k_s^b(x_2, \alpha) - (\partial_y y - b)k_s^b(x_2, \beta) \right] (g(x_2, t-s) - g(x_1, t-s))ds. \tag{6.131}
$$

As usual we use the estimate

$$
|g(x_2, t-s) - g(x_1, t-s)| \leq 2|\sqrt{x_2} - \sqrt{x_1}|^{\gamma}. \tag{6.132}
$$

Lemma 6.1.16 therefore completes this step; it shows that there is a constant C_b so that

$$
|I_3| \leq C_b \|g\|_{WF,0,\gamma} |\sqrt{x_1} - \sqrt{x_2}|^{\gamma}. \tag{6.133}
$$

We now turn to the compactly supported terms I_1 and I_2. These terms are estimated by

$$
\begin{aligned}
&\|g\|_{WF,0,\gamma} \int_0^t \int_\alpha^\beta |L_b k_s^b(x_2, y)| |\sqrt{y} - \sqrt{x_2}|^{\gamma} dy ds \\
&\|g\|_{WF,0,\gamma} \int_0^t \int_\alpha^\beta |L_b k_s^b(x_2, y)| |\sqrt{y} - \sqrt{x_1}|^{\gamma} dy ds.
\end{aligned}
\tag{6.134}
$$

The needed bounds are given in Lemma 6.1.17. This lemma shows that the terms I_1 and I_2 are estimated by

$$C_b\|g\|_{WF,0,\gamma}|\sqrt{x_2}-\sqrt{x_1}|^{\gamma}. \tag{6.135}$$

This leaves only the non-compact term, I_4. Recall that

$$I_4=\int_0^t\int_{J^c}(L_bk_s^b(x_2,y)-L_bk_s^b(x_1,y))(g(y,t-s)-g(x_1,t-s))dyds, \tag{6.136}$$

and that $J^c=[0,\alpha)\cup(\beta,\infty)$. We use (5.32) to estimate the difference $|g(y,t-s)-g(x_1,t-s)|$, hence Lemma 6.1.18 completes this case. Using Lemma 6.1.18 we see that I_4 also satisfies the bound in (6.135), which therefore completes the proof of the spatial part of the Hölder estimate. To complete the $k=0$ case all that remains is to estimate $|L_bu(x,t_1)-L_bu(x,t_2)|$.

The time estimate begins very much as the estimate for $|v(x,t_2)-v(x,t_1)|$; we first show that

$$|L_bu(x,t)|\leq C_b\|g\|_{WF,0,\gamma}t^{\frac{\gamma}{2}}. \tag{6.137}$$

This implies that for any $M>1$, there is a $C_{M,b}$ so that if $t_2>Mt_1$, then

$$|L_bu(x,t_2)-L_bu(x,t_1)|\leq C_{M,b}\|g\|_{WF,0,\gamma}|t_2-t_1|^{\frac{\gamma}{2}}, \tag{6.138}$$

which leaves only the case that $1<t_2/t_1<M$.

To prove (6.137) we use Lemma 6.1.10 and Lemma 6.1.15. The estimate in (6.110) shows that to prove (6.137) it suffices to show that

$$|x\partial_x^2u(x,t)|\leq C_b\|g\|_{WF,0,\gamma}t^{\frac{\gamma}{2}}. \tag{6.139}$$

To prove this we write

$$x\partial_x^2u(x,t)=\int_0^t\int_0^\infty x\partial_x^2k_s(x,y)[g(y,t-s)-g(x,t-s)]dyds, \tag{6.140}$$

which implies that

$$\begin{aligned}|x\partial_x^2u(x,t)|&\leq 2\|g\|_{WF,0,\gamma}\int_0^t\int_0^\infty|x\partial_x^2k_s^b(x,y)||\sqrt{x}-\sqrt{y}|^{\gamma}dyds\\&\leq C\|g\|_{WF,0,\gamma}\int_0^t s^{\frac{\gamma}{2}-1}\left(\frac{x/s}{1+x/s}\right)ds.\end{aligned} \tag{6.141}$$

The second line follows from Lemma 6.1.15; an elementary argument shows that the last integral is bounded by a constant times $t^{\frac{\gamma}{2}}$, completing the proof of (6.137).

To complete the time estimate we need to show that, for $M > 1$, there is a constant $C_{M,b}$, so that $x_1 < x_2 < Mx_1$ implies that

$$|L_b u(x, t_2) - L_b u(x, t_1)| \leq C_{M,b} \|g\|_{WF,0,\gamma} |t_2 - t_1|^{\frac{\gamma}{2}}. \tag{6.142}$$

The proof of (6.142), for the remaining cases, is broken into several parts, where we observe that, for $t_1 < t_2 < 2t_1$ we have:

$$L_b u(x, t_2) - L_b u(x, t_1) = \int_0^{t_2-t_1} \int_0^{\infty} L_b k_s^b(x, y)[g(y, t_2 - s) - g(y, t_1 - s)]dyds +$$
$$\int_0^{2t_1-t_2} \int_0^{\infty} [L_b k_{t_2-s}^b(x, y) - L_b k_{t_1-s}^b(x, y)](g(y, s) - g(x, s))dyds +$$
$$\int_{2t_1-t_2}^{t_1} \int_0^{\infty} L_b k_{t_2-s}^b(x, y)(g(y, s) - g(x, s))dyds. \tag{6.143}$$

We denote these terms by I_1, I_2 and I_3.

We start by estimating I_1, which we split into two parts, $I_{11} - I_{12}$; each part we rewrite as

$$I_{1j} = \int_0^{t_2-t_1} \int_0^{\infty} L_b k_s^b(x, y)[g(x, t_j - s) - g(y, t_j - s)]dyds \quad j = 1, 2. \tag{6.144}$$

Indeed this really explains the meaning of this term as a convergent integral. These are estimated, using the same argument, after we employ the estimate:

$$|(g(y, t_j - s) - g(x, t_j - s))| \leq 2\|g\|_{WF,0,\gamma} |\sqrt{x} - \sqrt{y}|^{\gamma}. \tag{6.145}$$

This shows, using Lemma 6.1.15, that

$$\begin{aligned} |I_{1j}| &\leq 2\|g\|_{WF,0,\gamma} | \int_0^{t_2-t_1} \int_0^{\infty} |L_b k_s^b(x, y)| |\sqrt{x} - \sqrt{y}|^{\gamma} dyds \\ &\leq C\|g\|_{WF,0,\gamma} | \int_0^{t_2-t_1} s^{\frac{\gamma}{2}-1} \left(\frac{x/s}{1 + x/s}\right) ds. \end{aligned} \tag{6.146}$$

An elementary argument now applies to show that this is bounded by

$$C\|g\|_{WF,0,\gamma} |t_2 - t_1|^{\frac{\gamma}{2}}.$$

Essentially the same argument works to estimate I_3, which, using (6.145), satisfies:

$$\begin{aligned}|I_3| &\leq 2\|g\|_{WF,0,\gamma} \int\limits_{2t_1-t_2}^{t_1} \int\limits_0^{\infty} |L_b k^b_{t_2-s}(x,y)||\sqrt{x}-\sqrt{y}|^{\gamma}\, dyds \\ &= 2\|g\|_{WF,0,\gamma} \int\limits_{t_2-t_1}^{2(t_2-t_1)} \int\limits_0^{\infty} |L_b k^b_s(x,y)||\sqrt{x}-\sqrt{y}|^{\gamma}\, dyds.\end{aligned} \tag{6.147}$$

The last line is again estimated using Lemma 6.1.15.

To complete the proof in the $k = 0$ case all that remains is to estimate I_2. This term is bounded by applying Lemma 6.1.19. This completes the proof of the Hölder estimates for $L_b u(x,t)$ in the $k = 0$ case. As

$$\partial_t u = L_b u + g, \tag{6.148}$$

the estimates on $\partial_t u$ are now an immediate consequence. Using equations (6.117) and (6.118) we apply Lemma 5.2.3 to conclude that $u \in \mathscr{C}^{0,2+\gamma}_{WF}(\mathbb{R}_+ \times [0,T])$, which completes the proof of (6.6) in the $k = 0$ case.

For the $k > 0$ cases, we need to add the assumption that $\operatorname{supp} g \subset \{(x,t) : x \in [0,R]\}$. To prove the higher order estimates we use Lemma 4.1.2, which is an extension of Corollaries 7.6 and 7.7 of [15], and Proposition 5.2.9 to reduce the higher order estimates to the $k = 0$ case and elementary estimates for functions in $\mathscr{C}^{k,\gamma}_{WF}(\mathbb{R} \times [0,T])$.

All that remains is to consider what happens as $b \to 0$. The estimates proved above hold uniformly for $b > 0$, with uniform bounds on the constants C_b for b in bounded subsets of $[0,\infty)$. If $g \in \mathscr{C}^{0,\gamma}_{WF}(\mathbb{R}_+ \times [0,T])$, then the solutions u^b are uniformly bounded in $\mathscr{C}^{0,2+\gamma}_{WF}(\mathbb{R}_+ \times [0,T])$. Proposition 5.2.8 implies that if $0 < \gamma' < \gamma$, then there is a subsequence $< u^{b_n} >$, with $b_n \to 0$, that converges to some function u^* in $\mathscr{C}^{0,2+\gamma'}_{WF}(\mathbb{R}_+ \times [0,T])$. Since u^* solves

$$(\partial_t - L_0)u^* = g \text{ and } u^*(x,0) = 0, \tag{6.149}$$

the uniqueness of the solution implies that in fact $u^* = u^0$, and that

$$\lim_{b \to 0^+} u^b = u^0 \text{ in } \mathscr{C}^{0,2+\gamma'}_{WF}(\mathbb{R}_+ \times [0,T]). \tag{6.150}$$

We can therefore take limits in the estimates satisfied by u^b for $b > 0$, to conclude that, in fact, $u^0 \in \mathscr{C}^{0,2+\gamma}_{WF}(\mathbb{R}_+ \times [0,T])$, and satisfies (6.6), with $k = 0$. The higher order estimates for the $b = 0$ case follow from this argument and (4.16).

The final claims in Proposition 6.0.11, asserting the convergence of $u(x,t)$ to 0, as $t \to 0^+$, with respect to the $\mathscr{C}^{k,2+\gamma'}$-norms, follow easily from these estimates, Proposition 5.2.8, and Lemma 6.0.12. This completes the proof of Proposition 6.0.11.

□

6.3 PROPERTIES OF THE RESOLVENT OPERATOR

We conclude this section by proving the estimates for $R(\mu)f$ stated in Proposition 6.0.13.

PROOF OF PROPOSITION 6.0.13. As in the proofs of the previous results, we begin by establishing these results for the $k = 0$ case, and arbitrary $0 < b$. The cases of arbitrary $k \in \mathbb{N}$ are obtained using Lemma 4.1.1. As noted earlier, no assumption about the support of the data is needed for the resolvent operator, since we do not have to estimate time derivatives. Hence we only require (6.88) with $q = 0$. We fix an angle $0 < \phi \leq \frac{\pi}{2}$.

We begin by showing that if $f \in \mathscr{C}^{0,\gamma}_{WF}(\mathbb{R}_+)$, then $R(\mu)f \in \mathscr{C}^2_{WF}(\mathbb{R}_+)$. First we see that Lemma 6.1.3 implies that

$$\|R(\mu)f\|_{L^\infty} \leq C_{b,\phi}\frac{\|f\|_{L^\infty}}{|\mu|}. \tag{6.151}$$

The argument in the proof of Proposition 6.0.10 between (6.69) and (6.78) applies mutatis mutandis to show that there is a constant $C_{b,\phi}$, so that if $t \in S_\phi$, then

$$|v(x,t) - v(y,t)| \leq C_{b,\phi}\|f\|_{WF,0,\gamma}|\sqrt{x} - \sqrt{y}|^\gamma. \tag{6.152}$$

Integrating this shows that there is a constant $C_{b,\alpha,\gamma}$ so that

$$\int_0^\infty [v(\cdot,t)]_{WF,0,\gamma}\,|e^{-\mu t}dt| \leq C_{b,\alpha,\gamma}\frac{\|f\|_{WF,0,\gamma}}{|\mu|}, \tag{6.153}$$

completing the proof of (6.12).

Next observe that, for $t \in S_0$, we have the formulæ:

$$\begin{aligned}\partial_x v(x,t) &= \int_0^\infty \partial_x k_t^b(x,y)(f(y) - f(x))dy\\ x\partial_x^2 v(x,t) &= \int_0^\infty x\partial_x^2 k_t^b(x,y)(f(y) - f(x))dy.\end{aligned} \tag{6.154}$$

Using the first formula and the estimate in Lemma 6.1.10 we see that, for $t \in S_\phi$

$$|\partial_x v(x,t)| \leq C_{b,\phi}\|f\|_{WF,0,\gamma}\frac{|t|^{\frac{\gamma}{2}-1}}{1 + \sqrt{x/|t|}}. \tag{6.155}$$

If $|\arg\mu| < \frac{\pi}{2} + \phi$, then, by choosing an appropriate ray in the right half plane we see that there is a positive constant m_ϕ, so that

$$\begin{aligned}|\partial_x R(\mu)f(x)| &\leq C_{b,\phi}\|f\|_{WF,0,\gamma}\int_0^\infty \frac{e^{-m_\phi|\mu|s}s^{\frac{\gamma}{2}-1}ds}{1 + \sqrt{x/s}}\\ &\leq C'_{b,\phi}\frac{\|f\|_{WF,0,\gamma}}{(m_\phi|\mu|)^{\frac{\gamma}{2}}}.\end{aligned} \tag{6.156}$$

Using the estimate in Lemma 6.1.15 we can show that

$$
\begin{aligned}
|x\partial_x^2 R(\mu) f(x)| &\leq C_{b,\phi} \|f\|_{WF,0,\gamma} \int_0^\infty \frac{e^{-m_\phi|\mu|s} x s^{\frac{\gamma}{2}-1} ds}{x+s} \\
&\leq C'_{b,\phi} \frac{\|f\|_{WF,0,\gamma}}{(m_\phi|\mu|)^{\frac{\gamma}{2}}}.
\end{aligned} \tag{6.157}
$$

It is useful to note that by splitting this integral into a part from 0 to x and the rest, we can also show that

$$
|x\partial_x^2 R(\mu) f(x)| \leq C'_{b,\phi} \|f\|_{WF,0,\gamma} x^{\frac{\gamma}{2}}. \tag{6.158}
$$

These estimates show that $R(\mu) f \in \mathscr{C}^2_{WF}(\mathbb{R}_+)$ and, by integrating (6.158), establish the Hölder estimate on the first derivative:

$$
[\partial_x R(\mu) f]_{WF,0,\gamma} \leq C_{b,\phi} \|f\|_{WF,0,\gamma}. \tag{6.159}
$$

With these estimates in hand, we can integrate by parts to establish that

$$
(\mu - L_b) R(\mu) f = f \text{ for } \mu \in \mathbb{C} \setminus (-\infty, 0]. \tag{6.160}
$$

Below we show that $R(\mu) f \in \mathscr{C}^{0,2+\gamma}_{WF}(\mathbb{R}_+)$. By the open mapping theorem, to show that $R(\mu)$ is also a left inverse for $(\mu - L_b)$ it suffices to show that the null-space of $\mu - L_b$ is trivial. For $\mu \in S_0$, this follows immediately from the estimate in (6.2), and the uniqueness of the solution to the Cauchy problem. If, for some $\mu \in S_0$, there were a solution $f \in \mathscr{C}^{0,2+\gamma}_{WF}(\mathbb{R}_+)$ to $L_b f = \mu f$, then the solution to the Cauchy problem with this initial data would be $v(x,t) = e^{\mu t} f$. This solution grows exponentially, contradicting (6.2). Thus for $\mu \in S_0$, and $f \in \mathscr{C}^{0,2+\gamma}_{WF}(\mathbb{R}_+)$ we also have the identity

$$
R(\mu)(\mu - L_b) f = f. \tag{6.161}
$$

The permanence of functional relations implies that this holds for $\mu \in \mathbb{C} \setminus (-\infty, 0]$.

We can also apply the observation in (6.72), along with the estimate in (6.158), to see that if any $0 < c < 1$ is fixed, then there is a $C_{b,c,\phi}$ so that if $y < cx$, then

$$
|x\partial_x^2 R(\mu) f(x) - y\partial_x^2 R(\mu) f(y)| \leq C_{b,c,\phi} \|f\|_{WF,0,\gamma} |\sqrt{x} - \sqrt{y}|^\gamma. \tag{6.162}
$$

To complete the proof that $R(\mu) f \in \mathscr{C}^{0,2+\gamma}_{WF}(\mathbb{R}_+)$ we only need to show that there is a $0 < c < 1$, so that a similar estimate holds for $cx < y < x$.

This is accomplished exactly as in the proof of Proposition 6.0.11. It suffices to estimate $[L_bR(\mu)f]_{WF,0,\gamma}$, and use a decomposition like that given in (6.129):

$$L_bR(\mu)f(x_2) - L_bR(\mu)f(x_1) =$$

$$\int_0^\infty e^{\mu s e^{i\theta}}\Bigg[\int_J L_b k^b_{se^{i\theta}}(x_2,y)(f(y)-f(x_2))dy -$$

$$\int_J L_b k^b_{se^{i\theta}}(x_1,y)(f(y)-f(x_1))dy -$$

$$\int_{J^c} L_b k^b_{se^{i\theta}}(x_2,y)(f(x_2)-f(x_1))dy +$$

$$\int_{J^c} [L_b k^b_{se^{i\theta}}(x_2,y) - L_b k^b_{se^{i\theta}}(x_2,y)](f(y)-f(x_1))dy\Bigg] e^{i\theta}ds$$

$$= I_1 + I_2 + I_3 + I_4. \quad (6.163)$$

Here we select $\theta \in (-\frac{\pi}{2}, \frac{\pi}{2})$, so that

$$|\arg \mu e^{i\theta}| < \frac{\pi}{2}. \quad (6.164)$$

Fix a positive constant $1/3 < c < 1$, and assume that $cx_2 < x_1 < x_2$, so that we can apply Lemmas 6.1.16–6.1.18 as in the earlier argument. We use the fact that $L_{b,x}k^b_t(x,y) = L^t_{b,y}k^b_t(x,y)$, to perform the y-integral in I_3. As the estimate in Lemma 6.1.16 holds uniformly for all $|t|$, it applies to show that

$$|I_3| \leq C_{b,\phi}\|f\|_{WF,0,\gamma}|\sqrt{x_2}-\sqrt{x_1}|^\gamma. \quad (6.165)$$

The estimates in Lemmas 6.1.17 and 6.1.18 also apply uniformly, for all $|t|$, and show that $|I_1|$, $|I_2|$, and $|I_4|$ each satisfy an estimate of the same form, thereby completing the proof that

$$\left[\!\left[x\partial_x^2 R(\mu)f\right]\!\right]_{WF,0,\gamma} \leq C_{b,\phi}\|f\|_{WF,0,\gamma}. \quad (6.166)$$

This completes the $k = 0$ case for $b > 0$. The case of $b = 0$ is obtained by using the fact that the constants in the estimates are uniformly bounded for $0 < b < 1$, and Proposition 5.2.8. This shows that if we let $f \in \mathscr{C}^{0,\gamma}_{WF}(\mathbb{R}_+)$ and set $w_b = (\mu - L_b)^{-1}f$ for $0 < b$, then $\{w_b : 0 < b < 1\}$ are uniformly bounded in $\mathscr{C}^{0,2+\gamma}_{WF}(\mathbb{R}_+)$. Proposition 5.2.8 shows that for any $0 < \tilde{\gamma} < \gamma$, this sequence has a subsequence that converges in $\mathscr{C}^{0,2+\tilde{\gamma}}_{WF}(\mathbb{R}_+)$. Any such limit $w_0 \in \mathscr{C}^{0,2+\gamma}_{WF}(\mathbb{R}_+)$ and satisfies

$$(\mu - L_0)w_0 = f. \quad (6.167)$$

The uniqueness result, Proposition 3.3.4, shows that w_0 is uniquely determined, which implies that $\{w_b\}$ itself converges in $\mathscr{C}^{0,2+\tilde{\gamma}}_{WF}(\mathbb{R}_+)$ to w_0, and that w_0 therefore satisfies the estimates in the statement of the proposition. Finally we use Lemma 4.1.1 to commute the x-derivatives past $k^b_t(x,y)$ and follow the argument above to establish this theorem for arbitrary $k \in \mathbb{N}$. □

Remark. The solution to the Cauchy problem can be expressed as contour integral involving $R(\mu)$:

$$v(x,t) = \frac{1}{2\pi i} \int\limits_{\Gamma_{\alpha,R}} e^{\mu t} R(\mu) f d\mu, \tag{6.168}$$

where

$$\Gamma_{\alpha,R} := -b\{\mu : |\arg \mu| < \pi - \alpha \text{ and } |\mu| > R\}. \tag{6.169}$$

Here the $-$ sign indicates that Γ_α is taken with the opposite orientation to that it inherits as the boundary of the region on the right-hand side in (6.169). Using this formula we easily establish the analytic continuation of v to $t \in H_+$ as well as estimates of the form

$$\|v(\cdot,t)\|_{WF,k,2+\gamma} \leq C(t)\|f\|_{WF,k,\gamma}. \tag{6.170}$$

The constant $C(t)$ tends to infinity as $t \to 0$ at a rate that depends on γ. As $e^{\mu t}$ is exponentially decreasing along $\Gamma_{\alpha,R}$, we can also estimate the time derivatives $\partial_t^j v(\cdot,t)$, for $t > 0$.

Chapter Seven

Hölder Estimates for Higher Dimensional Corner Models

The estimates proved in the previous chapter form a solid foundation for proving analogous results in higher dimensions for model operators of the form

$$L_{\boldsymbol{b},m} = \sum_{j=1}^{n}[x_j\partial_{x_j}^2 + b_j\partial_{x_j}] + \sum_{k=1}^{m}\partial_{y_k}^2, \tag{7.1}$$

here $\boldsymbol{b} \in \overline{\mathbb{R}}_+^n$. In this context we exploit the fact that the solution operator for $L_{\boldsymbol{b},m}$ is a product of solution operators for 1-dimensional problems.

In 2-dimensions we can write

$$u(x_1, x_2, t) - u(y_1, y_2, t) = [u(x_1, x_2, t) - u(x_1, y_2, t)] + [u(x_1, y_2, y) - u(y_1, y_2, t)], \tag{7.2}$$

and in $n > 2$ dimensions we rewrite $u(\boldsymbol{x}, t) - u(\boldsymbol{y}, t)$ as

$$u(\boldsymbol{x}, t) - u(\boldsymbol{y}, t) = \sum_{j=0}^{n-1}[u(\boldsymbol{x}'_j, x_{j+1}, \boldsymbol{y}''_j, t) - u(\boldsymbol{x}'_j, y_{j+1}, \boldsymbol{y}''_j, t)], \tag{7.3}$$

where:

$$\begin{aligned} \boldsymbol{x}'_j &= (x_1, \dots, x_j) \text{ if } 1 \leq j \text{ and } \emptyset \text{ if } j \leq 0, \\ \boldsymbol{x}''_j &= (x_{j+2}, \dots, x_n) \text{ if } j < n-1 \text{ and } \emptyset \text{ if } j \geq n-1. \end{aligned} \tag{7.4}$$

In this way we are reduced to estimating these differences 1-variable-at-a-time, which, in light of Lemma 6.1.1 suffices.

In the proofs of the 1-dimensional estimates the only facts about the data we use are contained in the estimates in (5.21) and (5.32). This makes it possible to use these arguments to prove estimates in higher dimensions "one variable at a time." The only other fact we use is that if $f(x_1, \dots, x_n)$ is an absolutely integrable function, such that for some j we know that, for any $\boldsymbol{x}'_j$, $\boldsymbol{x}''_j$, the 1-dimensional integral

$$\int_0^\infty f(\boldsymbol{x}'_j, z_{j+1}, \boldsymbol{x}''_j)dz_j = 0, \tag{7.5}$$

then Fubini's theorem implies that

$$\int_0^\infty \cdots \int_0^\infty f(z) dz_1 \cdots dz_n = 0. \tag{7.6}$$

While we cannot simply quote the 1-dimensional estimates, using formulæ like that in (7.3), we can reduce the proof of an estimate in higher dimensions to the estimation of a product of 1-dimensional integrals. These integrals are in turn estimated in the lemmas stated here in the previous chapter.

Using the "1-variable-at-a-time" approach we prove the higher dimensional estimates in several stages; we begin by considering the "pure corner" case where $m = 0$, and then turn to the Euclidean case, where $n = 0$. The Euclidean case is of course classical. In the next chapter we state the results we need for the case of general $(0, m)$ and the estimates on the 1-dimensional solution Euclidean heat kernel needed to prove them. Finally, in Chapter 9 we do the general case, where n and m can assume arbitrary non-negative values.

We first consider the homogeneous Cauchy problem

$$L_{\boldsymbol{b},0} v(\boldsymbol{x}, t) = 0 \text{ in } \mathbb{R}_+^n \times (0, \infty) \text{ and } v(\boldsymbol{x}, 0) = f(\boldsymbol{x}). \tag{7.7}$$

Here $\boldsymbol{b}$ is a vector in $\mathbb{R}_+^n$. If f is bounded and continuous, then the unique bounded solution is given by

$$v(\boldsymbol{x}, t) = \int_0^\infty \cdots \int_0^\infty \prod_{j=1}^n k_t^{b_j}(x_j, z_j) f(z) dz, \tag{7.8}$$

from which it is clear that

$$|v(\boldsymbol{x}, t)| \leq \|f\|_{\mathscr{C}^0(\mathbb{R}_+^n)}. \tag{7.9}$$

For fixed $\boldsymbol{x}$, $v(\boldsymbol{x}, t)$ extends analytically in t to define a function in S_0.

We next turn to estimating the solution, u, of the inhomogeneous problem:

$$\left[\partial_t - \sum_{j=1}^N [x_j \partial_{x_j}^2 + b_j \partial_{x_j}] \right] u = g, \tag{7.10}$$

vanishing at $t = 0$. Proposition 4.2.4 shows that the unique bounded solution is given by the integral:

$$u(\boldsymbol{x}, t) = \int_0^t \int_0^\infty \cdots \int_0^\infty \prod_{j=1}^n k_s^{b_j}(x_j, z_j) g(z, t-s) dz ds. \tag{7.11}$$

It is quite easy to see that, for any $k \in \mathbb{N}_0$ and $0 < \gamma < 1$, the operator $\partial_t - L_{\boldsymbol{b},0}$ maps data with compact support in $C_{WF}^{k,2+\gamma}(\mathbb{R}_+^n \times [0, T])$ to $C_{WF}^{k,\gamma}(\mathbb{R}_+^n \times [0, T])$. Our aim, once again, is to prove that, for data with compact support in $C_{WF}^{k,\gamma}(\mathbb{R}_+^n \times [0, T])$, the solution belongs to $C_{WF}^{k,2+\gamma}(\mathbb{R}_+^n \times [0, T])$. As in the 1-dimensional case, when $k = 0$ we do not need to assume that the data has compact support.

7.1 THE CAUCHY PROBLEM

We begin with the somewhat simpler homogeneous Cauchy problem.

PROPOSITION 7.1.1. *Fix* $k \in \mathbb{N}_0$, $0 < R$, *and* $\boldsymbol{b} \in \mathbb{R}_+^n$. *Let* $f \in \mathscr{C}_{WF}^{k,\gamma}(\mathbb{R}_+^n)$, *and let* v *be the unique solution, given in* (7.8), *to*

$$(\partial_t - L_{\boldsymbol{b},0})v = 0 \quad v(\boldsymbol{x}, 0) = f(\boldsymbol{x}). \tag{7.12}$$

If $k > 0$, *then assume that* f *is supported in*

$$B_R^+(\mathbf{0}) = \{\boldsymbol{x} : \boldsymbol{x} \in \mathbb{R}_+^N \text{ and } \|\boldsymbol{x}\| \leq R\}.$$

For $0 < \gamma < 1$ *there a constant* $C_{k,\gamma,b,R}$ *so that*

$$\|v\|_{WF,k,\gamma} \leq C_{k,\gamma,b,R}\|f\|_{WF,k,\gamma}, \tag{7.13}$$

and, if $f \in \mathscr{C}_{WF}^{k,2+\gamma}(\mathbb{R}_+^n)$, *then*

$$\|v\|_{WF,k,2+\gamma} \leq C_{k,\gamma,b,R}\|f\|_{WF,k,2+\gamma}. \tag{7.14}$$

For fixed γ, *the constants* $C_{k,\gamma,b,R}$ *are uniformly bounded for* $0 \leq b \leq B$. *If* $k = 0$, *then the constants are independent of* R.

PROOF. Suppose that we have proved the estimates above with constants C which, for any B, are uniformly bounded for $0 < b_j < B$. As shown in the proof of Proposition 6.0.10, the case where $b_j = 0$, for one or more values j, can be treated by choosing a sequence $< \boldsymbol{b}_n >$ so that

$$b_{n,j} > 0 \text{ for all } n \text{ and } \lim_{n\to\infty} b_{n,j} = b_j. \tag{7.15}$$

We let $v_{\boldsymbol{b}_n}(\boldsymbol{x}, t)$ denote the solutions with the given initial data f. Given that the estimates in the lemma have been proved for each $\boldsymbol{b}_n$, Proposition 5.2.8 shows that the sequence $< v_{\boldsymbol{b}_n} >$ contains subsequences convergent with respect to the topology on $\mathscr{C}_{WF}^{0,\gamma'}$, for any $0 < \gamma' < \gamma$. If $t > \epsilon > 0$, then these solutions also converge uniformly in $\mathscr{C}^m$, for any $m > 0$. Hence the limit satisfies the limiting diffusion equation with the given initial data; the uniqueness of such solutions shows that any convergent subsequence has the same limit. Thus $< v_{\boldsymbol{b}_n} >$ itself converges in $\mathscr{C}_{WF}^{0,\gamma'}$ to $v_{\boldsymbol{b}}$, the solution in the limiting case. This implies that $v_{\boldsymbol{b}}$ also satisfies the estimates in the proposition. This reasoning applies equally well to all the function spaces under consideration. Thus it suffices to consider the case where $b_j > 0$ for $j = 1, \ldots, n$, which we henceforth assume. In the sequel we use C to denote positive constants that may depend on γ and $\boldsymbol{b}$, which are uniformly bounded so long as $0 < \gamma < 1$ is fixed and, for $j = 1, \ldots, n$, $0 < b_j \leq B$, for any fixed B.

The solution is given by formula (7.8). We observe that

$$\begin{aligned} v(x_1, \ldots, x_n, t) - v(y_1, \ldots, y_n, t) = v(x_1, \ldots, x_n, t) - v(x_1, \ldots, x_{n-1}, y_n, t) + \\ v(x_1, \ldots, x_{n-1}, y_n, t) - v(x_1, \ldots, x_{n-2}, y_{n-1}, y_n, t) \\ + \cdots + v(x_1, y_2, \ldots, y_n, t) - v(y_1, \ldots, y_n, t). \end{aligned} \tag{7.16}$$

Hence it is enough to show that for each $1 < k \leq n$ we have:

$$|v(x_1, \ldots, x_{k-1}, x_k, y_{k+1}, \ldots, y_n, t) - v(x_1, \ldots, x_{k-1}, y_k, y_{k+1}, \ldots, y_n, t)| \leq C\|f\|_{WF,0,\gamma}\rho_s(\boldsymbol{x}, \boldsymbol{y})^{\gamma}. \quad (7.17)$$

It suffices to assume that $\boldsymbol{x}$ and $\boldsymbol{y}$ differ in exactly one coordinate, which we can choose to be n. For an n-vector $\boldsymbol{x}$ we let

$$\boldsymbol{x}' = (x_1, \ldots, x_{n-1}). \quad (7.18)$$

The proof is simply a matter of recapitulating the steps in the 1-dimensional case, and showing how the n-dimensional case can be reduced to this case. We first do the $k = 0$ case, which does not require additional hypotheses on f, and then do the $k > 0$ case assuming that f has bounded support.

The first step is to consider the special case $v(\boldsymbol{x}', x_n, t) - v(\boldsymbol{x}', 0, t)$. Using the fact that

$$\int_0^\infty k_t^b(x, y)dy = 1, \quad (7.19)$$

we see that

$$v(\boldsymbol{x}', x_n, t) - v(\boldsymbol{x}', 0, t) = \int_0^\infty \cdots \int_0^\infty \prod_{j=1}^{n-1} k_t^{b_j}(x_j, z_j) \times \Big[(k_t^{b_n}(x_n, z_n) - k_t^{b_n}(0, z_n))(f(z', z_n) - f(z', 0))\Big] dz_n dz'. \quad (7.20)$$

This follows because, for any z' we have:

$$\int_0^\infty (k_t^{b_n}(x_n, z_n) - k_t^{b_n}(0, z_n)) f(z', 0) dz_n = 0. \quad (7.21)$$

Using the triangle inequality, the positivity of the kernels, the obvious estimate

$$|f(\boldsymbol{x}) - f(\boldsymbol{y})| \leq 2\|f\|_{WF,0,\gamma}\rho_s(\boldsymbol{x}, \boldsymbol{y})^{\gamma}, \quad (7.22)$$

and (7.19), we obtain the estimate

$$|v(\boldsymbol{x}', x_n, t) - v(\boldsymbol{x}', 0, t)| \leq 2\|f\|_{WF,0,\gamma} \int_0^\infty |k_t^{b_n}(x_n, z_n) - k_t^{b_n}(0, z_n)| z_n^{\frac{\gamma}{2}} dz_n. \quad (7.23)$$

Lemma 6.1.4 shows that integral is bounded by $Cx_n^{\frac{\gamma}{2}}$, showing as before that for any $c < 1$, there is a C so that if $y_n < cx_n$, then

$$|v(\boldsymbol{x}', x_n, t) - v(\boldsymbol{x}', y_n, t)| \leq C\|f\|_{WF,0,\gamma}|\sqrt{x_n} - \sqrt{y_n}|^{\gamma}. \quad (7.24)$$

For the second step we show that

$$|\partial_{x_n} v(\boldsymbol{x}', x_n, t)| \leq C x_n^{\frac{\gamma}{2}-1} \|f\|_{WF,0,\gamma} \frac{\lambda^{1-\frac{\gamma}{2}}}{1+\lambda^{\frac{1}{2}}} = C t^{\frac{\gamma}{2}-1} \frac{\|f\|_{WF,0,\gamma}}{1+\lambda^{\frac{1}{2}}}, \tag{7.25}$$

where $\lambda = x_n/t$. Taking advantage of the fact that, for all $x_n \geq 0$ and $t > 0$, we have

$$\int_0^\infty \partial_{x_n} k_t^{b_n}(x_n, z_n) dz_n = 0, \tag{7.26}$$

it follows that

$$\partial_{x_n} v(\boldsymbol{x}', x_n, t) = \int_0^\infty \cdots \int_0^\infty \prod_{j=1}^{n-1} k_t^{b_j}(x_j, z_j) \times$$
$$\int_0^\infty \left[\partial_{x_n} k_t^{b_n}(x_n, z_n)(f(z', z_n) - f(z', x_n))\right] dz_n dz'. \tag{7.27}$$

As before it follows easily that

$$|\partial_{x_n} v(\boldsymbol{x}', x_n, t)| \leq C \|f\|_{WF,0,\gamma} \int_0^\infty |\partial_{x_n} k_t^{b_n}(x_n, z_n)| |\sqrt{z_n} - \sqrt{x_n}|^\gamma dz_n. \tag{7.28}$$

An application of Lemma 6.1.10 suffices to complete the proof of (7.25). Integrating (7.25), we can now verify that (7.24) holds so long as λ is bounded.

For the last step we fix $0 < c < 1$, and consider $cx_n < y_n < x_n$, and $\lambda \to \infty$. For this case we need to find an analogue of the rather complicated formula in (6.77), which is again straightforward:

$$v(\boldsymbol{x}', x_n, t) - v(\boldsymbol{x}', y_n, t) = \int_0^\infty \cdots \int_0^\infty \prod_{j=1}^{n-1} k_t^{b_j}(x_j, z_j) \times$$

$$\Bigg[\int_J k_t^{b_n}(x_n, z_n)(f(z', z_n) - f(z', x_n))dz_n +$$

$$\int_J k_t^{b_n}(y_n, z_n)(f(z', y_n) - f(z', z_n))dz_n +$$

$$\int_{J^c} k_t^{b_n}(x_n, z_n)(f(z', y_n) - f(z', x_n))dz_n +$$

$$\int_0^\infty k_t^{b_n}(x_n, z_n)(f(z', x_n) - f(z', y_n))dz_n +$$

$$\int_{J^c} [k_t^{b_n}(x_n, z_n) - k_t^{b_n}(y_n, z_n)](f(z', z_n) - f(z', y_n))dz_n \Bigg] dz'. \quad (7.29)$$

Recall that $J = [\alpha, \beta]$, where

$$\sqrt{\alpha} = \frac{3\sqrt{y_n} - \sqrt{x_n}}{2} \quad \sqrt{\beta} = \frac{3\sqrt{x_n} - \sqrt{y_n}}{2}. \quad (7.30)$$

Using the triangle inequality repeatedly, and (7.19), we see that each term reduces to one appearing in the 1-d argument multiplied by $\prod_{j=1}^{n-1} k_t^{b_j}(x_j, z_j)$, from the fact that

$$|f(z', x) - f(z', y)| \leq 2\|f\|_{WF,0,\gamma} |\sqrt{x} - \sqrt{y}|^{\gamma}. \quad (7.31)$$

It follows immediately that the first four z_n-integrals contribute terms bounded by a constant times $\|f\|_{WF,0,\gamma} |\sqrt{x_n} - \sqrt{y_n}|^{\gamma}$. This leaves just the last integral over J^c. This term is estimated by

$$2\|f\|_{WF,0,\gamma} \int_{J^c} |k_t^{b_n}(x_n, z_n) - k_t^{b_n}(y_n, z_n)||\sqrt{z_n} - \sqrt{y_n}|^{\gamma} dz_n. \quad (7.32)$$

For $c > 1/9$ we may apply Lemma 6.1.7 to this term, and the estimate in (7.24) follows once again. This completes the spatial part of the $k = 0$ case.

We now turn to the estimate of $v(\boldsymbol{x}, t) - v(\boldsymbol{x}, s)$; we begin with the case $ct < s < t$, for a $0 < c < 1$. By definition we have:

$$v(\boldsymbol{x}, t) - v(\boldsymbol{x}, s) = \int_0^\infty \cdots \int_0^\infty \left[\prod_{j=1}^{n} k_t^{b_j}(x_j, z_j) - \prod_{j=1}^{n} k_s^{b_j}(x_j, z_j) \right] f(z)dz. \quad (7.33)$$

The difference of products can be represented as a telescoping sum:

$$\prod_{j=1}^{n} k_t^{b_j}(x_j, z_j) - \prod_{j=1}^{n} k_s^{b_j}(x_j, z_j) = \sum_{l=1}^{n} \left\{ \prod_{j=1}^{n-l} k_t^{b_j}(x_j, z_j) \times \prod_{j=n-l+2}^{n} k_s^{b_j}(x_j, z_j) \times \left[k_t^{b_{n-l+1}}(x_{n-l+1}, z_{n-l+1}) - k_s^{b_{n-l+1}}(x_{n-l+1}, z_{n-l+1}) \right] \right\}, \tag{7.34}$$

with the convention that, if $q < p$, then $\prod_p^q = 1$. Recall that

$$\begin{cases} z'_l = (z_1, \dots, z_{n-l}) \text{ for } 0 \leq l \leq n-1 \\ z''_l = (z_{n-l+2}, \dots, z_n) \text{ for } 2 \leq l \leq n+1, \end{cases} \tag{7.35}$$

with z'_l and z''_l equal to the empty set outside the stated ranges. For $1 \leq l \leq n$, we have:

$$\int_0^\infty \left[k_t^{b_{n-l+1}}(x_{n-l+1}, z_{n-l+1}) - k_s^{b_{n-l+1}}(x_{n-l+1}, z_{n-l+1}) \right] \times f(z'_l, x_{n-l+1}, z''_l) dz_{n-l+1} = 0. \tag{7.36}$$

Using these observations we can re-express $v(\boldsymbol{x}, t) - v(\boldsymbol{x}, s)$ as

$$v(\boldsymbol{x}, t) - v(\boldsymbol{x}, s) = \int_0^\infty \cdots \int_0^\infty \sum_{l=1}^{n} \left\{ \prod_{j=1}^{n-l} k_t^{b_j}(x_j, z_j) \times \prod_{j=n-l+2}^{n} k_s^{b_j}(x_j, z_j) \times \left[k_t^{b_{n-l+1}}(x_{n-l+1}, z_{n-l+1}) - k_s^{b_{n-l+1}}(x_{n-l+1}, z_{n-l+1}) \right] \times [f(\boldsymbol{z}) - f(z'_l, x_{n-l+1}, z''_l)] \right\} d\boldsymbol{z}. \tag{7.37}$$

Inserting absolute values, and using (7.19) repeatedly, we see that

$$|v(\boldsymbol{x}, t) - v(\boldsymbol{x}, s)| \leq 2\|f\|_{WF,0,\gamma} \int_0^\infty \sum_{l=1}^{n} \left| k_t^{b_{n-l+1}}(x_{n-l+1}, z_{n-l+1}) - k_s^{b_{n-l+1}}(x_{n-l+1}, z_{n-l+1}) \right| \times |\sqrt{z_{n-l+1}} - \sqrt{x_{n-l+1}}|^\gamma dz_{n-l+1}. \tag{7.38}$$

Applying Lemma 6.1.8 it follows that for any $0 < c < 1$, there is a C, so that if $ct < s < t$, then

$$|v(\boldsymbol{x},t) - v(\boldsymbol{x},s)| \leq C\|f\|_{WF,0\gamma}|t-s|^{\frac{\gamma}{2}}. \tag{7.39}$$

To complete the proof of the proposition we need to consider only $v(\boldsymbol{x},t) - v(\boldsymbol{x},0)$; using (7.19) this can be expressed as

$$v(\boldsymbol{x},t) - v(\boldsymbol{x},0) = \int_0^\infty \cdots \int_0^\infty \prod_{j=1}^n k_t^{b_j}(x_j,z_j)[f(\boldsymbol{z}) - f(\boldsymbol{x})]d\boldsymbol{z}. \tag{7.40}$$

We rewrite

$$f(\boldsymbol{z}) - f(\boldsymbol{x}) = \sum_{l=0}^{n-1}[f(\boldsymbol{z}_l', \boldsymbol{x}_{l+1}'') - f(\boldsymbol{z}_{l+1}', \boldsymbol{x}_{l+2}'')]. \tag{7.41}$$

Putting this expression into the integral above and repeatedly using (7.19), we obtain the estimate:

$$|v(\boldsymbol{x},t) - v(\boldsymbol{x},0)| \leq 2\|f\|_{WF,0,\gamma}\sum_{j=1}^n \int_0^\infty k_t^{b_j}(x_j,z_j)|\sqrt{z_j} - \sqrt{x_j}|^\gamma dx_j. \tag{7.42}$$

Applying Lemma 6.1.6 shows that there is a constant C so that

$$|v(\boldsymbol{x},t) - v(\boldsymbol{x},0)| \leq C\|f\|_{WF,0,\gamma}t^{\frac{\gamma}{2}}. \tag{7.43}$$

As in the 1-dimensional case, this completes the proof that (7.39) holds for all $\boldsymbol{x}, s, t$, and thereby the proof of (7.13) in the $k = 0$ case. If we assume that f is supported in $B_R^+(\boldsymbol{0})$, then the estimates in (7.13) for $k > 0$ follow from the $k = 0$ case by repeatedly applying Propositions 4.2.2 and 5.2.9.

Many of the estimates needed to prove (7.14) with $k = 0$ follow from (7.13), Lemma 4.2.3 and applications of Propositions 4.2.2 and 5.2.9. We can use these results to show that

$$\|\nabla_{\boldsymbol{x}}v\|_{WF,0,\gamma} + \|\partial_t v\|_{WF,0,\gamma} + \sum_{i=1}^n \|x_i\partial_{x_i}^2 v\|_{WF,0,\gamma} \leq C\|f\|_{WF,0,2+\gamma}. \tag{7.44}$$

To complete the proof in this case we need to similarly estimate the derivatives

$$\sqrt{x_i x_j}\partial_{x_i}\partial_{x_j}v \text{ with } i \neq j \tag{7.45}$$

in $\mathscr{C}_{WF}^{0,\gamma}(\mathbb{R}_+^n \times [0,\infty))$. We can relabel so that $i = 1$ and $j = 2$. Using Proposition 4.2.2 we can express these derivatives as

$$\sqrt{x_1x_2}\partial_{x_1}\partial_{x_2}v(\boldsymbol{x},t) = \int_{\mathbb{R}_+^{n-2}} \prod_{j=3}^n k_t^{b_j}(x_j,z_j)$$

$$\int_0^\infty\int_0^\infty \left(\frac{x_1}{z_1}\right)^{\frac{1}{2}} k_t^{b_1+1}(x_1,z_1)\left(\frac{x_2}{z_2}\right)^{\frac{1}{2}} k_t^{b_2+1}(x_2,z_2)\sqrt{z_1z_2}\partial_{z_1}\partial_{z_2}f(\boldsymbol{z})d\boldsymbol{z}. \tag{7.46}$$

Since $f \in \mathscr{C}^{0,2+\gamma}_{WF}$ it is not immediately obvious that this is true, but can be obtained by a simple limiting argument. We let

$$f_\epsilon = f(x_1 + \epsilon, x_2 + \epsilon, x_3, \ldots, x_n), \tag{7.47}$$

and v_ϵ be the solution of the Cauchy problem with this initial data. For $\epsilon > 0$ it follows easily from Proposition 4.2.2 that

$$\sqrt{x_1x_2}\partial_{x_1}\partial_{x_2}v_\epsilon(\boldsymbol{x}, t) = \int\limits_{\mathbb{R}_+^{n-2}} \prod_{j=3}^{n} k_t^{b_j}(x_j, z_j)$$

$$\int\limits_0^\infty\int\limits_0^\infty \left(\frac{x_1}{z_1}\right)^{\frac{1}{2}} k_t^{b_1+1}(x_1, z_1) \left(\frac{x_2}{z_2}\right)^{\frac{1}{2}} k_t^{b_2+1}(x_2, z_2)\sqrt{z_1z_2}\partial_{z_1}\partial_{z_2} f_\epsilon(\boldsymbol{z})d\boldsymbol{z}. \tag{7.48}$$

For $t > 0$, the left-hand side converges uniformly to $\sqrt{x_1x_2}\partial_{x_1}\partial_{x_2}v$. Since the scaled derivative $\sqrt{(z_1+\epsilon)(z_2+\epsilon)}\partial_{z_1}\partial_{z_2}f_\epsilon(\boldsymbol{z})$ is uniformly bounded and converges to the limit $\sqrt{z_1z_2}\partial_{z_1}\partial_{z_2}f(\boldsymbol{z})$, and the kernel

$$\left(\frac{x_1}{z_1}\right)^{\frac{1}{2}} k_t^{b_1+1}(x_1, z_1) \left(\frac{x_2}{z_2}\right)^{\frac{1}{2}} k_t^{b_2+1}(x_2, z_2) \tag{7.49}$$

is absolutely integrable, we see that the limit can be taken inside the integral to give (7.46).

The following lemma is used to bound these integrals.

LEMMA 7.1.2. *If* $0 \le \gamma \le 1$, *and* $b > \nu - \frac{\gamma}{2} > 0$, *then there is a constant* $C_{b,\phi}$, *bounded for* $b \le B$, *and* $B^{-1} < b + \frac{\gamma}{2} - \nu$, *so that, for* $t \in S_\phi$, *where* $0 < \phi < \frac{\pi}{2}$, *we have the estimate*

$$\int\limits_0^\infty \left(\frac{x}{y}\right)^\nu |k_t^b(x, y)| y^{\frac{\gamma}{2}} dy \le C_{b,\phi} x^{\frac{\gamma}{2}}. \tag{7.50}$$

The lemma is proved in Appendix A.

Since $f \in \mathscr{C}^{0,2+\gamma}_{WF}(\mathbb{R}_+^n)$ it follows that

$$|\sqrt{x_1x_2}\partial_{x_1}\partial_{x_2}f(\boldsymbol{x})| \le \|f\|_{WF,0,2+\gamma} \min\{x_1^{\frac{\gamma}{2}}, x_2^{\frac{\gamma}{2}}, 1\}; \tag{7.51}$$

applying this lemma shows that

$$|\sqrt{x_ix_j}\partial_{x_i}\partial_{x_j}v(\boldsymbol{x}, t)| \le C\|f\|_{WF,0,2+\gamma}. \tag{7.52}$$

We need to now establish the Hölder continuity of these derivatives. The argument used above for v applies directly to show the Hölder continuity in the $(x_3, \ldots, x_n)$ variables, leaving only x_1 and x_2. It clearly suffices to do the x_1-case. We have the estimate:

$$|\sqrt{x_1x_2}\partial_{x_1}\partial_{x_2}v(\boldsymbol{x}, t)| \le$$

$$\|f\|_{WF,0,2+\gamma} \int\limits_0^\infty\int\limits_0^\infty \left(\frac{x_1}{z_1}\right)^{\frac{1}{2}} k_t^{b_1+1}(x_1, z_1) \left(\frac{x_2}{z_2}\right)^{\frac{1}{2}} k_t^{b_2+1}(x_2, z_2) z_1^{\frac{\gamma}{2}} dz_1 dz_2. \tag{7.53}$$

Lemma 7.1.2 bounds both the z_1- and z_2-integrals and therefore

$$|\sqrt{x_1x_2}\partial_{x_1}\partial_{x_2}v(\boldsymbol{x},t)| \leq C\min\{x_1^{\frac{\gamma}{2}}, x_2^{\frac{\gamma}{2}}\}\|f\|_{WF,0,2+\gamma}. \tag{7.54}$$

Note that this implies that

$$\lim_{x_i\vee x_j\to 0+}|\sqrt{x_ix_j}\partial_{x_i}\partial_{x_j}v(\boldsymbol{x},t)| = 0. \tag{7.55}$$

If $c < 1$, then this implies the estimate

$$|\sqrt{x_1x_2}\partial_{x_1}\partial_{x_2}v(\boldsymbol{x},t) - \sqrt{x_1'x_2}\partial_{x_1}\partial_{x_2}v(\boldsymbol{x}',t)| \leq C|\sqrt{x_1}-\sqrt{x_1'}|^{\gamma}\,\|f\|_{WF,0,2+\gamma} \tag{7.56}$$

for $x_1' < cx_1$.

Thus we are left to consider $cx_1 < x_1' < x_1$. To simplify the notation in the following calculations, we let $\tilde{z} = (z_3,\dots,z_n)$, $\tilde{\boldsymbol{b}} = (b_3,\dots,b_n)$,

$$f_{12}(z) = \partial_{z_1}\partial_{z_2}f(z), \tag{7.57}$$

and

$$k_t^{\tilde{\boldsymbol{b}}}(\tilde{x},\tilde{z}) = \prod_{j=3}^{n}k_t^{b_j}(x_j,z_j). \tag{7.58}$$

We have the formula

$$\begin{aligned}
&\sqrt{x_1x_2}\partial_{x_1}\partial_{x_2}v(\boldsymbol{x},t) - \sqrt{x_1'x_2}\partial_{x_1}\partial_{x_2}v(\boldsymbol{x}',t) = \\
&\qquad\int\limits_{\mathbb{R}_+^{n-2}}\int_0^\infty\int_0^\infty k_t^{\tilde{\boldsymbol{b}}}(\tilde{x},\tilde{z})k_t^{b_2+1}(x_2,z_2)\left(\frac{x_2}{z_2}\right)^{\frac{1}{2}}\times \\
&\quad\left[\sqrt{x_1}k_t^{b_1+1}(x_1,z_1) - \sqrt{x_1'}k_t^{b_1+1}(x_1',z_1)\right]\sqrt{z_2}f_{12}(z_1,z_2,\tilde{z})dz_1dz_2d\tilde{z}.
\end{aligned} \tag{7.59}$$

We assume that

$$\frac{1}{4} \leq \frac{x_1'}{x_1} \leq 1, \tag{7.60}$$

and let

$$\sqrt{\alpha} = \max\left\{\frac{3\sqrt{x_1'}-\sqrt{x_1}}{2}, 0\right\} \text{ and } \sqrt{\beta} = \frac{3\sqrt{x_1}-\sqrt{x_1'}}{2}. \tag{7.61}$$

Note that (7.60) implies that $\alpha > x_1'/4$. We let $J = [\alpha,\beta]$, and observe that

$$|\sqrt{x_1z_2}f_{12}(x_1,z_2,\tilde{z}) - \sqrt{x_1'z_2}f_{12}(x_1',z_2,\tilde{z})| \leq 2\|f\|_{WF,0,2+\gamma}|\sqrt{x_1}-\sqrt{x_1'}|^{\gamma}. \tag{7.62}$$

This estimate implies that

$$|\sqrt{z_2} f_{12}(x_1, z_2, \tilde{z})| \leq 2C\|f\|_{WF,0,2+\gamma}|x_1^{\frac{\gamma-1}{2}}. \tag{7.63}$$

To estimate the difference in(7.62) we dissect the z_1-integral in a manner similar to that used in (6.77):

$$\begin{aligned}
&\sqrt{x_1 x_2}\partial_{x_1}\partial_{x_2} v(\boldsymbol{x}, t) - \sqrt{x_1' x_2}\partial_{x_1}\partial_{x_2} v(\boldsymbol{x}', t) = \\
&\qquad \int_{\mathbb{R}_+^{n-2}} \int_0^\infty k_t^{\tilde{\boldsymbol{b}}}(\tilde{\boldsymbol{x}}, \tilde{z}) k_t^{b_2+1}(x_2, z_2)\left(\frac{x_2}{z_2}\right)^{\frac{1}{2}} \times \\
&\qquad \Bigg[\int_J k_t^{b_1+1}(x_1, z_1)\sqrt{x_1 z_2}[f_{12}(z_1, z_2, \tilde{z}) - f_{12}(x_1, z_2, \tilde{z})]dz_1 - \\
&\qquad \int_J k_t^{b_1+1}(x_1', z_1)\sqrt{x_1' z_2}[f_{12}(z_1, z_2, \tilde{z}) - f_{12}(x_1', z_2, \tilde{z})]dz_1 + \\
&\qquad \int_{J^c} k_t^{b_1+1}(x_1, z_1)\sqrt{x_1 z_2}[f_{12}(x_1', z_2, \tilde{z}) - f_{12}(x_1, z_2, \tilde{z})]dz_1 + \\
&\int_{J^c} [k_t^{b_1+1}(x_1, z_1)\sqrt{x_1 z_2} - k_t^{b_1+1}(x_1', z_1)\sqrt{x_1' z_2}][f_{12}(z_1, z_2, \tilde{z}) - f_{12}(x_1', z_2, \tilde{z})]dz_1 + \\
&\qquad [\sqrt{x_1 z_2} f_{12}(x_1, z_2, \tilde{z}) - \sqrt{x_1' z_2} f_{12}(x_1', z_2, \tilde{z})]\Bigg] dz_2 d\tilde{z}.
\end{aligned} \tag{7.64}$$

We denote the terms on the right-hand side by I, II, III, IV, and V. The terms I, II, III, and V can be estimated fairly easily. For V we apply the estimate in (7.62) and Lemma 7.1.2 to conclude that

$$|V| \leq 2\|f\|_{WF0,2+\gamma}|\sqrt{x_1} - \sqrt{x_1'}|^\gamma. \tag{7.65}$$

To handle I we observe that

$$\begin{aligned}
&|\sqrt{x_1 z_2}[f_{12}(z_1, z_2, \tilde{z}) - f_{12}(x_1, z_2, \tilde{z})]| \\
&\leq |\sqrt{z_1 z_2} f_{12}(z_1, z_2, \tilde{z}) - \sqrt{x_1 z_2} f_{12}(x_1, z_2, \tilde{z})| + |\sqrt{z_1 z_2} - \sqrt{x_1 z_2}||f_{12}(z_1, z_2, \tilde{z})| \\
&\leq 2\|f\|_{WF0,2+\gamma}\left[|\sqrt{x_1} - \sqrt{z_1}|^\gamma + z_1^{\frac{\gamma-1}{2}}|\sqrt{z_1} - \sqrt{x_1}|\right].
\end{aligned} \tag{7.66}$$

We note that

$$z_1^{\frac{\gamma-1}{2}}|\sqrt{z_1} - \sqrt{x_1}| = |\sqrt{z_1} - \sqrt{x_1}|^\gamma \left|1 - \sqrt{\frac{x_1}{z_1}}\right|^{1-\gamma}. \tag{7.67}$$

For $z_1 \in J$ the ratio

$$\frac{x_1}{z_1} \leq 16, \tag{7.68}$$

and the differences $|\sqrt{z_1} - \sqrt{x_1}|$ and $|\sqrt{z_1} - \sqrt{x_1'}|$ are bounded above by a multiple of $|\sqrt{x_1} - \sqrt{x_1'}|$. Once again we can use Lemma 7.1.2 to see that

$$|I| \leq 2C\|f\|_{WF0,2+\gamma}|\sqrt{x_1} - \sqrt{x_1'}|^{\gamma}. \tag{7.69}$$

The same argument applies with minor modifications to show that

$$|II| \leq 2C\|f\|_{WF0,2+\gamma}|\sqrt{x_1} - \sqrt{x_1'}|^{\gamma}. \tag{7.70}$$

This argument also shows that

$$|\sqrt{x_1 z_2}[f_{12}(x_1', z_2, \tilde{z}) - f_{12}(x_1, z_2, \tilde{z})]| \leq 2\|f\|_{WF,0,2+\gamma}|\sqrt{x_1} - \sqrt{x_1'}|^{\gamma}, \tag{7.71}$$

which implies that

$$|III| \leq 2C\|f\|_{WF0,2+\gamma}|\sqrt{x_1} - \sqrt{x_1'}|^{\gamma}. \tag{7.72}$$

This leaves only the term of type IV. We rewrite this term as

$$IV = \int\limits_{J^c} \left| k_t^{b_1+1}(x_1, z_1)\sqrt{\frac{x_1}{z_1}} - k_t^{b_1+1}(x_1', z_1)\sqrt{\frac{x_1'}{z_1}} \right| \times \\ \sqrt{z_1 z_2}|f_{12}(z_1, z_2, \tilde{z}) - f_{12}(x_1', z_2, \tilde{z})|dz_1. \tag{7.73}$$

Arguing as in (7.67) we see that

$$|\sqrt{z_1 z_2}[f_{12}(z_1, z_2, \tilde{z}) - f_{12}(x_1', z_2, \tilde{z})]| \leq \\ 2\|f\|_{WF,0,2+\gamma}\left[|\sqrt{z_1} - \sqrt{x_1'}|^{\gamma} + (x_1')^{\frac{\gamma-1}{2}}|\sqrt{z_1} - \sqrt{x_1'}|\right]. \tag{7.74}$$

We complete the estimate of IV with the following lemma:

LEMMA 7.1.3. *If $J = [\alpha, \beta]$, with α, β are given by* (7.61), *assuming that x_1', x_1 satisfy* (7.60), *and $b > 0$, $0 < \gamma \leq 1$, and $0 < \phi < \frac{\pi}{2}$, there is a $C_{b,\phi}$ so that if $t \in S_{\phi}$, then,*

$$\int\limits_{J^c} \left| k_t^{b+1}(x_1, z_1)\sqrt{\frac{x_1}{z_1}} - k_t^{b+1}(x_1', z_1)\sqrt{\frac{x_1'}{z_1}} \right| |\sqrt{z_1} - \sqrt{x_1'}|^{\gamma} dz_1 \leq C_{b,\phi}|\sqrt{x_1} - \sqrt{x_1'}|^{\gamma}. \tag{7.75}$$

Applying the lemma to the expression in (7.73) completes the proof that

$$|\sqrt{x_i x_j}\partial_{x_i}\partial_{x_j}v(\boldsymbol{x},t) - \sqrt{x_i' x_j'}\partial_{x_i}\partial_{x_j}v(\boldsymbol{x}',t)| \leq C\|f\|_{WF,0,2+\gamma}\rho_s(\boldsymbol{x},\boldsymbol{x}')^{\gamma}. \quad (7.76)$$

We now consider the Hölder continuity in time for these derivatives. We re-write this difference as

$$\sqrt{x_1x_2}\partial_{x_1}\partial_{x_2}[v(\boldsymbol{x},t) - v(\boldsymbol{x},0)] =$$
$$\sqrt{x_1x_2}\int\limits_{\mathbb{R}_+^{n-2}}\int\limits_0^{\infty}\int\limits_0^{\infty} k_t^{\tilde{\boldsymbol{b}}}(\tilde{\boldsymbol{x}},\tilde{\boldsymbol{z}})k_t^{b_1+1}(x_1,z_1)k_t^{b_2+1}(x_2,z_2)[f_{12}(\boldsymbol{z}) - f_{12}(\boldsymbol{x})]dz_1dz_2d\tilde{\boldsymbol{z}}, \quad (7.77)$$

where $f_{12} = \partial_{z_1}\partial_{z_2}f$. We re-write the difference $f_{12}(\boldsymbol{z}) - f_{12}(\boldsymbol{x})$ as

$$f_{12}(\boldsymbol{z}) - f_{12}(\boldsymbol{x}) = \sqrt{\frac{z_1z_2}{z_1z_2}}\sum_{j=3}^{n}[f_{12}(\boldsymbol{z}_l', z_l, \boldsymbol{x}_l'') - f_{12}(\boldsymbol{z}_l', x_l, \boldsymbol{x}_l'')] +$$
$$[f_{12}(z_1,z_2,\boldsymbol{x}_n'') - f_{12}(z_1,x_2,\boldsymbol{x}_n'')] + [f_{12}(z_1,x_2,\boldsymbol{x}_n'') - f_{12}(x_1,x_2,\boldsymbol{x}_n'')]. \quad (7.78)$$

Substituting from the sum in (7.78) into (7.77), we see that the terms for $l = 3, \ldots, n$ are each bounded by:

$$\|f\|_{WF,0,2+\gamma}\int\limits_0^{\infty}\int\limits_0^{\infty}\int\limits_0^{\infty} k_t^{b_1+1}(x_1,z_1)k_t^{b_2+1}(x_2,z_2)\times$$
$$\sqrt{\frac{x_1x_2}{z_1z_2}}k_t^{b_l}(x_l,z_l)|\sqrt{x_l} - \sqrt{z_l}|^{\gamma}dz_ldz_1dz_2. \quad (7.79)$$

From Lemma 7.1.2 and Lemma 6.1.6 we see that these terms are bounded by a constant times $\|f\|_{WF,0,2+\gamma}t^{\frac{\gamma}{2}}$.

We re-write $[f_{12}(z_1,z_2,\boldsymbol{x}_n'') - f_{12}(z_1,x_2,\boldsymbol{x}_n'')]$ as

$$[f_{12}(z_1,z_2,\boldsymbol{x}_n'') - f_{12}(z_1,x_2,\boldsymbol{x}_n'')] = \frac{1}{\sqrt{z_1x_2}}[(\sqrt{z_1x_2} - \sqrt{z_1z_2})f_{12}(z_1,z_2,\boldsymbol{x}_n'') +$$
$$(\sqrt{z_1z_2}f_{12}(z_1,z_2,\boldsymbol{x}_n'') - \sqrt{z_1x_2}f_{12}(z_1,x_2,\boldsymbol{x}_n''))]. \quad (7.80)$$

The right-hand side is estimated by

$$\frac{\|f\|_{WF,0,2+\gamma}}{\sqrt{z_1x_2}}[|\sqrt{x_2} - \sqrt{z_2}|z_2^{\frac{\gamma-1}{2}} + |\sqrt{x_2} - \sqrt{z_2}|^{\gamma}]. \quad (7.81)$$

The contribution of this term is therefore bounded by

$$\|f\|_{WF,0,2+\gamma}\int\limits_0^{\infty}\int\limits_0^{\infty}\sqrt{\frac{x_1}{z_1}}k_t^{b_1+1}(x_1,z_1)k_t^{b_2+1}(x_2,z_2)\times$$
$$[|\sqrt{x_2} - \sqrt{z_2}|^{\gamma} + |\sqrt{x_2} - \sqrt{z_2}|z_2^{\frac{\gamma-1}{2}}]dz_1dz_2. \quad (7.82)$$

Lemmas 7.1.2 and 6.1.6 show that the $|\sqrt{x_2}-\sqrt{z_2}|^\gamma$ -term is bounded by

$$C\|f\|_{WF,0,2+\gamma}t^{\frac{\gamma}{2}}.$$

To bound the contribution of the other term we apply

LEMMA 7.1.4. *For $0\le b$, $0\le\gamma<1$, there is a constant $C_{b,\phi}$, bounded for $b\le B$, so that for $t\in S_\phi$,*

$$\int\limits_0^\infty |k_t^{b+1}(x,y)||\sqrt{y}-\sqrt{x}|y^{\frac{\gamma-1}{2}}dy\le C_{b,\phi}|t|^{\frac{\gamma}{2}}. \tag{7.83}$$

The last term is re-written as

$$\sqrt{x_1x_2}[f_{12}(z_1,x_2,\boldsymbol{x}_n'')-f_{12}(x_1,x_2,\boldsymbol{x}_n'')]=$$
$$[\sqrt{z_1x_2}f_{12}(z_1,x_2,\boldsymbol{x}_n'')-\sqrt{x_1x_2}f_{12}(x_1,x_2,\boldsymbol{x}_n'')]+(\sqrt{x_1x_2}-\sqrt{z_1x_2})f_{12}(z_1,x_2,\boldsymbol{x}_n''), \tag{7.84}$$

which is estimated by

$$\|f\|_{WF0,2+\gamma}[|\sqrt{z_1}-\sqrt{x_1}|z_1^{\frac{\gamma-1}{2}}+|\sqrt{z_1}-\sqrt{x_1}|^\gamma]. \tag{7.85}$$

These terms are estimated as in the previous case, showing that altogether there is a C so that

$$|\sqrt{x_ix_j}\partial_{x_i}\partial_{x_j}v(\boldsymbol{x},t)-\sqrt{x_ix_j}\partial_{x_i}\partial_{x_j}v(\boldsymbol{x},0)|\le C\|f\|_{WF,0,2+\gamma}t^{\frac{\gamma}{2}}, \tag{7.86}$$

which implies that for a $c<1$, there is a C so that, if $s<ct$, then

$$|\sqrt{x_ix_j}\partial_{x_i}\partial_{x_j}v(\boldsymbol{x},t)-\sqrt{x_ix_j}\partial_{x_i}\partial_{x_j}v(\boldsymbol{x},s)|\le C\|f\|_{WF,0,2+\gamma}|t-s|^{\frac{\gamma}{2}}. \tag{7.87}$$

This leaves only the case $ct<s<t$, for a $c<1$. We begin with the analogue of (7.37),

$$\sqrt{x_1x_2}\partial_{x_1}\partial_{x_2}[v(\boldsymbol{x},t)-v(\boldsymbol{x},s)]=\int\limits_0^\infty\cdots\int\limits_0^\infty\sqrt{x_1x_2}\sum_{l=1}^n\Bigg\{\prod_{j=1}^{n-l}k_t^{\widetilde{b}_j}(x_j,z_j)\times$$
$$\prod_{j=n-l+2}^n k_s^{\widetilde{b}_j}(x_j,z_j)\left[k_t^{\widetilde{b}_{n-l+1}}(x_{n-l+1},z_{n-l+1})-k_s^{\widetilde{b}_{n-l+1}}(x_{n-l+1},z_{n-l+1})\right]\times$$
$$[f_{12}(\boldsymbol{z})-f_{12}(\boldsymbol{z}_l',x_{n-l+1},\boldsymbol{z}_l'')]\Bigg\}d\boldsymbol{z}, \tag{7.88}$$

where

$$\widetilde{b}_1=b_1+1,\quad \widetilde{b}_2=b_2+1,\text{ and }\widetilde{b}_j=b_j\text{ for }j>2. \tag{7.89}$$

Each term in this sum with $1 \leq l \leq n-2$ is estimated by

$$\|f\|_{WF,0,2+\gamma} \int_0^\infty \int_0^\infty \int_0^\infty \left(\frac{x_1}{z_1}\right)^{\frac{1}{2}} k_t^{b_1+1}(x_1, z_1) \left(\frac{x_2}{z_2}\right)^{\frac{1}{2}} k_t^{b_2+1}(x_2, z_2)$$
$$\times |k_t^{\widetilde{b}_{n-l+1}}(x_{n-l+1}, z_{n-l+1}) - k_s^{\widetilde{b}_{n-l+1}}(x_{n-l+1}, z_{n-l+1})|$$
$$\times |\sqrt{z_{n-l+1}} - \sqrt{x_{n-l+1}}|^\gamma dz_{n-l+1} dz_2 dz_1. \quad (7.90)$$

Lemmas 7.1.2 and 6.1.8 show that these terms are bounded by

$$C\|f\|_{WF,0,2+\gamma} |t-s|^{\frac{\gamma}{2}}. \quad (7.91)$$

We now turn to $l = n-1$, and n. These cases are essentially identical; we give the details for $l = n-1$. The contribution of this term is bounded by

$$\|f\|_{WF,0,2+\gamma} \int_{\mathbb{R}_+^{n-2}} \prod_{j=3}^n k_s^{b_j}(x_j, z_j) \int_0^\infty \int_0^\infty \sqrt{x_1 x_2} k_t^{b_1+1}(x_1, z_1) \times$$
$$\left| k_t^{b_2+1}(x_2, z_2) - k_s^{b_2+1}(x_2, z_2) \right| |f_{12}(z) - f_{12}(z_1, x_2, z''_{n-1})| dz_2 dz_1 dz''_{n-1}. \quad (7.92)$$

Proceeding as in (7.84) and (7.85), we see that

$$\sqrt{x_1 x_2} |f_{12}(z) - f_{12}(z_1, x_2, z''_{n-1})| \leq$$
$$\|f\|_{WF,0,2+\gamma} \sqrt{\frac{x_1}{z_1}} [|\sqrt{x_2} - \sqrt{z_2}| z_2^{\frac{\gamma-1}{2}} + |\sqrt{x_2} - \sqrt{z_2}|^\gamma]. \quad (7.93)$$

Applying Lemma 7.1.2 shows that we are left to estimate

$$\|f\|_{WF,0,2+\gamma} \int_0^\infty \left| k_t^{b_2+1}(x_2, z_2) - k_s^{b_2+1}(x_2, z_2) \right|$$
$$\times [|\sqrt{x_2} - \sqrt{z_2}| z_2^{\frac{\gamma-1}{2}} + |\sqrt{x_2} - \sqrt{z_2}|^\gamma] dz_2. \quad (7.94)$$

Lemma 6.1.8 shows that

$$\int_0^\infty \left| k_t^{b_2+1}(x_2, z_2) - k_s^{b_2+1}(x_2, z_2) \right| |\sqrt{x_2} - \sqrt{z_2}|^\gamma dz_2 \leq C|t-s|^{\frac{\gamma}{2}}, \quad (7.95)$$

leaving only

$$\int_0^\infty \left| k_t^{b_2+1}(x_2, z_2) - k_s^{b_2+1}(x_2, z_2) \right| |\sqrt{x_2} - \sqrt{z_2}| z_2^{\frac{\gamma-1}{2}} dz_2. \quad (7.96)$$

This term is bounded in the following lemma:

LEMMA 7.1.5. *For* $0 \leq \gamma < 1$, $1 \leq b$, *and* $0 < c < 1$, *there is a constant* C_b, *so that if* $ct < s < t$, *then*

$$\int_0^\infty \left| k_t^b(x,z) - k_s^b(x,z) \right| |\sqrt{x} - \sqrt{z}| z^{\frac{\gamma-1}{2}} dz \leq C_b |t-s|^{\frac{\gamma}{2}}. \tag{7.97}$$

The proof of the lemma is in Appendix A. Applying this result completes the proof that

$$|\sqrt{x_1 x_2} \partial_{x_1} \partial_{x_2} [v(\boldsymbol{x},t) - v(\boldsymbol{x},s)]| \leq C \|f\|_{WF,0,2+\gamma} |t-s|^{\frac{\gamma}{2}}. \tag{7.98}$$

The fact that $\partial_t v = L_{\boldsymbol{b},0} v$ allows us to deduce that $\partial_t v \in \mathscr{C}^{0,\gamma}_{WF}(\mathbb{R}^n_+ \times [0,\infty))$ and satisfies the same estimates as the spatial derivative. This finishes the proof of (7.14) in the $k = 0$ case.

We can now proceed as we did in the proof of (7.13) for $k > 0$, applying Proposition 4.2.2 to commute derivatives past the kernel functions. We now assume that f has support in $B_R^+(\mathbf{0})$, which allows the use of Proposition 5.2.9 to estimate the resultant data. This reduces the proof of (7.14) for $k > 0$, to the $k = 0$ case, which thereby completes the proof of the proposition. □

7.2 THE INHOMOGENEOUS CASE

We now turn to estimating the solution of the inhomogeneous problem in a n-dimensional corner. Let $g \in \mathscr{C}_{WF,0,\gamma}(\mathbb{R}^n_+ \times [0,T])$, and let u denote the solution to

$$\partial_t u - \sum_{j=1}^N [x_j \partial_{x_j}^2 + b_j \partial_{x_j}] u = g, \tag{7.99}$$

which vanishes at $t = 0$. According to Proposition 4.2.4, it is given by the integral:

$$u(\boldsymbol{x},t) = \int_0^t \int_0^\infty \cdots \int_0^\infty \prod_{j=1}^n k_s^{b_j}(x_j,z_j) g(\boldsymbol{z},t-s) d\boldsymbol{z} ds. \tag{7.100}$$

PROPOSITION 7.2.1. *Fix* $0 < \gamma < 1$, $0 < R$, $k \in \mathbb{N}_0$, *and* $(b_1,\ldots,b_n) \in \mathbb{R}^n_+$. *Let* $g \in \mathscr{C}^{k,\gamma}_{WF}(\mathbb{R}^n_+ \times [0,T])$, *and let* u *be the unique solution, given in* (7.11) *to* (7.99), *with* $u(\boldsymbol{x},0) = 0$. *If* $k > 0$, *then assume that* g *is supported in* $B_R^+(\mathbf{0}) \times [0,T]$. *The solution*

$$u \in \mathscr{C}^{k,2+\gamma}_{WF}(\mathbb{R}^n_+ \times [0,T]); \tag{7.101}$$

there is a constant $C_{k,\gamma,b,R}$ *so that*

$$\|u\|_{WF,k,2+\gamma,T} \leq C_{k,\gamma,b,R}(1+T)\|g\|_{WF,k,\gamma,T}. \tag{7.102}$$

For fixed γ, *the constants* $C_{k,\gamma,b,R}$ *are uniformly bounded for* $0 \leq b < B$. *If* $k = 0$, *then the constants are independent of* R.

PROOF. As before it suffices to assume that $b_j > 0$ for $j = 1, \dots, n$. The estimates we prove below have constants C, which, for any B, are uniformly bounded if $0 < b_j < B$. The case where $b_j = 0$, for one or more values j, is again treated by choosing a sequence $< \boldsymbol{b}_n >$ so that

$$b_{n,j} > 0 \text{ for all } n \text{ and } \lim_{n\to\infty} b_{n,j} = b_j. \tag{7.103}$$

We let $< u_{\boldsymbol{b}_n} >$ denote the solutions with the given data g. Given that the estimates in the lemma have been proved for each $\boldsymbol{b}_n$, we see that Proposition 5.2.8 and uniqueness imply that for $0 < \gamma' < \gamma$, the sequence $< u_{\boldsymbol{b}_n} >$ converges, in $\mathscr{C}^{0,2+\gamma'}_{WF}$, to $u_{\boldsymbol{b}}$, the solution in the limiting case. This implies that $u_{\boldsymbol{b}}$ also satisfies the estimates in the proposition. This reasoning applies equally well to all the function spaces under consideration. It therefore suffices to consider the case where $b_j > 0$ for $j = 1, \dots, n$, which we henceforth assume.

As in the proof of Proposition 6.0.11 we note that with

$$u_\epsilon(\boldsymbol{x}, t) = \int_0^{t-\epsilon}\int_0^\infty \cdots \int_0^\infty \prod_{j=1}^n k_s^{b_j}(x_j, z_j) g(\boldsymbol{z}, t-s) d\boldsymbol{z} ds, \tag{7.104}$$

the solution u is the uniform limit of u_ϵ. The functions u_ϵ are smooth where $t > 0$, and we can show as before that, for $0 < \epsilon < t$, we have

$$\begin{gathered}\partial_{x_k} u_\epsilon(\boldsymbol{x}, t) = \\ \int_0^{t-\epsilon}\int_0^\infty \cdots \int_0^\infty \partial_{x_k} \prod_{j=1}^n k_s^{b_j}(x_j, z_j)[g(\boldsymbol{z}, t-s) - g(z_k', x_k, z_k'')] d\boldsymbol{z} ds \\ \sqrt{x_k x_l}\partial_{x_l}\partial_{x_k} u_\epsilon(\boldsymbol{x}, t) = \\ \int_0^{t-\epsilon}\int_0^\infty \cdots \int_0^\infty \sqrt{x_k x_l}\partial_{x_l}\partial_{x_k} \prod_{j=1}^n k_s^{b_j}(x_j, z_j)[g(\boldsymbol{z}, t-s) - g(z_k', x_k, z_k'')] d\boldsymbol{z} ds.\end{gathered} \tag{7.105}$$

Assume that $g \in \mathscr{C}^{0,\gamma}_{WF}(\mathbb{R}^n_+ \times [0, T])$, for a $0 < \gamma < 1$. Using Lemma 6.1.10 for the first derivatives, the mixed derivatives where $k \neq l$, and Lemma 6.1.15, when $k = l$, we can again show that these derivatives converge, as $\epsilon \to 0^+$, uniformly on $\mathbb{R}^n_+ \times [0, T]$. This shows that u has continuous first partial $\boldsymbol{x}$-derivatives on $\mathbb{R}^n_+ \times [0, T]$, and continuous second $\boldsymbol{x}$-derivatives on $(0, \infty)^n \times [0, T]$, with

$$\lim_{x_l \vee x_k \to 0^+} \sqrt{x_k x_l}\partial_{x_l}\partial_{x_k} u(\boldsymbol{x}, t) = 0. \tag{7.106}$$

This also shows that we can allow $\epsilon \to 0^+$, in the expressions for these derivatives in (7.105), to obtain absolutely convergent expressions for the corresponding derivatives of u. Finally we argue as before to show that

$$\partial_t u = L_{\boldsymbol{b},0} u + g, \tag{7.107}$$

and therefore the t-derivative of u is continuous and u satisfies the desired equation.

Note that

$$u(\boldsymbol{x},t)=\int_0^t v^s(\boldsymbol{x},t-s)ds, \tag{7.108}$$

where $v^s(\boldsymbol{x},t)$ is the solution to

$$[\partial_t - L_{\boldsymbol{b},0}]v^s = 0 \text{ with } v^s(\boldsymbol{x},0)=g(\boldsymbol{x},s). \tag{7.109}$$

This relation allows us to use estimates on the solution to the Cauchy problem to derive bounds on u.

From the positivity of the heat kernel and (7.19) it is immediate that

$$|u(\boldsymbol{x},t)| \leq \|g\|_{WF,0,\gamma}t. \tag{7.110}$$

To establish the Lipschitz continuity of u we integrate (7.25) to conclude that, for each $1 \leq j \leq n$, we have the estimate:

$$|\partial_{x_j}u(\boldsymbol{x},t)| \leq C\|g\|_{WF,0,\gamma}t^{\frac{\gamma}{2}} \tag{7.111}$$

and therefore

$$|u(\boldsymbol{x}'_j,x_j,\boldsymbol{x}''_j,t)-u(\boldsymbol{x}'_j,y_j,\boldsymbol{x}''_j,t)| \leq C\|g\|_{WF,0,\gamma}t^{\frac{\gamma}{2}}|x_j-y_j|. \tag{7.112}$$

Thus we can also integrate the estimate in (7.24) with respect to t to see that

$$|u(\boldsymbol{x}'_j,x_j,\boldsymbol{x}''_j,t)-u(\boldsymbol{x}'_j,y_j,\boldsymbol{x}''_j,t)| \leq C\|g\|_{WF,0,\gamma}t|\sqrt{x_j}-\sqrt{y_j}|^{\gamma}. \tag{7.113}$$

The estimates, proved below on the first and second derivatives, show that

$$|L_{\boldsymbol{b},0}u(\boldsymbol{x},t)| \leq C\|g\|_{WF,0,\gamma}t^{\frac{\gamma}{2}}, \tag{7.114}$$

thus the equation $[\partial_t - L_{\boldsymbol{b},0}]u(\boldsymbol{x},t)=g(\boldsymbol{x},t)$ implies that, for $t_1 < t_2$, we have

$$|u(\boldsymbol{x},t_2)-u(\boldsymbol{x},t_1)| \leq C(1+t_2^{\frac{\gamma}{2}})\|g\|_{WF,0,\gamma}|t_2-t_1|. \tag{7.115}$$

Our next task is to establish the Hölder estimate for the first spatial-derivatives. There is a small twist in the higher dimensional case: we use one argument to estimate $|\partial_{x_j}u(\boldsymbol{x}'_j,x_j,\boldsymbol{x}''_j,t)-\partial_{x_j}u(\boldsymbol{x}'_j,y_j,\boldsymbol{x}''_j,t)|$, and a rather different argument to estimate $|\partial_{x_l}u(\boldsymbol{x}'_j,x_j,\boldsymbol{x}''_j,t)-\partial_{x_l}u(\boldsymbol{x}'_j,y_j,\boldsymbol{x}''_j,t)|$, for $l \neq j$. The former follows exactly as in the 1-dimensional case; we show that there is a constant so that for $1 \leq l \leq n$,

$$|x_l\partial_{x_l}^2u(\boldsymbol{x},t)| \leq C\|g\|_{WF,0,\gamma}\min\{t^{\frac{\gamma}{2}},x_l^{\frac{\gamma}{2}}\}. \tag{7.116}$$

This estimate implies that

$$\lim_{x_l\to 0^+}x_l\partial_{x_l}^2u(\boldsymbol{x},t)=0. \tag{7.117}$$

The proof of (7.116) follows simply from

$$x_l\partial_{x_l}^2 u(\boldsymbol{x},t) = \int_0^t\int_0^\infty\cdots\int_0^\infty \prod_{j\neq l}^n k_s^{b_j}(x_j,z_j)\times$$
$$x_l\partial_{x_l}^2 k_s^{b_l}(x_l,z_l)[g(\boldsymbol{z},t-s)-g(\boldsymbol{z}_l',x_l,\boldsymbol{z}_l'',t-s)]d\boldsymbol{z}ds. \quad (7.118)$$

Putting absolute values inside the integral, using the estimates

$$|g(\boldsymbol{x},t)-g(\boldsymbol{y},s)| \leq 2\|g\|_{WF,0,\gamma}[\rho_s(\boldsymbol{x},\boldsymbol{y})+\sqrt{|t-s|}]^\gamma, \quad (7.119)$$

and (7.19), we see that (7.116) follows from Lemma 6.1.14.

Integrating $\partial_{x_j}^2 u$ and applying (7.116), we see that

$$|\partial_{x_j}u(\boldsymbol{x}_j',x_j,\boldsymbol{x}_j'',t)-\partial_{x_j}u(\boldsymbol{x}_j',y_j,\boldsymbol{x}_j'',t)| \leq C\|g\|_{WF,0,\gamma}|\sqrt{x_j}-\sqrt{y_j}|^\gamma. \quad (7.120)$$

To do the "off-diagonal" case we use Lemma 6.1.5. To estimate

$$|\partial_{x_l}u(\boldsymbol{x}_m',x_m,\boldsymbol{x}_m'',t)-\partial_{x_l}u(\boldsymbol{x}_m',y_m,\boldsymbol{x}_m'',t)|, \text{ for } l\neq m,$$

we observe that

$$\partial_{x_l}u(\boldsymbol{x}_m',x_m,\boldsymbol{x}_m'',t)-\partial_{x_l}u(\boldsymbol{x}_m',y_m,\boldsymbol{x}_m'') = \int_0^t\int_0^\infty\cdots\int_0^\infty \prod_{j\neq l,m}^n k_s^{b_j}(x_j,z_j)\times$$
$$[k_s^{b_m}(x_m,z_m)-k_s^{b_m}(y_m,z_m)]\partial_{x_l}k_s^{b_l}(x_l,z_l)[g(\boldsymbol{z},t-s)-g(\boldsymbol{z}_l',x_l,\boldsymbol{z}_l'',t-s)]d\boldsymbol{z}ds. \quad (7.121)$$

Putting absolute values into the integral and using (7.119), and (7.19), gives:

$$|\partial_{x_l}u(\boldsymbol{x}_m',x_m,\boldsymbol{x}_m'',t)-\partial_{x_l}u(\boldsymbol{x}_m',y_m,\boldsymbol{x}_m'')| \leq \|g\|_{WF,0,\gamma}\times$$
$$\int_0^t\int_0^\infty\int_0^\infty |k_s^{b_m}(x_m,z_m)-k_s^{b_m}(y_m,z_m)||\partial_{x_l}k_s^{b_l}(x_l,z_l)||\sqrt{z_l}-\sqrt{x_l}|^\gamma dz_m dz_l ds. \quad (7.122)$$

If $y_m = 0$, then applying (6.32) we see that this is estimated by

$$C\|g\|_{WF,0,\gamma}\int_0^t s^{\frac{\gamma}{2}-1}\frac{x_m/s}{1+x_m/s}ds, \quad (7.123)$$

which is easily seen to be bounded by $C\|g\|_{WF,0,\gamma}x_m^{\frac{\gamma}{2}}$. Applying (6.72) we see that, if $0<c<1$, then there is a constant C so that for $y_m < cx_m$, we have

$$|\partial_{x_l}u(\boldsymbol{x}_m',x_m,\boldsymbol{x}_m'',t)-\partial_{x_l}u(\boldsymbol{x}_m',y_m,\boldsymbol{x}_m'')| \leq C\|g\|_{WF,0,\gamma}|\sqrt{x_m}-\sqrt{y_m}|^\gamma. \quad (7.124)$$

We are therefore reduced to considering $cx_m < y_m < x_m$, for a $c < 1$. If we use (6.33) it follows that

$$|\partial_{x_l} u(\boldsymbol{x}'_m, x_m, \boldsymbol{x}''_m, t) - \partial_{x_l} u(\boldsymbol{x}'_m, y_m, \boldsymbol{x}''_m)| \leq C\|g\|_{WF,0,\gamma} \times \int_0^t s^{\frac{\gamma}{2}-1} \left(\frac{\frac{\sqrt{x_m}-\sqrt{y_m}}{\sqrt{s}}}{1+\frac{\sqrt{x_m}-\sqrt{y_m}}{\sqrt{s}}} \right) ds. \quad (7.125)$$

We split this into an integral from 0 to $(\sqrt{x_m} - \sqrt{y_m})^2$ and the rest, to obtain:

$$|\partial_{x_l} u(\boldsymbol{x}'_m, x_m, \boldsymbol{x}''_m, t) - \partial_{x_l} u(\boldsymbol{x}'_m, y_m, \boldsymbol{x}''_m, t)| \leq C\|g\|_{WF,0,\gamma} \times \left[\int_0^{(\sqrt{x_m}-\sqrt{y_m})^2} s^{\frac{\gamma}{2}-1} + \int_{(\sqrt{x_m}-\sqrt{y_m})^2}^t s^{\frac{\gamma}{2}-1} \left(\frac{\sqrt{x_m}-\sqrt{y_m}}{\sqrt{s}} \right) ds \right]. \quad (7.126)$$

Performing these integrals shows that (7.124) holds in this case as well.

To complete the analysis of $\partial_{x_j} u(\boldsymbol{x}, t)$ we need to show that there is a constant so that

$$|\partial_{x_j} u(\boldsymbol{x}, t_2) - \partial_{x_j} u(\boldsymbol{x}, t_1)| \leq C\|g\|_{WF,0,\gamma} |t_2 - t_1|^{\frac{\gamma}{2}}. \quad (7.127)$$

This follows immediately from the 1-dimensional argument. Using (7.111), we see that for any $c < 1$, there is a C so that this estimate holds for $t_1 < ct_2$. As in the 1-dimensional case, we now assume that $t_1 < t_2 < 2t_1$. Without loss of generality we can take $j = n$; use (6.111) to re-express $\partial_{x_n} u(\boldsymbol{x}, t_2) - \partial_{x_n} u(\boldsymbol{x}, t_1)$ as

$$\begin{aligned}
\partial_{x_n} u(\boldsymbol{x}, t_2) - \partial_{x_n} u(\boldsymbol{x}, t_1) = \\
&\int_0^{t_2-t_1} \int_0^\infty \cdots \int_0^\infty \prod_{j=1}^{n-1} k_s^{b_j}(x_j, z_j) \partial_{x_n} k_s^{b_n}(x_n, z_n) \times \\
&\quad [g(z'_n, z_n, t_2 - s) - g(z'_n, z_n, t_1 - s)] dz_n dz'_n ds + \\
&\int_0^{2t_1-t_2} \int_0^\infty \cdots \int_0^\infty \partial_{x_n} \left[\prod_{j=1}^n k_{t_2-s}^{b_j}(x_j, z_j) - \prod_{j=1}^n k_{t_1-s}^{b_j}(x_j, z_j) \right] \times \\
&\quad [g(z'_n, z_n, s) - g(z'_n, x_n, s)] dz_n dz'_n ds + \\
&\int_{2t_1-t_2}^{t_1} \int_0^\infty \cdots \int_0^\infty \prod_{j=1}^{n-1} k_{t_2-s}^{b_j}(x_j, z_j) \partial_{x_n} k_{t_2-s}^{b_n}(x_n, z_n) \times \\
&\quad [g(z'_n, z_n, s) - g(z'_n, x_n, s)] dz_n dz'_n ds. \quad (7.128)
\end{aligned}$$

In the first integral we replace $g(z'_n, z_n, t_j - s)$ with $g(z'_n, z_n, t_j - s) - g(z'_n, x_n, t_j - s)$, for $j = 1, 2$, and then apply Lemma 6.1.10, as in the 1-dimensional case, to show that this term is bounded by the right-hand side of (7.127). A similar argument is applied to estimate the third integral.

To handle the second term we use formula (7.34) to conclude that

$$\partial_{x_n}\left[\prod_{l=1}^{n} k_{t_2}^{b_l}(x_l,z_l)-\prod_{l=1}^{n} k_{t_1}^{b_l}(x_l,z_l)\right]=$$
$$\sum_{m=1}^{n}\partial_{x_n}\left\{\prod_{l=1}^{n-m} k_{t_2}^{b_l}(x_l,z_l)\times\prod_{l=n-m+2}^{n} k_{t_1}^{b_l}(x_l,z_l)\times\right.$$
$$\left.\left[k_{t_2}^{b_{n-m+1}}(x_{n-m+1},z_{n-m+1})-k_{t_1}^{b_{n-m+1}}(x_{n-m+1},z_{n-m+1})\right]\right\}. \tag{7.129}$$

To estimate the contribution to the second integral coming from the term in (7.129) with $m=1$, we observe that

$$|g(z_n',z_n,s)-g(z_n',x_n,s)|\leq 2\|g\|_{WF,0,\gamma}|\sqrt{x_n}-\sqrt{z_n}|^{\gamma}, \tag{7.130}$$

and apply Lemma 6.1.12. The contributions of the other terms are bounded by

$$C\|g\|_{WF,0,\gamma}\int_0^{2t_1-t_2}\int_0^{\infty}\int_0^{\infty}|k_{t_2-s}^{b_j}(x_j,z_j)-k_{t_1-s}^{b_j}(x_j,z_j)|\times$$
$$|\partial_{x_n}k_{t_1-s}^{b_n}(x_n,z_n)||\sqrt{x_n}-\sqrt{z_n}|^{\gamma}dz_jdz_nds. \tag{7.131}$$

We use Lemma 6.1.10 and Lemma 6.1.9 to see that, upon setting $\sigma=t_1-s$, this integral is bounded by

$$C\|g\|_{WF,0,\gamma}\int_{t_2-t_1}^{t_1}\frac{\sigma^{\frac{\gamma-1}{2}}(t_2-t_1)}{(\sqrt{\sigma}+\sqrt{x_n})(t_2-t_1+\sigma)}\leq C\|g\|_{WF,0,\gamma}|t_2-t_1|^{\frac{\gamma}{2}}. \tag{7.132}$$

This completes the proof that there is a constant C so that

$$|\partial_{x_j}u(\boldsymbol{x},t_1)-\partial_{x_j}u(\boldsymbol{x},t_2)|\leq C\|g\|_{WF,0,\gamma}|t_2-t_1|^{\frac{\gamma}{2}}. \tag{7.133}$$

The fact that there is a constant C so that

$$|\nabla_{\boldsymbol{x}}u(\boldsymbol{x}_1,t_1)-\nabla_{\boldsymbol{x}}u(\boldsymbol{x}_2,t_2)|\leq C\|g\|_{WF,0,\gamma}[\rho_s(\boldsymbol{x}_1,\boldsymbol{x}_2)+|t_2-t_1|^{\frac{1}{2}}]^{\gamma} \tag{7.134}$$

now follows from the foregoing estimates and Lemma 6.1.1.

An estimate showing the boundedness of $|x_j\partial_{x_j}^2u(\boldsymbol{x},t)|$ is given in (7.116). We can use Lemma 6.1.10 to prove an analogous estimate for the mixed partial derivatives. Arguing as above, we easily establish that, for $j\neq k$, x_j and x_k both positive, we have

$$|\partial_{x_j}\partial_{x_k}u(\boldsymbol{x},t)|\leq\|g\|_{WF,0,\gamma}\int_0^{t}\int_0^{\infty}\int_0^{\infty}|\partial_{x_k}k_s^{b_j}(x_k,z_k)|\times$$
$$|\partial_{x_j}k_s^{b_j}(x_j,z_j)||g(z_j',z_j,z_j'',t-s)-g(z_j',x_j,z_j'',t-s)|dz_jdz_kds. \tag{7.135}$$

Lemma 6.1.10 applies to show that this quantity is bounded by

$$C\|g\|_{WF,0,\gamma}\int\limits_0^t \frac{s^{\frac{\gamma}{2}-1}ds}{\sqrt{x_jx_k}}, \tag{7.136}$$

which implies that

$$|\sqrt{x_jx_k}\partial_{x_j}\partial_{x_k}u(\boldsymbol{x},t)| \leq C\|g\|_{WF,0,\gamma}t^{\frac{\gamma}{2}}. \tag{7.137}$$

All that remains to complete the estimate of spatial derivatives is the proof of the Hölder continuity of the second derivatives of u.

We begin by proving the Hölder continuity of $x_j\partial_{x_j}^2 u$. The estimates in (7.116) and (6.72) show that for any $c < 1$, there is a C so that if $y_j < cx_j$, then

$$|x_j\partial_{x_j}^2 u(\boldsymbol{x}_j', x_j, \boldsymbol{x}_j'', t) - y_j\partial_{x_j}^2 u(\boldsymbol{x}_j', y_j, \boldsymbol{x}_j'', t)| \leq C\|g\|_{WF,0,\gamma}|\sqrt{x_j}-\sqrt{y_j}|^{\gamma}. \tag{7.138}$$

Thus in the "diagonal" case we only need to consider $cx_j < y_j < x_j$. The proof in this case follows exactly as in the 1-dimensional case; we establish the Hölder continuity of $L_{b_j,x_j}u$, which is sufficient, as we have already done so for the first derivatives. To do this we express the difference:

$$L_{b_j,x_j}u(\boldsymbol{x}_j', x_j, \boldsymbol{x}_j'', t) - L_{b_j,x_j}u(\boldsymbol{x}_j', y_j, \boldsymbol{x}_j'', t), \tag{7.139}$$

using (6.129) in the j-variable, much like the formula in (7.29). The estimate for each term in (6.129) carries over to the present situation to immediately establish that (7.138) holds for a suitable C, for all pairs (x_j, y_j).

To finish the spatial estimate in this case we need to consider the "non-diagonal" situation. With $j \neq m$, we express this difference as

$$\begin{aligned} &x_j\partial_{x_j}^2 u(\boldsymbol{x}_m', x_m, \boldsymbol{x}_m'', t) - x_j\partial_{x_j}^2 u(\boldsymbol{x}_m', y_m, \boldsymbol{x}_m'', t) = \\ &\int\limits_0^t\int\limits_0^\infty\cdots\int\limits_0^\infty \prod_{l\neq j,m} k_s^{b_l}(x_l,z_l)x_j\partial_{x_j}^2 k_s^{b_j}(x_j,z_j)[k_s^{b_m}(x_m,z_m) - k_s^{b_m}(y_m,z_m)]\times \\ &\qquad [g(\boldsymbol{z}_j', z_j, \boldsymbol{z}_j'', t-s) - g(\boldsymbol{z}_j', x_j, \boldsymbol{z}_j'', t-s)]d\boldsymbol{z}ds. \end{aligned} \tag{7.140}$$

From this formula it follows that

$$\begin{aligned} &|x_j\partial_{x_j}^2 u(\boldsymbol{x}_m', x_m, \boldsymbol{x}_m'', t) - x_j\partial_{x_j}^2 u(\boldsymbol{x}_m', y_m, \boldsymbol{x}_m'', t)| \leq \\ &2\|g\|_{WF,0,\gamma}\int\limits_0^t\int\limits_0^\infty\int\limits_0^\infty |x_j\partial_{x_j}^2 k_s^{b_j}(x_j,z_j)||\sqrt{x_j}-\sqrt{z_j}|^{\gamma}\times \\ &\qquad |k_s^{b_m}(x_m,z_m) - k_s^{b_m}(y_m,z_m)]|dz_jdz_mds. \end{aligned} \tag{7.141}$$

This case is completed by employing Lemmas 6.1.5 and 6.1.15.

We begin with the case that $y_m = 0$; Lemmas 6.1.5 and 6.1.15 in (7.141) show that

$$|x_j\partial_{x_j}^2 u(\boldsymbol{x}'_m, x_m, \boldsymbol{x}''_m, t) - x_j\partial_{x_j}^2 u(\boldsymbol{x}'_m, 0, \boldsymbol{x}''_m, t)| \leq \\ 2\|g\|_{WF,0,\gamma} \int_0^t s^{\frac{\gamma}{2}-1} \frac{x_m/s}{1 + x_m/s} ds. \quad (7.142)$$

This is easily seen to be bounded by $C\|g\|_{WF,0,\gamma} x_m^{\frac{\gamma}{2}}$. We are therefore left to consider the case $cx_m < y_m < x_m$, for any $c < 1$. We now use the second estimate in Lemma 6.1.5 to see that

$$|x_j\partial_{x_j}^2 u(\boldsymbol{x}'_m, x_m, \boldsymbol{x}''_m, t) - x_j\partial_{x_j}^2 u(\boldsymbol{x}'_m, y_m, \boldsymbol{x}''_m, t)| \leq \\ \|g\|_{WF,0,\gamma} \int_0^t s^{\frac{\gamma}{2}-1} \frac{\frac{\sqrt{x_m}-\sqrt{y_m}}{\sqrt{s}}}{1 + \frac{\sqrt{x_m}-\sqrt{y_m}}{\sqrt{s}}} ds. \quad (7.143)$$

An elementary argument shows that the right-hand side is bounded by

$$C\|g\|_{WF,0,\gamma} |\sqrt{x_m} - \sqrt{y_m}|^{\gamma}. \quad (7.144)$$

This completes the proof of the spatial part of the Hölder estimates for $x_j\partial_{x_j}^2 u(\boldsymbol{x}, t)$. We next turn to the time estimate.

From (7.116) it follows that

$$|x_j\partial_{x_j}^2 u(\boldsymbol{x}, t)| \leq C\|g\|_{WF,0,\gamma} t^{\frac{\gamma}{2}}. \quad (7.145)$$

This shows that $x_j\partial_{x_j}^2 u(\boldsymbol{x}, t)$ tends uniformly to zero like $t^{\frac{\gamma}{2}}$. Applying (6.72) we see that for $c < 1$, there is a C so that, if $s < ct$, then

$$|x_j\partial_{x_j}^2 u(\boldsymbol{x}, t) - x_j\partial_{x_j}^2 u(\boldsymbol{x}, s)| \leq C\|g\|_{WF,0,\gamma} |t - s|^{\frac{\gamma}{2}}. \quad (7.146)$$

We are therefore left to consider the case $t_1 < t_2 < 2t_1$. To handle this case we begin with the formula from (6.143):

$$L_{b_j}u(\boldsymbol{x},t_2)-L_{b_j}u(\boldsymbol{x},t_1)=$$
$$\int_0^{t_2-t_1}\int_0^{\infty}\cdots\int_0^{\infty}\prod_{l\neq j}k_s^{b_l}(x_l,z_l)L_{b_j}k_s^{b_j}(x_j,z_j)[g(z,t_2-s)-g(z,t_1-s)]dzds+$$
$$\int_0^{2t_1-t_2}\int_0^{\infty}\cdots\int_0^{\infty}L_{b_j,x_j}\left[\prod_{l=1}^{n}k_{t_2-s}^{b_l}(x_l,z_l)-\prod_{l=1}^{n}k_{t_1-s}^{b_l}(x_l,z_l)\right]$$
$$\times[g(z,s)-g(z_j',x_j,z_j'',s)]dzds+$$
$$\int_{2t_1-t_2}^{t_1}\int_0^{\infty}\cdots\int_0^{\infty}\prod_{l\neq j}k_{t_2-s}^{b_l}(x_l,z_l)L_{b_j}k_{t_2-s}^{b_j}(x_j,z_j)[g(z,s)-g(z_j',x_j,z_j'',s)]dzds. \tag{7.147}$$

In the first integral, as in the 1-dimensional cases, we use the estimates

$$|g(z,t_q-s)-g(z_j',x_j,z_j'',t_q-s)|\leq 2\|g\|_{WF,0,\gamma}|\sqrt{x_j}-\sqrt{z_j}|^{\gamma}; \tag{7.148}$$

here $q=1,2$. In the last integral in (7.147) we use the estimate

$$|g(z,s)-g(z_j',x_j,z_j'',s)|\leq 2\|g\|_{WF,0,\gamma}|\sqrt{x_j}-\sqrt{z_j}|^{\gamma}. \tag{7.149}$$

This immediately reduces these cases to 1-dimensional cases, and these terms are therefore bounded by $C\|g\|_{WF,0,\gamma}|t_2-t_1|^{\frac{\gamma}{2}}$.

To handle the second term we use formula (7.34) to conclude that

$$L_{b_j,x_j}\left[\prod_{l=1}^{n}k_{t_2}^{b_l}(x_l,z_l)-\prod_{l=1}^{n}k_{t_1}^{b_l}(x_l,z_l)\right]=$$
$$\sum_{m=1}^{n}L_{b_j,x_j}\left\{\prod_{l=1}^{n-m}k_{t_2}^{b_l}(x_l,z_l)\times\prod_{l=n-m+2}^{n}k_{t_1}^{b_l}(x_l,z_l)\times\right.$$
$$\left.\left[k_{t_2}^{b_{n-m+1}}(x_{n-m+1},z_{n-m+1})-k_{t_1}^{b_{n-m+1}}(x_{n-m+1},z_{n-m+1})\right]\right\}. \tag{7.150}$$

There are now two types of terms: those with $n-m+1\neq j$, and the term with $n-m+1=j$. In all cases we use the estimate in (7.149). With this understood, the term with $n-m+1=j$ immediately reduces to the 1-dimensional case. Terms where $n-m+1\neq j$ are bounded by

$$I=\|g\|_{WF,0,\gamma}\int_0^{2t_1-t_2}\int_0^{\infty}\int_0^{\infty}|k_{t_2-s}^{b_l}(x_l,z_l)-k_{t_1-s}^{b_l}(x_l,z_l)|\times$$
$$|L_{b_j}k_{t_q-s}^{b_j}(x_j,z_j)||\sqrt{x_j}-\sqrt{z_j}|^{\gamma}dz_ldz_jds; \tag{7.151}$$

here $q = 1$ or 2. Using Lemmas 6.1.10, 6.1.9 and 6.1.15 we see that

$$I \leq C\|g\|_{WF,0,\gamma} \int_{t_2-t_1}^{t_1} (s+(q-1)\tau)^{\frac{\gamma}{2}-1} \frac{\tau ds}{s+\tau}, \tag{7.152}$$

with $\tau = t_2 - t_1$. The case $q = 1$ clearly produces a larger value. In this case we set $w = s/\tau$, obtaining

$$I \leq C\|g\|_{WF,0,\gamma} \tau^{\frac{\gamma}{2}} \int_{1}^{\frac{t_1}{\tau}} w^{\frac{\gamma}{2}-1} \frac{dw}{1+w}, \tag{7.153}$$

which completes the proof that, for $j = 1, \dots, n$, we have:

$$|L_{b_j}u(\boldsymbol{x}, t_2) - L_{b_j}u(\boldsymbol{x}, t_1)| \leq C\|g\|_{WF,0,\gamma} |t_2 - t_1|^{\frac{\gamma}{2}}. \tag{7.154}$$

To finish the proof of the proposition we need to show that the mixed derivatives $\sqrt{x_j x_l}\partial_{x_j}\partial_{x_l} u$ are Hölder continuous. There are two cases depending upon whether the variable that is allowed to vary is one of x_j, x_l or not. The latter case is immediate from lemmas we have already proved. Let $m \neq j$ or l, then we easily see that

$$\begin{aligned}|\sqrt{x_j x_l}\partial_{x_j}\partial_{x_l} u(\boldsymbol{x}'_m, x_m, \boldsymbol{x}''_m, t) - \sqrt{x_j x_l}\partial_{x_j}\partial_{x_l} u(\boldsymbol{x}'_m, y_m, \boldsymbol{x}''_m, t)| \leq \\ \|g\|_{WF,0,\gamma} \int_0^t \int_0^\infty \int_0^\infty \int_0^\infty |k_s^{b_m}(x_m, z_m) - k_s^{b_m}(y_m, z_m)| \times \\ |\sqrt{x_j}\partial_{x_j} k_s^{b_j}(x_j, z_j)||\sqrt{x_l}\partial_{x_l} k_s^{b_l}(x_l, z_l)||\sqrt{z_l} - \sqrt{y_l}|^\gamma dz_m dz_l dz_j ds.\end{aligned} \tag{7.155}$$

We first let $y_m = 0$ and use the first estimate in Lemma 6.1.5 to bound the z_m-integral, and Lemma 6.1.10 to estimate the other two. This shows that this expression is bounded by

$$\|g\|_{WF,0,\gamma} \int_0^t \left(\frac{x_m}{s+x_m}\right) s^{\frac{\gamma}{2}-1} ds. \tag{7.156}$$

This is bounded by $C\|g\|_{WF,0,\gamma} x_m^{\frac{\gamma}{2}}$, which allows us to restrict to the case that, for a $c < 1$, we have the estimate $cx_m < y_m < x_m$. Applying the other estimate in Lemma 6.1.5 we easily deduce that

$$\begin{aligned}|\sqrt{x_j x_l}\partial_{x_j}\partial_{x_l} u(\boldsymbol{x}'_m, x_m, \boldsymbol{x}''_m, t) - \sqrt{x_j x_l}\partial_{x_j}\partial_{x_l} u(\boldsymbol{x}'_m, y_m, \boldsymbol{x}''_m, t)| \leq \\ C\|g\|_{WF,0,\gamma} |\sqrt{x_m} - \sqrt{y_m}|^\gamma.\end{aligned} \tag{7.157}$$

Now suppose that $m = l$ and $y_l = 0$. In this case we see that

$$|\sqrt{x_j x_l}\partial_{x_j}\partial_{x_l} u(\boldsymbol{x}_l', x_l, \boldsymbol{x}_l'', t)| \leq$$
$$\|g\|_{WF,0,\gamma} \int_0^t \int_0^\infty \int_0^\infty |\sqrt{x_j}\partial_{x_j} k_s^{b_j}(x_j, z_j)||\sqrt{x_l}\partial_{x_l} k_s^{b_l}(x_l, z_l)||\sqrt{z_l} - \sqrt{x_l}|^\gamma \, dz_l dz_j ds. \tag{7.158}$$

We apply Lemma 6.1.10 to see that this is bounded by

$$\int_0^t \left(\frac{\sqrt{x_l} s^{\frac{\gamma-1}{2}}}{\sqrt{s} + \sqrt{x_l}}\right)\left(\frac{\sqrt{x_j} s^{-\frac{1}{2}}}{\sqrt{s} + \sqrt{x_j}}\right) ds. \tag{7.159}$$

An elementary argument shows that

$$|\sqrt{x_j x_l}\partial_{x_j}\partial_{x_l} u| \leq C\|g\|_{WF,0,\gamma} \min\{x_j^{\frac{\gamma}{2}}, x_l^{\frac{\gamma}{2}}, t^{\frac{\gamma}{2}}\}. \tag{7.160}$$

This estimate implies that

$$\lim_{x_j \vee x_l \to 0^+} \sqrt{x_j x_l}\partial_{x_j}\partial_{x_l} u(\boldsymbol{x}, t) = 0. \tag{7.161}$$

In light of (6.72) all that remains is to consider $cx_l < y_l < x_l$, for a $0 < c < 1$, for which we require an estimate of the quantity:

$$\int_0^\infty |\sqrt{x_1}\partial_x k_t^b(x_1, y) - \sqrt{x_2}\partial_x k_t^b(x_2, y)||\sqrt{x_1} - \sqrt{y}|^\gamma \, dy. \tag{7.162}$$

We now show how to use (6.40), and the estimate in Lemma 6.1.11, to prove the spatial Hölder estimate for $\sqrt{x_j x_l}\partial_{x_j}\partial_{x_l} u$, with respect to x_j and x_l.

$$|\sqrt{x_j x_l}\partial_{x_j}\partial_{x_l} u(\boldsymbol{x}_l', x_l, \boldsymbol{x}_l'', t) - \sqrt{x_j y_l}\partial_{x_j}\partial_{x_l} u(\boldsymbol{x}_l', y_l, \boldsymbol{x}_l'', t)| \leq$$
$$2\|g\|_{WF,0,\gamma} \int_0^t \int_0^\infty \int_0^\infty |\sqrt{x_j}\partial_{x_j} k_s^{b_j}(x_j, z_j)| \times$$
$$|\sqrt{x_l}\partial_{x_l} k_s^{b_l}(x_l, z_l) - \sqrt{y_l}\partial_{x_l} k_s^{b_l}(y_l, z_l)||\sqrt{z_l} - \sqrt{y_l}|^\gamma \, dz_l dz_j ds. \tag{7.163}$$

We apply Lemmas 6.1.10 and 6.1.11 to see that this integral is bounded by

$$C\int_0^t s^{\frac{\gamma}{2}-1} \frac{\left(\frac{|\sqrt{x_l}-\sqrt{y_l}|}{s}\right)}{1 + \left(\frac{|\sqrt{x_l}-\sqrt{y_l}|}{s}\right)} \leq$$
$$C\left[\int_0^{|\sqrt{x_l}-\sqrt{y_l}|^2} s^{\frac{\gamma}{2}-1} + \int_{|\sqrt{x_l}-\sqrt{y_l}|^2}^t s^{\frac{\gamma-3}{2}}|\sqrt{x_l} - \sqrt{y_l}| ds\right]. \tag{7.164}$$

Here we implicitly assume that $t > |\sqrt{x_l} - \sqrt{y_l}|^2$; if this is not the case then only the first integral on the right side of (7.164) is needed. In either case we easily see that the right-hand side is bounded by $C|\sqrt{x_l} - \sqrt{y_l}|^\gamma$.

To finish the proof of the proposition all that remains is to show that these derivatives are Hölder continuous with respect to time. The estimates in (7.137) and (6.72) show that if $0 < c < 1$, then there is a C so that for $0 < t_1 < ct_2$, we have the estimate

$$|\sqrt{x_j x_l}\partial_{x_j}\partial_{x_l}u(\boldsymbol{x}, t_2) - \sqrt{x_j x_l}\partial_{x_j}\partial_{x_l}u(\boldsymbol{x}, t_1)| \leq C\|g\|_{WF,0,\gamma}|t_2 - t_1|^{\frac{\gamma}{2}}. \quad (7.165)$$

Thus we are left to consider the case $t_1 < t_2 < 2t_1$. To that end we express

$$\sqrt{x_j x_l}\partial_{x_j}\partial_{x_l}[u(\boldsymbol{x}, t_2) - u(\boldsymbol{x}, t_1)] =$$
$$\int_0^{t_2-t_1}\int_0^\infty\cdots\int_0^\infty \prod_{m\neq j,l} k_s^{b_m}(x_m, z_m)\sqrt{x_j}\partial_{x_j}k_s^{b_j}(x_j, z_j)\sqrt{x_l}\partial_{x_l}k_s^{b_l}(x_l, z_l)\times$$
$$[(g(z, t_2 - s) - g(z'_j, x_j, z''_j, t_2 - s)) + (g(z'_j, x_j, z''_j, t_1 - s) - g(z, t_1 - s))]dzds+$$
$$\int_0^{2t_1-t_2}\int_0^\infty\cdots\int_0^\infty \sqrt{x_j x_l}\partial_{x_j}\partial_{x_l}\left[\prod_{m=1}^n k_{t_2-s}^{b_m}(x_m, z_m) - \prod_{m=1}^n k_{t_1-s}^{b_m}(x_m, z_m)\right]\times$$
$$g(z, s)dzds+$$
$$\int_{2t_1-t_2}^{t_1}\int_0^\infty\cdots\int_0^\infty \prod_{m\neq j,l} k_{t_2-s}^{b_m}(x_m, z_m)\sqrt{x_j}\partial_{x_j}k_{t_2-s}^{b_j}(x_j, z_j)\sqrt{x_l}\partial_{x_l}k_{t_2-s}^{b_l}(x_l, z_l)\times$$
$$[g(z, s) - g(z'_j, x_j, z''_j, s)]dzds. \quad (7.166)$$

In the first integral, as in the 1-dimensional case, we use the estimates

$$|g(z, t_q - s) - g(z'_j, x_j, z''_j, t_q - s)| \leq 2\|g\|_{WF,0,\gamma}|\sqrt{x_j} - \sqrt{z_j}|^\gamma; \quad (7.167)$$

here $q = 1, 2$. In the last integral in (7.166) we use the estimate

$$|g(z, s) - g(z'_j, x_j, z''_j, s)| \leq 2\|g\|_{WF,0,\gamma}|\sqrt{x_j} - \sqrt{z_j}|^\gamma. \quad (7.168)$$

Applying Lemma 6.1.10 we see that these terms are bounded by

$$C\int_0^{2(t_2-t_1)} \frac{\sqrt{x_j x_l}s^{\frac{\gamma}{2}-1}}{(\sqrt{x_j} + \sqrt{s})(\sqrt{x_l} + \sqrt{s})} \leq C|t_2 - t_1|^{\frac{\gamma}{2}}. \quad (7.169)$$

All that remains is to estimate the second integral in (7.166), where once again we employ formula (7.34). In each of the terms which arise, we can replace $g(z, s)$ with either $[g(z, s) - g(z'_j, x_j, z''_j, s)]$, or $[g(z, s) - g(z'_l, x_l, z''_l, s)]$, without changing the values of these integrals. With this understood, there are only five essentially different cases to consider, depending upon which terms in the products on the right-hand side of (7.34) are differentiated. We let $\tau = t_2 - t_1$; the cases requiring consideration are integrands with terms of the form:

I.

$$|\sqrt{x_j}\partial_{x_j}k_s^{b_j}(x_j,z_j)\sqrt{x_l}\partial_{x_l}[k_{\tau+s}^{b_l}(x_l,z_l)-k_s^{b_l}(x_l,z_l)]||\sqrt{x_j}-\sqrt{z_j}|^{\gamma}, \quad (7.170)$$

II.

$$|\sqrt{x_j}\partial_{x_j}k_{\tau+s}^{b_j}(x_j,z_j)\sqrt{x_l}\partial_{x_l}[k_{\tau+s}^{b_l}(x_l,z_l)-k_s^{b_l}(x_l,z_l)]||\sqrt{x_j}-\sqrt{z_j}|^{\gamma}, \quad (7.171)$$

III. With $m \neq j$ or l

$$|\sqrt{x_j}\partial_{x_j}k_s^{b_j}(x_j,z_j)\sqrt{x_l}\partial_{x_l}k_s^{b_l}(x_l,z_l)\times \\ [k_{\tau+s}^{b_m}(x_m,z_m)-k_s^{b_m}(x_m,z_m)]||\sqrt{x_j}-\sqrt{z_j}|^{\gamma}, \quad (7.172)$$

IV. With $m \neq j$ or l

$$|\sqrt{x_j}\partial_{x_j}k_{\tau+s}^{b_j}(x_j,z_j)\sqrt{x_l}\partial_{x_l}k_s^{b_l}(x_l,z_l)\times \\ [k_{\tau+s}^{b_m}(x_m,z_m)-k_s^{b_m}(x_m,z_m)]||\sqrt{x_j}-\sqrt{z_j}|^{\gamma}, \quad (7.173)$$

V. With $m \neq j$ or l

$$|\sqrt{x_j}\partial_{x_j}k_{\tau+s}^{b_j}(x_j,z_j)\sqrt{x_l}\partial_{x_l}k_{\tau+s}^{b_l}(x_l,z_l)\times \\ [k_{\tau+s}^{b_m}(x_m,z_m)-k_s^{b_m}(x_m,z_m)]||\sqrt{x_j}-\sqrt{z_j}|^{\gamma}. \quad (7.174)$$

Applying Lemmas 6.1.10 and 6.1.9 we see that the integrals of types III, IV and V are all bounded by

$$C\int_{\tau}^{\infty}\frac{s^{\frac{\gamma}{2}-1}\tau ds}{\tau+s}=C\tau^{\frac{\gamma}{2}}\int_{1}^{\infty}\frac{\sigma^{\frac{\gamma}{2}-1}d\sigma}{1+\sigma}\leq C\tau^{\frac{\gamma}{2}}, \quad (7.175)$$

leaving just the terms of types I and II. These are estimated using Lemma 6.1.10 and Lemma 6.1.13. Both of these terms are bounded by

$$C\int_{\tau}^{\infty}\frac{\sqrt{x_jx_l}s^{\frac{\gamma}{2}-1}\tau ds}{(\sqrt{x_j}+\sqrt{s})(\sqrt{x_l}+\sqrt{s})(\tau+s)}\leq C\int_{\tau}^{\infty}s^{\frac{\gamma}{2}-2}\tau ds\leq C\tau^{\frac{\gamma}{2}}. \quad (7.176)$$

This completes the proof that

$$|\sqrt{x_jx_l}\partial_{x_j}\partial_{x_l}u(\boldsymbol{x},t_2)-\sqrt{x_jx_l}\partial_{x_j}\partial_{x_l}u(\boldsymbol{x},t_1)|\leq C\|g\|_{WF,0,\gamma}|t_2-t_1|^{\frac{\gamma}{2}}. \quad (7.177)$$

Using the estimates on the spatial derivatives, and the differential equation (7.99), we easily establish that $\partial_t u \in \mathscr{C}^{0,\gamma}_{WF}(\mathbb{R}^n_+ \times \mathbb{R}_+)$ satisfies the desired estimates. The estimates in (7.117) and (7.161) show that the appropriate scaled second derivatives tend to zero along portions of $b\mathbb{R}^n_+$. The argument applied in the 1-dimensional case to show that (see equations (6.118) to (6.126))

$$\lim_{x\to\infty}[|u(x,t)| + |\partial_x u(x,t)| + |x\partial_x^2 u(x,t)|] = 0, \tag{7.178}$$

applies mutatis mutandis to show that

$$\lim_{\|\boldsymbol{x}\|\to\infty}[|u(\boldsymbol{x},t)| + |\nabla_{\boldsymbol{x}} u(\boldsymbol{x},t)| + \sum_{k,l}|\sqrt{x_k x_l}\partial_{x_k}\partial_{x_l} u(\boldsymbol{x},t)|] = 0. \tag{7.179}$$

One merely needs to observe that, if

$$\Phi_R(\boldsymbol{x}) = \prod_{j=1}^{n} \varphi_{R,R}(x_j), \tag{7.180}$$

then:

1. $[1 - \Phi_R(\boldsymbol{x})]g$ tends to zero in $\mathscr{C}^{0,\gamma}_{WF}(\mathbb{R}^n_+ \times [0,T])$, as $R \to \infty$.
2. For any fixed R the solution

$$u^0_R(\boldsymbol{x},t) = \int_0^t\int_0^\infty \cdots \int_0^\infty \prod_{j=1}^{n} k_s^{b_j}(x_j,y_j)\Phi_R(\boldsymbol{y})g(\boldsymbol{y},t-s)d\boldsymbol{y}ds \tag{7.181}$$

along with all derivatives, tends rapidly to zero as $\|\boldsymbol{x}\| \to \infty$.

These observations and the various Hölder estimates established above imply that we can apply Lemma 5.2.7 to conclude that

$$u \in \mathscr{C}^{0,2+\gamma}_{WF}(\mathbb{R}^n_+ \times [0,T]). \tag{7.182}$$

This completes the proof of the proposition in the $k = 0$ case. As in the 1-dimensional case, we can use Proposition 4.2.4 to commute derivatives $\partial_t^j \partial_{\boldsymbol{x}}^{\boldsymbol{\alpha}}$ past the kernel function in the integral representation. Assuming that g has support in $B^+_R(\mathbf{0}) \times [0,T]$ allows us to apply Proposition 5.2.9 to bound the resultant data in terms of $\|g\|_{WF,k,\gamma}$. Hence we can apply the estimates in the $k = 0$ case to establish the estimates in (7.102) for all $k \in \mathbb{N}$. □

7.3 THE RESOLVENT OPERATOR

As in the 1-dimensional case we can define the resolvent operator as the Laplace transform of the heat kernel. For $\mu \in S_0$, we have the formula

$$R(\mu)f(\boldsymbol{x}) = \lim_{\epsilon\to 0^+}\int_\epsilon^{\frac{1}{\epsilon}}\int_0^\infty \cdots \int_0^\infty \prod_{j=1}^{n} k_t^{b_j}(x_j,z_j)f(\boldsymbol{z})d\boldsymbol{z}e^{-t\mu}dt. \tag{7.183}$$

Using the asymptotic expansion for the 1-dimensional factors it follows easily that for each fixed $\boldsymbol{x} \in \mathbb{R}_+^n$, $R(\mu)f(\boldsymbol{x})$ is an analytic function of μ. Applying Cauchy's theorem we can easily show that, so long as $\mathrm{Re}[\mu e^{i\theta} > 0]$, we can rewrite this as:

$$R(\mu)f(\boldsymbol{x}) = \lim_{\epsilon \to 0+} \int_{\epsilon}^{\frac{1}{\epsilon}} \int_0^{\infty} \cdots \int_0^{\infty} \prod_{j=1}^{n} k_{se^{i\theta}}^{b_j}(x_j, z_j) f(z) dz e^{-e^{i\theta}\mu s} e^{i\theta} ds. \tag{7.184}$$

This shows that $R(\mu)f(\boldsymbol{x})$ extends analytically to $\mathbb{C}\setminus(-\infty, 0]$. We close this section by stating a proposition summarizing the properties of $R(\mu)$ as an operator on the Hölder spaces $\mathscr{C}_{WF}^{k,\gamma}(\mathbb{R}_+^n)$. The proof is deferred to the end of Chapter 9 where the analogous result covering all model operators is proved.

PROPOSITION 7.3.1. *The resolvent operator $R(\mu)$ is analytic in the complement of $(-\infty, 0]$, and is given by the integral in* (7.184) *provided that* $\mathrm{Re}(\mu e^{i\theta}) > 0$. *For $\alpha \in (0, \pi]$, there are constants $C_{b,\alpha}$ so that if*

$$\alpha - \pi \leq \arg \mu \leq \pi - \alpha, \tag{7.185}$$

then for $f \in \mathscr{C}_b^0(\mathbb{R}_+^n)$ we have

$$\|R(\mu)f\|_{L^\infty} \leq \frac{C_{b,\alpha}}{|\mu|} \|f\|_{L^\infty}, \tag{7.186}$$

with $C_{b,\pi} = 1$. Moreover, for $0 < \gamma < 1$, there is a constant $C_{b,\alpha,\gamma}$ so that if $f \in \mathscr{C}_{WF}^{0,\gamma}(\mathbb{R}_+^n)$, then

$$\|R(\mu)f\|_{WF,0,\gamma} \leq \frac{C_{b,\alpha,\gamma}}{|\mu|} \|f\|_{WF,0,\gamma}. \tag{7.187}$$

If for a $k \in \mathbb{N}_0$, and $0 < \gamma < 1$, $f \in \mathscr{C}_{WF}^{k,\gamma}(\mathbb{R}_+^n)$, then $R(\mu)f \in \mathscr{C}_{WF}^{k,2+\gamma}(\mathbb{R}_+^n)$, and we have

$$(\mu - L_b)R(\mu)f = f. \tag{7.188}$$

If $f \in \mathscr{C}_{WF}^{0,2+\gamma}(\mathbb{R}_+^n)$, then

$$R(\mu)(\mu - L_b)f = f. \tag{7.189}$$

There are constants $C_{b,k,\alpha}$ so that, for μ satisfying (7.185), *we have*

$$\|R(\mu)f\|_{WF,k,2+\gamma} \leq C_{b,k,\alpha} \left[1 + \frac{1}{|\mu|}\right] \|f\|_{WF,k,\gamma}. \tag{7.190}$$

For any $B > 0$, these constants are uniformly bounded for $0 \leq b < B$.

Chapter Eight

Hölder Estimates for Euclidean Models

The Euclidean model problems are given by

$$\partial_t u(\mathbf{y}, t) - \sum_{j=1}^{m} \partial_{y_j}^2 u(\mathbf{y}, t) = g(\mathbf{y}, t) \text{ with } u(\mathbf{y}, 0) = f(\mathbf{y}). \tag{8.1}$$

The 1-dimensional solution kernel is

$$k_t^e(x, y) = \frac{e^{\frac{|x-y|^2}{4t}}}{\sqrt{4\pi t}}, \tag{8.2}$$

and the solution to the equation in (8.1), vanishing at $t = 0$, is given by

$$u(\mathbf{y}, t) = \int_0^t \int_{-\infty}^{\infty} \cdots \int_{-\infty}^{\infty} \prod_{j=1}^{m} k_{t-s}^e(y_j, z_j) g(\mathbf{z}, s) d\mathbf{z} ds. \tag{8.3}$$

The solution to the homogeneous initial value problem with $v(\mathbf{y}, 0) = f(\mathbf{y})$ is given by

$$v(\mathbf{y}, t) = \int_{-\infty}^{\infty} \cdots \int_{-\infty}^{\infty} \prod_{j=1}^{m} k_t^e(y_j, z_j) f(\mathbf{z}) d\mathbf{z}. \tag{8.4}$$

For fixed $\mathbf{y}$, $v(\mathbf{y}, t)$ extends analytically in t to define a function in S_0. The Hölder estimates for the solutions of this problem are, of course, classical. In this chapter we state the estimates and the 1-dimensional kernel estimates needed to prove them.

8.1 HÖLDER ESTIMATES FOR SOLUTIONS IN THE EUCLIDEAN CASE

The solutions of the problems

$$\partial_t v(\mathbf{y}, t) - \sum_{j=1}^{m} \partial_{y_j}^2 v(\mathbf{y}, t) = 0 \text{ with } v(\mathbf{y}, t) = f(\mathbf{y}) \in \mathscr{C}^{k,\gamma}(\mathbb{R}^m) \tag{8.5}$$

and

$$\partial_t u(\mathbf{y}, t) - \sum_{j=1}^{m} \partial_{y_j}^2 u(\mathbf{y}, t) = g(\mathbf{y}, t) \in \mathscr{C}^{k,\gamma}(\mathbb{R}^m \times \mathbb{R}_+) \text{ with } u(\mathbf{y}, t) = 0, \tag{8.6}$$

are well known to satisfy Hölder estimates. These can easily be derived from the 1-dimensional kernel estimates, which are stated in the following subsection, much as in the degenerate case, though with considerably less effort.

For the homogeneous Cauchy problem we have:

PROPOSITION 8.1.1. *Let $k \in \mathbb{N}_0$ and $0 < \gamma < 1$. The solution v to* (8.5) *with initial data $f \in \mathscr{C}^{k,\gamma}(\mathbb{R}^m)$, given in* (8.4), *belongs to $\mathscr{C}^{k,\gamma}(\mathbb{R}^m \times \mathbb{R}_+)$. There are constants C so that*

$$\|v\|_{k,\gamma} \leq C\|f\|_{k,\gamma}. \tag{8.7}$$

For the inhomogeneous problem, with zero initial data, we have:

PROPOSITION 8.1.2. *Let $k \in \mathbb{N}_0$ and $0 < \gamma < 1$. The solution, u, to* (8.6) *with $g \in \mathscr{C}^{k,\gamma}(\mathbb{R}^m \times \mathbb{R}_+)$, is given in* (8.3)*; it belongs to $\mathscr{C}^{k+2,\gamma}(\mathbb{R}^m \times \mathbb{R}_+)$. There are constants C so that*

$$\|u\|_{k+2,\gamma,T} \leq C(1+T)\|g\|_{k,\gamma,T}. \tag{8.8}$$

The proofs of these propositions are in all essential ways identical to the proofs of Propositions 7.1.1 and 7.2.1, respectively, where the 1-dimensional kernel estimates from Chapter 6.1 are replaced by those given below in Chapter 8.2. The Euclidean arguments are a bit simpler, as there is no spatial boundary, and hence the special arguments needed, in the degenerate case, as $x_j \to 0$ are not necessary. The $k > 0$ estimates follow easily from the $k = 0$ estimates using Proposition 4.2.2, in the $n = 0$ case. The details of these arguments are left to the interested reader. As noted above, these results are classical, and complete proofs can be found in [29].

We can also define the resolvent operator $R(\mu)f$, as the Laplace transform of the heat kernel. For $\mu \in S_0$, we have the formula

$$R(\mu)f(\mathbf{y}) = \lim_{\epsilon \to 0^+} \int_{\epsilon}^{\frac{1}{\epsilon}} \int_0^{\infty} \cdots \int_0^{\infty} \prod_{j=1}^{n} k_t^e(y_j, w_j) f(\mathbf{w}) d\mathbf{w} e^{-t\mu} dt. \tag{8.9}$$

Using the asymptotic expansion for the 1-dimensional factors it follows easily that for each fixed $\mathbf{y} \in \mathbb{R}^m$, $R(\mu)f(\mathbf{y})$ is an analytic function of μ. Applying Cauchy's theorem we can easily show that, so long as $\mathrm{Re}[\mu e^{i\theta} > 0]$, we can rewrite this as:

$$R(\mu)f(\mathbf{y}) = \lim_{\epsilon \to 0^+} \int_{\epsilon}^{\frac{1}{\epsilon}} \int_0^{\infty} \cdots \int_0^{\infty} \prod_{j=1}^{n} k_{se^{i\theta}}^e(y_j, w_j) f(\mathbf{w}) d\mathbf{w} e^{-e^{i\theta}\mu s} e^{i\theta} ds. \tag{8.10}$$

This shows that $R(\mu)f(\mathbf{y})$ extends analytically to $\mathbb{C}\setminus(-\infty, 0]$. We close this section by stating a proposition summarizing the properties of $R(\mu)$ as an operator on the Hölder spaces $\mathscr{C}^{k,\gamma}(\mathbb{R}^m)$. The proof is deferred to the end of Chapter 9 where the analogous result covering all model operators is proved.

PROPOSITION 8.1.3. *The resolvent operator $R(\mu)$ is analytic in the complement of $(-\infty, 0]$, and is given by the integral in* (8.10) *provided that* $\mathrm{Re}(\mu e^{i\theta}) > 0$. *For $\alpha \in (0, \pi]$, there are constants C_α so that if*

$$\alpha - \pi \leq \arg\mu \leq \pi - \alpha, \tag{8.11}$$

then for $f \in \mathscr{C}_b^0(\mathbb{R}^m)$ we have

$$\|R(\mu)f\|_{L^\infty} \leq \frac{C_\alpha}{|\mu|}\|f\|_{L^\infty}, \tag{8.12}$$

with $C_\pi = 1$. Moreover, for $0 < \gamma < 1$, there is a constant $C_{\alpha,\gamma}$ so that if $f \in \mathscr{C}^{0,\gamma}(\mathbb{R}^m)$, then

$$\|R(\mu)f\|_{0,\gamma} \leq \frac{C_{\alpha,\gamma}}{|\mu|}\|f\|_{0,\gamma}. \tag{8.13}$$

For $k \in \mathbb{N}_0$, and $0 < \gamma < 1$, if $f \in \mathscr{C}^{k,\gamma}(\mathbb{R}^m)$, then $R(\mu)f \in \mathscr{C}^{2+k,\gamma}(\mathbb{R}^m)$, and, we have

$$(\mu - L_{\mathbf{0},m})R(\mu)f = f. \tag{8.14}$$

If $f \in \mathscr{C}^{2,\gamma}(\mathbb{R}^m)$, then

$$R(\mu)(\mu - L_{\mathbf{0},m})f = f. \tag{8.15}$$

There are constants $C_{k,\alpha}$ so that, for μ satisfying (8.11), *we have*

$$\|R(\mu)f\|_{k+2,\gamma} \leq C_{k,\alpha}\left[1 + \frac{1}{|\mu|}\right]\|f\|_{k,\gamma}. \tag{8.16}$$

Remark. As before the solution $v(\mathbf{y}, t)$ to the Cauchy problem can be expressed as a contour integral:

$$v(\mathbf{y}, t) = \frac{1}{2\pi i}\int\limits_{\Gamma_{\alpha,R}} [R(\mu)f](\mathbf{y})e^{\mu t}d\mu. \tag{8.17}$$

From this representation it follows that v extends analytically in t to S_0. Moreover, for $t \in S_0$ we see that $v(\cdot, t)$ belongs to $\mathscr{C}^{2,\gamma}(\mathbb{R}^m)$. Hence by the semi-group property $v(\cdot, t) \in \mathscr{C}^\infty(\mathbb{R}^m)$.

8.2 1-DIMENSIONAL KERNEL ESTIMATES

The 1-dimensional kernel estimates can easily be used to prove the Hölder estimates stated in the previous subsection. They form essential components of the proofs of the Hölder estimates for the general model problems, considered in the next chapter. The proofs of these estimates are elementary, largely following from the facts that the kernel, $k_t^e(x, y)$, is a function of $(x - y)^2/t$, which extends analytically to $\mathbb{R}^2 \times S_0$. As in the degenerate case, we have

$$\int\limits_{-\infty}^{\infty} k_t^e(x, y)dy = 1 \text{ for } x \in \mathbb{R}, t \in S_0. \tag{8.18}$$

The proofs of the following classical results are left to the reader.

8.2.1 Basic Kernel Estimates

LEMMA 8.2.1. *For $0 \le \gamma < 1$, $0 < \phi \le \frac{\pi}{2}$, there is a C_ϕ so that, for $t \in S_\phi$,*

$$\int_{-\infty}^{\infty} |k_t^e(x,y)||x-y|^\gamma \, dy \le C_\phi |t|^{\frac{\gamma}{2}}. \tag{8.19}$$

LEMMA 8.2.2. *For $0 < \phi \le \frac{\pi}{2}$, there is a constant C_ϕ so that for $t \in S_\phi$*

$$\int_{-\infty}^{\infty} |k_t^e(x_2,z) - k_t^e(x_1,z)| dz \le C_\phi \left(\frac{\frac{|x_2-x_1|}{\sqrt{|t|}}}{1 + \frac{|x_2-x_1|}{\sqrt{|t|}}} \right). \tag{8.20}$$

We set

$$\alpha_e = \frac{3x_1 - x_2}{2} \text{ and } \beta_e = \frac{3x_2 - x_1}{2}. \tag{8.21}$$

LEMMA 8.2.3. *Let $J = [\alpha_e, \beta_e]$, as defined in (8.21). For $0 < \gamma < 1$, $0 < \phi \le \frac{\pi}{2}$, there is a C_ϕ so that for $t = |t|e^{i\theta} \in S_\phi$*

$$\int_{J^c} |k_t^e(x_2,y) - k_t^e(x_1,y)||y - x_1|^\gamma \, dy \le C_\phi |x_2 - x_1|^\gamma e^{-\cos\theta \frac{(x_2-x_1)^2}{2|t|}}. \tag{8.22}$$

LEMMA 8.2.4. *For $0 < \gamma < 1$ and $c < 1$ there is a C such that if $c < s/t < 1$, then*

$$\int_{-\infty}^{\infty} \left| k_t^e(x,y) - k_s^e(x,y) \right| |x-y|^\gamma \, dy \le C|t-s|^{\frac{\gamma}{2}}. \tag{8.23}$$

Without an upper bound on $0 < t/s$, we have the estimate

$$\int_{-\infty}^{\infty} \left| k_t^e(x,y) - k_s^e(x,y) \right| dy \le C \left(\frac{t/s - 1}{1 + [t/s - 1]} \right). \tag{8.24}$$

8.2.2 First Derivative Estimates

LEMMA 8.2.5. *For $0 \le \gamma < 1$, and $0 < \phi \le \frac{\pi}{2}$, there is a C_ϕ so that for $t \in S_\phi$ we have*

$$\int_0^{\infty} |\partial_x k_t^e(x,y)||\sqrt{y} - \sqrt{x}|^\gamma \, dy \le C_\phi |t|^{\frac{\gamma-1}{2}}. \tag{8.25}$$

LEMMA 8.2.6. *For $0 < \gamma < 1$, $0 < \phi \leq \frac{\pi}{2}$, there is a constant C_ϕ so that for $t \in S_\phi$,*

$$\int_0^\infty |\partial_x k_t^e(x_1, y) - \partial_x k_t^e(x_2, y)||\sqrt{x_1} - \sqrt{y}|^\gamma \, dy \leq$$

$$C_\phi |t|^{\frac{\gamma-1}{2}} \frac{\left(\frac{|\sqrt{x_2}-\sqrt{x_1}|}{\sqrt{|t|}}\right)}{1 + \left(\frac{|\sqrt{x_2}-\sqrt{x_1}|}{\sqrt{|t|}}\right)}. \quad (8.26)$$

LEMMA 8.2.7. *For $0 \leq \gamma < 1$, and $0 < \tau$, we have for $s \in [\tau, \infty)$ that there is a constant C so that*

$$\int_{-\infty}^{\infty} |\partial_x k_{\tau+s}^e(x, y) - \partial_x k_s^e(x, y)||x - y|^\gamma \, dy < C \frac{\tau s^{\frac{\gamma-1}{2}}}{(\tau + s)}. \quad (8.27)$$

8.2.3 Second Derivative Estimates

LEMMA 8.2.8. *For $0 \leq \gamma < 1$, $0 < \phi \leq \frac{\pi}{2}$, there is a C_ϕ so that for $t \in S_\phi$ we have the estimate*

$$\int_{-\infty}^{\infty} |\partial_x^2 k_t^e(x, y)||x - y|^\gamma \, dy \leq C_\phi |t|^{\frac{\gamma}{2}-1}. \quad (8.28)$$

This implies that, if $\gamma > 0$, then

$$\int_0^{|t|} \int_{-\infty}^{\infty} |\partial_x^2 k_{se^{i\theta}}^e(x, y)||x - y|^\gamma \, dy ds \leq \left(\frac{2C_\phi}{\gamma}\right) |t|^{\frac{\gamma}{2}}. \quad (8.29)$$

LEMMA 8.2.9. *For $0 < \phi \leq \frac{\pi}{2}$, there is a C_ϕ so that for $t \in S_\phi$*

$$\int_0^{|t|} \left|\partial_y k_{se^{i\theta}}^e(x_2, \alpha_e) - \partial_y k_{se^{i\theta}}^e(x_2, \beta_e)\right| ds \leq C_\phi e^{-\cos\theta \frac{(x_2-x_1)^2}{|t|}}, \quad (8.30)$$

where α_e and β_e are defined in (8.21).

LEMMA 8.2.10. *For $0 < \gamma < 1$, $0 < \phi \leq \frac{\pi}{2}$, there is a C_ϕ so that for $|\theta| \leq \frac{\pi}{2} - \phi$ and $J = [\alpha_e, \beta_e]$, with the endpoints given by* (8.21)*, we have*

$$\begin{aligned} &\int_0^{|t|} \int_{\alpha_e}^{\beta_e} |\partial_x^2 k_{se^{i\theta}}^e(x_2, y)||y - x_2|^\gamma \, dy ds \leq C_\phi |x_2 - x_1|^\gamma \\ &\int_0^{|t|} \int_{\alpha_e}^{\beta_e} |\partial_x^2 k_{se^{i\theta}}^e(x_1, y)||y - x_1|^\gamma \, dy ds \leq C_\phi |x_2 - x_1|^\gamma. \end{aligned} \quad (8.31)$$

These estimates follow from the more basic

LEMMA 8.2.11. *For $0 < \gamma < 1$, $0 < \phi \leq \frac{\pi}{2}$, there is a C_ϕ so that for $|\theta| < \frac{\pi}{2} - \phi$ and $J = [\alpha_e, \beta_e]$, with the endpoints given by (8.21), we have*

$$\int_{\alpha_e}^{\beta_e} |\partial_x^2 k^e_{se^{i\theta}}(x_2, y)||y - x_2|^\gamma \, dy \leq C_\phi \begin{cases} s^{\frac{\gamma}{2}-1} & \text{if } s < (x_2 - x_1)^2 \\ \frac{|x_2 - x_1|^{\gamma+1}}{s^{\frac{3}{2}}} & \text{if } s \geq (x_2 - x_1)^2. \end{cases} \tag{8.32}$$

LEMMA 8.2.12. *For $0 < \gamma < 1$, $0 < \phi \leq \frac{\pi}{2}$, there is a C_ϕ so that for $|\theta| < \frac{\pi}{2} - \phi$ and $J = [\alpha_e, \beta_e]$, with the endpoints given by (8.21), we have*

$$\int_0^{|t|} \int_{J^c} |\partial_x^2 k^e_{se^{i\theta}}(x_2, y) - \partial_x^2 k^e_{se^{i\theta}}(x_1, y)||y - x_1|^\gamma \, dy ds \leq C_\phi |x_2 - x_1|^\gamma . \tag{8.33}$$

This follows from the more basic

LEMMA 8.2.13. *For $0 < \gamma < 1$, $0 < \phi \leq \frac{\pi}{2}$, there is a C_ϕ so that for $|\theta| < \frac{\pi}{2} - \phi$ and $J = [\alpha_e, \beta_e]$, with the endpoints given by (8.21), we have*

$$\int_{J^c} |\partial_x^2 k^e_{se^{i\theta}}(x_2, y) - \partial_x^2 k^e_{se^{i\theta}}(x_1, y)||y - x_1|^\gamma \, dy \leq C_\phi s^{\frac{\gamma}{2}-1} \left| \frac{x_2 - x_1}{\sqrt{s}} \right| e^{-\cos\theta \frac{(x_2-x_1)^2}{8s}} . \tag{8.34}$$

Finally we have

LEMMA 8.2.14. *For $0 \leq \gamma < 1$, $0 < \tau$ and $\tau < s$, there is a constant C so that*

$$\int_{-\infty}^{\infty} |\partial_x^2 k^e_{\tau+s}(x, y) - \partial_x^2 k^e_s(x, y)||x - y|^\gamma \, dy \leq C|\tau| s^{\frac{\gamma}{2}-2}. \tag{8.35}$$

8.2.4 Large t Behavior

To prove estimates on the resolvent, and to study the off-diagonal behavior of the heat kernel in many variables, it is useful to have estimates on the derivatives of $k^e_t(x, y)$ valid for t bounded away from zero.

LEMMA 8.2.15. *For $j \in \mathbb{N}$ and $0 < \phi \leq \frac{\pi}{2}$ there is a constant $C_{j,\phi}$ so that if $t \in S_\phi$, then*

$$\int_{-\infty}^{\infty} |\partial_x^j k^e_t(x, y)| dy \leq \frac{C_{j,\phi}}{|t|^{\frac{j}{2}}}. \tag{8.36}$$

The proof of this lemma is in the Appendix.

Chapter Nine

Hölder Estimates for General Models

We now turn to the task of estimating solutions to heat equations defined by the operators of the form:

$$L_{\boldsymbol{b},m} = \sum_{j=1}^{n} [x_j \partial_{x_j}^2 + b_j \partial_{x_j}] + \sum_{k=1}^{m} \partial_{y_k}^2. \tag{9.1}$$

The general model operator on $\mathbb{R}_+^n \times \mathbb{R}^m$, denoted $L_{\boldsymbol{b},m}$, is labeled by a non-negative n-vector $\boldsymbol{b} = (b_1, \ldots, b_n)$, and m, the dimension of the corner. We use x-variables to denote points in $\mathbb{R}_+^n$ and y-variables to denote points in $\mathbb{R}^m$. If we have a function of these variables $f(\boldsymbol{x}, \boldsymbol{y})$ then, as before, we estimate differences $f(\boldsymbol{x}^2, \boldsymbol{y}^2) - f(\boldsymbol{x}^1, \boldsymbol{y}^1)$ 1-variable-at-a-time. We first observe that

$$f(\boldsymbol{x}^2, \boldsymbol{y}^2) - f(\boldsymbol{x}^1, \boldsymbol{y}^1) = [f(\boldsymbol{x}^2, \boldsymbol{y}^2) - f(\boldsymbol{x}^1, \boldsymbol{y}^2)] + [f(\boldsymbol{x}^1, \boldsymbol{y}^2) - f(\boldsymbol{x}^1, \boldsymbol{y}^1)]; \tag{9.2}$$

each term in brackets can then be written as a telescoping sum:

$$\begin{aligned} f(\boldsymbol{x}^2, \boldsymbol{y}^2) - f(\boldsymbol{x}^1, \boldsymbol{y}^1) = \\ \left\{ \sum_{j=0}^{n-1} [f(\boldsymbol{x}_j^{2\prime}, x_{j+1}^2, \boldsymbol{x}_j^{1\prime\prime}, \boldsymbol{y}^2) - f(\boldsymbol{x}_j^{2\prime}, x_{j+1}^1, \boldsymbol{x}_j^{1\prime\prime}, \boldsymbol{y}^2)] \right\} + \\ \left\{ \sum_{l=0}^{m-1} [f(\boldsymbol{x}^1, \boldsymbol{y}_l^{2\prime}, y_{l+1}^2, \boldsymbol{y}_l^{1\prime\prime}) - f(\boldsymbol{x}^1, \boldsymbol{y}_l^{2\prime}, y_{l+1}^2, \boldsymbol{y}_l^{1\prime\prime})] \right\}, \end{aligned} \tag{9.3}$$

where $\boldsymbol{x}_j'$ and $\boldsymbol{x}_j''$ are defined in (7.4). We say that terms in the first sum have a "variation in an x-variable," and terms in the second have a "variation in a y-variable." We only need to deal with terms that have a variation in one or the other type of variable, and this simplifies the proofs for the general case considerably.

In this chapter we prove Hölder estimates for the solutions on $\mathbb{R}_+^n \times \mathbb{R}^m \times \mathbb{R}_+$ to the homogeneous Cauchy problem

$$[\partial_t - L_{\boldsymbol{b},m}] v(\boldsymbol{x}, \boldsymbol{y}, t) = 0 \text{ with } v(\boldsymbol{x}, \boldsymbol{y}, 0) = f(\boldsymbol{x}, \boldsymbol{y}), \tag{9.4}$$

and the inhomogeneous problem

$$[\partial_t - L_{\boldsymbol{b},m}] u(\boldsymbol{x}, \boldsymbol{y}, t) = g(\boldsymbol{x}, \boldsymbol{y}, t) \text{ with } u(\boldsymbol{x}, \boldsymbol{y}, 0) = 0. \tag{9.5}$$

The solution to the homogeneous initial value problem with $v(\boldsymbol{x}, \boldsymbol{y}, 0) = f(\boldsymbol{x}, \boldsymbol{y})$ is given by

$$\begin{aligned} v(\boldsymbol{x}, \boldsymbol{y}, t) &= \int_{-\infty}^{\infty} \cdots \int_{-\infty}^{\infty} \prod_{l=1}^{n} k_t^{b_l}(x_l, w_l) \prod_{j=1}^{m} k_t^{e}(y_j, z_j) f(\boldsymbol{w}, z) dz d\boldsymbol{w} \\ &= \kappa_t^{\boldsymbol{b},m} f. \end{aligned} \tag{9.6}$$

For fixed $(\boldsymbol{x}, \boldsymbol{y})$, $v(\boldsymbol{x}, \boldsymbol{y}, t)$ extends analytically in t to define a function in S_0.

The solution to the inhomogeneous problem is given by the operator $K_t^{\boldsymbol{b},m}$ defined by

$$\begin{aligned} u(\boldsymbol{x}, \boldsymbol{y}, t) &= \int_{0}^{t} \int_{-\infty}^{\infty} \cdots \int_{-\infty}^{\infty} \prod_{l=1}^{n} k_{t-s}^{b_l}(x_l, w_l) \prod_{j=1}^{m} k_{t-s}^{e}(y_j, z_j) g(\boldsymbol{w}, \boldsymbol{z}, s) d\boldsymbol{z} d\boldsymbol{w} ds \\ &= K_t^{\boldsymbol{b},m} g. \end{aligned} \tag{9.7}$$

As before, for $t > 0$, this expression should be understood as

$$u(\boldsymbol{x}, \boldsymbol{y}, t) = \lim_{\epsilon \to 0+} u_\epsilon(\boldsymbol{x}, \boldsymbol{y}, t), \tag{9.8}$$

where

$$u_\epsilon(\boldsymbol{x}, \boldsymbol{y}, t) = \int_{0}^{t-\epsilon} \int_{-\infty}^{\infty} \cdots \int_{-\infty}^{\infty} \prod_{l=1}^{n} k_{t-s}^{b_l}(x_l, w_l) \prod_{j=1}^{m} k_{t-s}^{e}(y_j, z_j) g(\boldsymbol{w}, \boldsymbol{z}, s) d\boldsymbol{z} d\boldsymbol{w} ds. \tag{9.9}$$

We also note that if we let $v^s(\boldsymbol{x}, \boldsymbol{y}, t)$ denote the solution to the Cauchy problem:

$$[\partial_t - L_{\boldsymbol{b},m}] v^s = 0 \text{ with } v^s(\boldsymbol{x}, \boldsymbol{y}, 0) = g(\boldsymbol{x}, \boldsymbol{y}, s), \tag{9.10}$$

then

$$u(\boldsymbol{x}, \boldsymbol{y}, t) = \int_{0}^{t} v^s(\boldsymbol{x}, \boldsymbol{y}, t - s) ds. \tag{9.11}$$

The resolvent operator $R(\mu) f$ is defined, for $f \in \dot{\mathscr{C}}^0(\mathbb{R}_+^n \times \mathbb{R}^m)$, and $\mu \in S_0$, by

$$R(\mu) f(\boldsymbol{x}, \boldsymbol{y}) = \lim_{\epsilon \to 0+} \int_{\epsilon}^{\frac{1}{\epsilon}} e^{-\mu t} v(\boldsymbol{x}, \boldsymbol{y}, t) dt. \tag{9.12}$$

As in the earlier cases, this is an analytic function of $\mu \in S_0$. By deforming the contour we can replace this representation with

$$R(\mu) f(\boldsymbol{x}, \boldsymbol{y}) = \lim_{\epsilon \to 0+} \int_{\epsilon}^{\frac{1}{\epsilon}} e^{-\mu s e^{i\theta}} v(\boldsymbol{x}, \boldsymbol{y}, s e^{i\theta}) e^{i\theta} ds, \tag{9.13}$$

which converges if $\mathrm{Re}[\mu e^{i\theta}] > 0$. This analytically extends $R(\mu)f(\boldsymbol{x}, \boldsymbol{y})$ to complex numbers $\mu \in \mathbb{C} \setminus (-\infty, 0]$. As in the previous two chapters, the estimates herein are all proved by reduction to 1-variable kernel estimates.

9.1 THE CAUCHY PROBLEM

We begin with estimates for the homogeneous Cauchy problem.

PROPOSITION 9.1.1. *Let $k \in \mathbb{N}_0$, $0 < R$, $(b_1, \dots, b_n) \in \mathbb{R}_+^n$, and $0 < \gamma < 1$. Suppose that the initial data f lies in $\mathscr{C}_{WF}^{k,\gamma}(\mathbb{R}_+^n \times \mathbb{R}^m)$; if $k > 0$, assume that f is supported in $B_R^+(\mathbf{0}) \times \mathbb{R}^m$. Then the solution v to (9.4), with initial data f, given in (9.6), belongs to $\mathscr{C}_{WF}^{k,\gamma}(\mathbb{R}_+^n \times \mathbb{R}^m \times \mathbb{R}_+)$. There are constants $C_{\boldsymbol{b},\gamma,R}$ so that*

$$\|v\|_{WF,k,\gamma} \le C_{\boldsymbol{b},\gamma,R}\|f\|_{WF,k,\gamma}. \tag{9.14}$$

If $f \in \mathscr{C}_{WF}^{k,2+\gamma}(\mathbb{R}_+^n \times \mathbb{R}^m)$, then v belongs to $\mathscr{C}_{WF}^{k,2+\gamma}(\mathbb{R}_+^n \times \mathbb{R}^m \times \mathbb{R}_+)$. There are constants $C_{\boldsymbol{b},\gamma,R}$ so that

$$\|v\|_{WF,k,2+\gamma} \le C_{\boldsymbol{b},\gamma,R}\|f\|_{WF,k,2+\gamma}. \tag{9.15}$$

For $B > 0$, these constants are uniformly bounded for $\boldsymbol{b} \le B\mathbf{1}$, and if $k = 0$, then the constants are independent of R.

PROOF. For the $k = 0, \boldsymbol{b} > \mathbf{0}$, the estimates in (9.14) follow as in the proof of Proposition 7.1.1, via the 1-variable-at-a-time method. The cases where two "x" (or parabolic) variables differ follow, essentially verbatim, as in the proof of Proposition 7.1.1, from the lemmas in Chapter 6.1. The new cases in the proof of this proposition are those involving the "y"- (or Euclidean) variables. As noted after the statement of Proposition 8.1.1, these cases follow, mutatis mutandis, via the arguments used in the proof of Proposition 7.1.1. The estimates for the kernel k_t^b must be replaced with estimates for the 1-dimensional, *Euclidean* heat kernel. These are stated in Section 8.2. As these cases also arise in the proof of Proposition 9.2.1, to avoid excessive repetition, we forego giving the details now, and leave them for the proof of the next proposition.

As before all the constants in these estimates are uniformly bounded for bounded $\mathbf{0} < \boldsymbol{b} \le \boldsymbol{B}$. Applying the compactness result in Proposition 5.2.8, we can allow entries of $\boldsymbol{b}$ to tend to zero, obtaining the unique limiting solution with all the desired estimates for these cases as well. If we assume that f is supported in $B_R^+(\mathbf{0}) \times \mathbb{R}^m$, then the estimates in (9.14) for the Hölder spaces with $k > 0$ follow from the $k = 0$ results, and Propositions 4.2.2 and 5.2.9.

As in the proof of Proposition 7.1.1, some additional estimates are needed to establish (9.15). We begin with the $k = 0$ case. Applying Proposition 4.2.2 to commute derivatives through the integral kernel, we see that estimates for $\mathscr{C}_{WF}^{0,\gamma}$-norm of $\nabla_{\boldsymbol{x}} v$, $\nabla_{\boldsymbol{y}} v$, $x_j \partial_{x_j}^2 v$, $j = 1, \dots, n$ and $\partial_{y_k}\partial_{y_l} v$, $1 \le k, l \le m$ follow from (9.14). To establish (9.15), we need only estimate the $\mathscr{C}_{WF}^{0,\gamma}$-norm of

$$\sqrt{x_i}\partial_{x_i}\partial_{y_k} v \text{ and } \sqrt{x_i x_j}\partial_{x_i}\partial_{x_j} v. \tag{9.16}$$

To estimate $\sqrt{x_i}\partial_{x_i}\partial_{y_k}v$, we can relabel so that $i = n$, and $k = m$; this derivative is given by

$$\sqrt{x_n}\partial_{x_n}\partial_{y_m}v = \int\limits_{\mathbb{R}_+^{n-1}}\int\limits_{\mathbb{R}^{m-1}} \kappa_t^{\boldsymbol{b}',m-1}(\boldsymbol{x}'_{n-1}, \boldsymbol{w}'_{n-1}; \boldsymbol{y}'_{m-1}, \boldsymbol{z}'_{m-1})\times$$

$$\int\limits_0^\infty\int\limits_{-\infty}^\infty \sqrt{\frac{x_n}{w_n}}k_t^{b_n+1}(x_n, w_n)k_t^e(y_m, z_m)\sqrt{w_n}\partial_{w_n}\partial_{z_m}f(\boldsymbol{w}, \boldsymbol{z})dw_n dz_m d\boldsymbol{w}'_{n-1}d\boldsymbol{z}'_{m-1}, \tag{9.17}$$

where

$$\kappa_t^{\boldsymbol{b}',m-1}(\boldsymbol{x}'_{n-1}, \boldsymbol{w}'_{n-1}; \boldsymbol{y}'_{m-1}, \boldsymbol{z}'_{m-1}) = \prod_{j=1}^{n-1} k_t^{b_j}(x_j, w_j)\prod_{k=1}^{m-1} k_t^e(y_k, z_k). \tag{9.18}$$

Applying Lemma 7.1.2 we see that

$$|\sqrt{x_n}\partial_{x_n}\partial_{y_m}v| \leq C\|f\|_{WF,0,2+\gamma}. \tag{9.19}$$

Since $f \in \mathscr{C}_{WF}^{0,2+\gamma}$, we know that

$$|\sqrt{x_n}\partial_{x_n}\partial_{y_m}f(\boldsymbol{x}, \boldsymbol{y})| \leq \|f\|_{WF,0,2+\gamma}x_n^{\frac{\gamma}{2}}. \tag{9.20}$$

Using this estimate in (9.17) and applying Lemma 7.1.2 shows that

$$|\sqrt{x_n}\partial_{x_n}\partial_{y_m}v| \leq C\|f\|_{WF,0,2+\gamma}x_n^{\frac{\gamma}{2}}, \tag{9.21}$$

establishing that

$$\lim_{x_n\to 0^+}\sqrt{x_n}\partial_{x_n}\partial_{y_m}v = 0. \tag{9.22}$$

The Hölder continuity of this derivative in the $\boldsymbol{y}$-variables follows by re-expressing the difference,

$$\sqrt{x_n}\partial_{x_n}\partial_{y_m}v(\boldsymbol{x}, \boldsymbol{y}, t) - \sqrt{x_n}\partial_{x_n}\partial_{y_m}v(\boldsymbol{x}, \widetilde{\boldsymbol{y}}, t), \tag{9.23}$$

as a sum of terms like those appearing in (7.29). We then apply estimates from Lemmas 8.2.2 and 8.2.3 to the terms in this sum, along with Lemma 7.1.2, to conclude that

$$|\sqrt{x_n}\partial_{x_n}\partial_{y_m}v(\boldsymbol{x}, \boldsymbol{y}, t) - \sqrt{x_n}\partial_{x_n}\partial_{y_m}v(\boldsymbol{x}, \widetilde{\boldsymbol{y}}, t)| \leq C\|f\|_{WF,0,2+\gamma}\rho_e(\boldsymbol{y}, \widetilde{\boldsymbol{y}})^\gamma. \tag{9.24}$$

The argument to establish the estimates

$$|\sqrt{x_n}\partial_{x_n}\partial_{y_m}v(\boldsymbol{x}'_j, x^1_{j+1}, \boldsymbol{x}''_j, \boldsymbol{y}, t) - \sqrt{x_n}\partial_{x_n}\partial_{y_m}v(\boldsymbol{x}'_j, x^2_{j+1}, \boldsymbol{x}''_j, \boldsymbol{y}, t)| \leq$$
$$C\|f\|_{WF,0,2+\gamma}\left|\sqrt{x^1_{j+1}} - \sqrt{x^2_{j+1}}\right|^\gamma, \text{ with } j \leq n-2, \tag{9.25}$$

is essentially identical, with Lemmas 8.2.2 and 8.2.3 replaced by Lemmas 6.1.5 and 6.1.7.

The only remaining spatial estimate is (9.25) with $j = n-1$. The estimate in (9.21) implies that for any $c < 1$, there is a C so that, if $x_n^1 < cx_n^2$, then

$$|\sqrt{x_n}\partial_{x_n}\partial_{y_m}v(\boldsymbol{x}'_{n-1}, x_n^1, \boldsymbol{y}, t) - \sqrt{x_n}\partial_{x_n}\partial_{y_m}v(\boldsymbol{x}'_{n-1}, x_n^2, \boldsymbol{y}, t)| \leq C\|f\|_{WF,0,2+\gamma}\left|\sqrt{x_n^1} - \sqrt{x_n^2}\right|^{\gamma}, \tag{9.26}$$

leaving only the case $cx_n^2 < x_n^1 < x_n^2$. Using an obvious modification of (7.64), and essentially the same argument as appears after (7.64), we can prove this estimate as well. To complete the estimates of this derivative we need to show that it is Hölder continuous in the time variable. The proof of this estimate is a small modification of that used to prove (7.98). We begin with

$$\sqrt{x_n}\partial_{x_n}\partial_{y_m}[v(\boldsymbol{x}, \boldsymbol{y}, t) - v(\boldsymbol{x}, \boldsymbol{y}, 0)] = \int\limits_{\mathbb{R}_+^{n-1}}\int\limits_{\mathbb{R}^{m-1}} k_t^{\boldsymbol{b}',m-1}(\boldsymbol{x}'_{n-1}, \boldsymbol{w}'_{n-1}; \boldsymbol{y}'_{m-1}, \boldsymbol{z}'_{m-1}) \times$$
$$\int\limits_0^\infty\int\limits_0^\infty \sqrt{x_n}k_t^{b_n+1}(x_n, w_n)k_t^e(y_m, z_m)[f_{nm}(\boldsymbol{w}, \boldsymbol{z}) - f_{nm}(\boldsymbol{x}, \boldsymbol{y})]dw_n dz_m d\boldsymbol{w}'_{n-1}d\boldsymbol{z}'_{m-1}, \tag{9.27}$$

where $f_{nm} = \partial_{w_n}\partial_{z_m} f$. We rewrite the difference, $[f_{nm}(\boldsymbol{w}, \boldsymbol{z}) - f_{nm}(\boldsymbol{x}, \boldsymbol{y})]$, as a telescoping sum like that in (9.3):

$$f_{nm}(\boldsymbol{w}, \boldsymbol{z}) - f_{nm}(\boldsymbol{x}, \boldsymbol{y}) = \left\{\sum_{j=0}^{n-1}[f_{nm}(\boldsymbol{w}'_j, w_{j+1}, \boldsymbol{x}''_j, \boldsymbol{z}) - f_{nm}(\boldsymbol{w}'_j, x_{j+1}, \boldsymbol{x}''_j, \boldsymbol{z})]\right\} + \left\{\sum_{l=0}^{m-1}[f_{nm}(\boldsymbol{x}, \boldsymbol{z}'_l, z_{l+1}, \boldsymbol{y}''_l) - f_{nm}(\boldsymbol{x}, \boldsymbol{z}'_l, y_{l+1}, \boldsymbol{y}''_l)]\right\}. \tag{9.28}$$

Each term in the second sum is estimated by

$$\sqrt{x_n}|f_{nm}(\boldsymbol{x}, \boldsymbol{z}'_l, z_{l+1}, \boldsymbol{y}''_l) - f_{nm}(\boldsymbol{x}, \boldsymbol{z}'_l, y_{l+1}, \boldsymbol{y}''_l)| \leq \|f\|_{WF,0,2+\gamma}|z_{l+1} - y_{l+1}|^{\gamma}. \tag{9.29}$$

Each term in the first sum, except for $j = n-1$, is estimated by

$$\sqrt{x_n}|f_{nm}(\boldsymbol{w}'_j, w_{j+1}, \boldsymbol{x}''_j, \boldsymbol{z}) - f_{nm}(\boldsymbol{w}'_j, x_{j+1}, \boldsymbol{x}''_j, \boldsymbol{z})| \leq \|f\|_{WF,0,2+\gamma}|\sqrt{x_{j+1}} - \sqrt{w_{j+1}}|^{\gamma}. \tag{9.30}$$

The remaining case is estimated by

$$\begin{aligned}\sqrt{x_n}|f_{nm}(\boldsymbol{w}'_{n-1}, w_n, \boldsymbol{z}) - f_{nm}(\boldsymbol{w}'_{n-1}, x_n, \boldsymbol{z})| &\leq |\sqrt{x_n} - \sqrt{w_n}||f_{nm}(\boldsymbol{w}'_{n-1}, w_n, \boldsymbol{z})| + \\ &\quad |\sqrt{w_n}f_{nm}(\boldsymbol{w}'_{n-1}, w_n, \boldsymbol{z}) - \sqrt{x_n}f_{nm}(\boldsymbol{w}'_{n-1}, x_n, \boldsymbol{z})| \\ &\leq \|f\|_{WF,0,2+\gamma}[|\sqrt{x_n} - \sqrt{w_n}|w_n^{\frac{\gamma-1}{2}} + |\sqrt{x_n} - \sqrt{w_n}|^{\gamma}].\end{aligned} \tag{9.31}$$

Using these estimates in (9.27) we apply Lemmas 6.1.6, 7.1.4 and 8.2.1 to deduce that

$$|\sqrt{x_n}\partial_{x_n}\partial_{y_m}[v(\boldsymbol{x},\boldsymbol{y},t)-v(\boldsymbol{x},\boldsymbol{y},0)]| \leq C\|f\|_{WF,0,2+\gamma}t^{\frac{\gamma}{2}}. \tag{9.32}$$

As usual, this shows that for $0 < c < 1$, there is a C so that if $s < ct$, then

$$|\sqrt{x_n}\partial_{x_n}\partial_{y_m}[v(\boldsymbol{x},\boldsymbol{y},t)-v(\boldsymbol{x},\boldsymbol{y},s)]| \leq C\|f\|_{WF,0,2+\gamma}|t-s|^{\frac{\gamma}{2}}. \tag{9.33}$$

We are left with the case $ct < s < t$, which again closely follows the pattern of the proof of (7.98). As before we use an analogue of (7.37):

$$\begin{aligned}
&\sqrt{x_n}\partial_{x_n}\partial_{y_m}[v(\boldsymbol{x},\boldsymbol{y},t)-v(\boldsymbol{x},\boldsymbol{y},s)] \\
&= \sqrt{x_n}\int_{\mathbb{R}^n_+}\int_{\mathbb{R}^m}\Bigg[\kappa_t^{e,m}(\boldsymbol{y},\boldsymbol{z})\sum_{l=1}^{n}\Bigg\{\prod_{j=1}^{n-l}k_t^{\widetilde{b}_j}(x_j,w_j)\times\prod_{j=n-l+2}^{n}k_s^{\widetilde{b}_j}(x_j,w_j)\times \\
&\qquad\Big[k_t^{\widetilde{b}_{n-l+1}}(x_{n-l+1},w_{n-l+1})-k_s^{\widetilde{b}_{n-l+1}}(x_{n-l+1},w_{n-l+1})\Big]\times \\
&\qquad\qquad[f_{nm}(\boldsymbol{w},\boldsymbol{z})-f_{nm}(\boldsymbol{w}'_l,x_{n-l+1},\boldsymbol{w}''_l,\boldsymbol{z})]\Bigg\}+ \\
&\qquad\kappa_s^{\widetilde{\boldsymbol{b}},0}(\boldsymbol{x},\boldsymbol{w})\sum_{q=1}^{m}\Bigg\{\prod_{j=1}^{m-q}k_t^{e}(y_j,z_j)\times\prod_{j=m-q+2}^{m}k_s^{e}(y_j,z_j)\times \\
&\qquad\Big[k_t^{e}(y_{m-q+1},z_{m-q+1})-k_s^{e}(y_{m-q+1},z_{m-q+1})\Big]\times \\
&\qquad\qquad[f_{nm}(\boldsymbol{w},\boldsymbol{z})-f_{nm}(\boldsymbol{w},\boldsymbol{z}'_q,y_{m-q+1},\boldsymbol{z}''_q)]\Bigg\}\Bigg]d\boldsymbol{w}d\boldsymbol{z},
\end{aligned} \tag{9.34}$$

where

$$\widetilde{b}_j = b_j \text{ for } j=1,\ldots,n \text{ and } \widetilde{b}_n = b_n+1. \tag{9.35}$$

Each term in the second sum is estimated by an integral of the form

$$\|f\|_{WF,0,2+\gamma}\int_0^\infty\int_0^\infty\sqrt{\frac{x_n}{w_n}}k_s^{b_n+1}(x_n,w_n)\times |k_t^e(y,z)-k_s^e(y,z)||y-z|^{\gamma}dzdw_n. \tag{9.36}$$

Lemmas 7.1.2 and 8.2.4 show that these terms are bounded by

$$C\|f\|_{WF,0,2+\gamma}|t-s|^{\frac{\gamma}{2}}. \tag{9.37}$$

Every term in the first sum, with $l \neq 1$, is bounded by an integral of the form

$$\|f\|_{WF,0,2+\gamma}\int_0^\infty\int_0^\infty\sqrt{\frac{x_n}{w_n}}k_s^{b_n+1}(x_n,w_n)\times |k_t^{b_j}(x_j,w_j)-k_s^{b_j}(x_j,w_j)||\sqrt{x_j}-\sqrt{w_j}|^{\gamma}dw_jdw_n. \tag{9.38}$$

Lemmas 7.1.2 and 6.1.8 show that these terms are bounded by

$$C\|f\|_{WF,0,2+\gamma}|t-s|^{\frac{\gamma}{2}}. \tag{9.39}$$

This leaves just the $l = 1$ case, which is bounded by

$$\|f\|_{WF,0,2+\gamma}\int_0^\infty |k_t^{b_n+1}(x_n,w_n)-k_s^{b_n+1}(x_n,w_n)|\times$$
$$[|\sqrt{x_n}-\sqrt{w_n}|^\gamma+|\sqrt{x_n}-\sqrt{w_n}|w_n^{\frac{1-\gamma}{2}}]dw_n. \tag{9.40}$$

Lemmas 6.1.8 and 7.1.5 show that this is also bounded by $C\|f\|_{WF,0,2+\gamma}|t-s|^{\frac{\gamma}{2}}$, thereby completing the proof that

$$|\sqrt{x_n}\partial_{x_n}\partial_{y_m}(v(\boldsymbol{x},\boldsymbol{y},t)-v(\boldsymbol{x},\boldsymbol{y},s)]|\leq C\|f\|_{WF,0,2+\gamma}|t-s|^{\frac{\gamma}{2}}. \tag{9.41}$$

This brings us to the Hölder estimates for $\sqrt{x_ix_j}\partial_{x_i}\partial_{x_j}v$. The proofs here are quite similar to the analogous result in Proposition 7.1.1. The proof that

$$|\sqrt{x_ix_j}\partial_{x_i}\partial_{x_j}v(\boldsymbol{x},\boldsymbol{y},t)-\sqrt{\widetilde{x}_i\widetilde{x}_j}\partial_{x_i}\partial_{x_j}v(\widetilde{\boldsymbol{x}},\boldsymbol{y},t)|\leq C\rho_s(\boldsymbol{x},\widetilde{\boldsymbol{x}})^\gamma, \tag{9.42}$$

follows exactly as before. Using Lemma 7.1.2, working 1-variable-at-a-time, we also easily establish

$$|\sqrt{x_ix_j}\partial_{x_i}\partial_{x_j}v(\boldsymbol{x},\boldsymbol{y},t)-\sqrt{x_ix_j}\partial_{x_i}\partial_{x_j}v(\boldsymbol{x},\widetilde{\boldsymbol{y}},t)|\leq C\rho_e(\boldsymbol{y},\widetilde{\boldsymbol{y}})^\gamma. \tag{9.43}$$

The Hölder continuity in time follows as in Proposition 7.1.1, while incorporating the Euclidean variables as in the previous case, i.e., $\sqrt{x_j}\partial_{x_j}\partial_{y_l}v$. Finally we observe that, as $\partial_t v = L_{\boldsymbol{b},m}v$, and we have established that $L_{\boldsymbol{b},m}v \in \mathscr{C}^{0,\gamma}_{WF}(\mathbb{R}_+^n\times\mathbb{R}^m\times[0,\infty))$, the same is true of $\partial_t v$. This completes the proof of (9.15) in the $k = 0$ case. Assuming that f is supported in $B_R^+(\mathbf{0})\times\mathbb{R}^m$, applying Propositions 4.2.2 and 5.2.9, we can easily deduce (9.15) when $k > 0$ from the $k = 0$ case. □

9.2 THE INHOMOGENEOUS PROBLEM

We now turn to the inhomogeneous problem.

PROPOSITION 9.2.1. *For any $k\in\mathbb{N}_0$, $(b_1,\ldots,b_n)\in\mathbb{R}_+^n$, $0<R$, and $0<\gamma<1$, we let $g\in\mathscr{C}^{k,\gamma}(\mathbb{R}_+^n\times\mathbb{R}^m\times\mathbb{R}_+)$. If $0<k$, then assume that g is supported in $B_R^+(\mathbf{0})\times\mathbb{R}^m\times[0,T]$. The solution u to (9.5), with right-hand side g, given in (9.7), belongs to $\mathscr{C}^{k,2+\gamma}(\mathbb{R}_+^n\times\mathbb{R}^m\times\mathbb{R}_+)$. There are constants $C_{k,\boldsymbol{b},\gamma,R}$ so that*

$$\|u\|_{WF,k+2,\gamma,T}\leq C_{k,\boldsymbol{b},\gamma,R}(1+T)\|g\|_{WF,k,\gamma,T}. \tag{9.44}$$

The tangential first derivatives satisfy a stronger estimate; there is a constant C so that, if $T\leq 1$, then

$$\|\nabla_y u\|_{WF,0,\gamma,T}\leq CT^{\frac{\gamma}{2}}\|g\|_{WF,0,\gamma,T}. \tag{9.45}$$

The constants are uniformly bounded for $\boldsymbol{b}\leq B\mathbf{1}$, and independent of R if $k=0$.

PROOF. As before we begin by assuming that $\mathbf{0} < \boldsymbol{b}$, and $k = 0$. Using the 1-variable-at-a-time method, any estimate of the variation in an x-variable of a derivative in the x-variables alone, or the variation in a y-variable of a derivative in the y-variables alone, follows easily from the lemmas in Sections 6.1 and 8.2.

The maximum principle and (9.11) show that

$$|u(\boldsymbol{x}, \boldsymbol{y}, t)| \leq t\|g\|_{L^\infty(\mathbb{R}_+^n \times \mathbb{R}^m \times [0,t])}. \tag{9.46}$$

We use the representation in (9.11) and Proposition 9.1.1 to deduce that, for $t \leq T$,

$$|u(\boldsymbol{x}^1, \boldsymbol{y}^1, t) - u(\boldsymbol{x}^2, \boldsymbol{y}^2, t)| \leq Ct\rho((\boldsymbol{x}^1, \boldsymbol{y}^1), (\boldsymbol{x}^2, \boldsymbol{y}^2))^\gamma \, \|g\|_{WF,0,\gamma,T}. \tag{9.47}$$

As before, estimates of the second derivatives (see equations (6.105) and (6.110) and Lemma 8.2.8) show that

$$|L_{\boldsymbol{b},m} u(\boldsymbol{x}, \boldsymbol{y}, t)| \leq C\|g\|_{WF,0,\gamma} t^{\frac{\gamma}{2}}. \tag{9.48}$$

Integrating the equation, $\partial_t u = L_{\boldsymbol{b},m} u + g$, in t we can therefore show that there is a constant C so that, if $t_1 < t_2 \leq T$, then

$$|u(\boldsymbol{x}, \boldsymbol{y}, t_2) - u(\boldsymbol{x}, \boldsymbol{y}, t_1)| \leq C\|g\|_{WF,0,\gamma,T} \left[|t_2^{\frac{\gamma}{2}+1} - t_1^{\frac{\gamma}{2}+1}| + |t_2 - t_1| \right]. \tag{9.49}$$

These results show that there is a constant C so that

$$\|u\|_{WF,0,\gamma,T} \leq CT^{1-\frac{\gamma}{2}}(1 + T^{\frac{\gamma}{2}})\|g\|_{WF,0,\gamma,T}. \tag{9.50}$$

9.2.1 First Derivative Estimates

Using the estimates proved above, we can easily show that

$$|\partial_{x_j} u(\boldsymbol{x}, \boldsymbol{y}, t)| \leq C\|g\|_{WF,0,\gamma} t^{\frac{\gamma}{2}} \text{ and } |\partial_{y_l} u(\boldsymbol{x}, \boldsymbol{y}, t)| \leq C\|g\|_{WF,0,\gamma} t^{\frac{\gamma+1}{2}}. \tag{9.51}$$

The first estimate follows by the argument used to prove (7.111). We indicate how the second estimate is proved. The standard limiting argument shows that

$$\partial_{y_m} u(\boldsymbol{x}, \boldsymbol{y}, t) = \int_0^t \int_0^\infty \cdots \int_0^\infty \int_{-\infty}^\infty \cdots \int_{-\infty}^\infty \prod_{j=1}^n k_s^{b_j}(x_j, w_j) \prod_{l \neq m} k_s^e(y_l, z_l) \times$$
$$\partial_{y_m} k_s^e(y_m, z_m)[g(\boldsymbol{w}, z, t-s) - g(\boldsymbol{w}, z'_{m-1}, y_m, z''_{m-1}, t-s)] d\boldsymbol{w} dz ds. \tag{9.52}$$

Putting in absolute values we see that

$$|\partial_{y_m} u(\boldsymbol{x}, \boldsymbol{y}, t)| \leq 2\|g\|_{WF,0,\gamma} \int_0^t \int_0^\infty |\partial_{y_m} k_s^e(y_m, z_m)||y_m - z_m|^\gamma \, dz_m ds. \tag{9.53}$$

Lemma 8.2.5 shows that

$$|\partial_{y_m} u(\boldsymbol{x}, \boldsymbol{y}, t)| \leq C\|g\|_{WF,0,\gamma} \int_0^t s^{\frac{\gamma-1}{2}} ds = C\|g\|_{WF,0,\gamma} \frac{2t^{\frac{\gamma+1}{2}}}{\gamma+1}. \tag{9.54}$$

We note that by integrating these estimates for $\nabla_{\boldsymbol{x},\boldsymbol{y}} u$ we obtain a Lipschitz estimate for u itself:

$$|u(\boldsymbol{x}^2, \boldsymbol{y}^2, t) - u(\boldsymbol{x}^1, \boldsymbol{y}^1, t)| \leq C\|g\|_{WF,0,\gamma} t^{\frac{\gamma}{2}} (1 + \sqrt{t}) \|(\boldsymbol{x}^2, \boldsymbol{y}^2) - (\boldsymbol{x}^1, \boldsymbol{y}^1)\|, \tag{9.55}$$

though these estimates are not directly relevant to estimating $\|u\|_{WF,0,2+\gamma,T}$.

The arguments used to prove (7.120) and (7.124) apply, essentially verbatim, to show that

$$|\partial_{x_j} u(\boldsymbol{x}^2, \boldsymbol{y}, t) - \partial_{x_j} u(\boldsymbol{x}^1, \boldsymbol{y}, t)| \leq C\|g\|_{WF,0,\gamma} \rho_s(\boldsymbol{x}^1, \boldsymbol{x}^2)^\gamma \tag{9.56}$$

provided $\rho_s(\boldsymbol{x}^1, \boldsymbol{x}^2)$ is bounded by 1. For $\rho_e(\boldsymbol{y}^1, \boldsymbol{y}^2) < 1$ we have

$$|\partial_{y_l} u(\boldsymbol{x}, \boldsymbol{y}^2, t) - \partial_{y_l} u(\boldsymbol{x}, \boldsymbol{y}^1, t)| \leq C t^{\frac{\gamma}{2}} \|g\|_{WF,0,\gamma} \rho_e(\boldsymbol{y}^1, \boldsymbol{y}^2)^\gamma. \tag{9.57}$$

To prove this we can assume that $\boldsymbol{y}^2 - \boldsymbol{y}^1$ has exactly one non-zero entry. If $\boldsymbol{y}^1$ and $\boldsymbol{y}^2$ differ in the lth entry, then

$$|\partial_{y_l} u(\boldsymbol{x}, \boldsymbol{y}^2, t) - \partial_{y_l} u(\boldsymbol{x}, \boldsymbol{y}^1, t)| \leq \\ \|g\|_{WF,0,\gamma} \int_0^t \int_{-\infty}^{\infty} |\partial_{y_l} [k_s^e(y_l^2, z_l) - k_s^e(y_l^1, z_l)]| |y_l^1 - z_l|^\gamma dz_l ds. \tag{9.58}$$

Applying Lemma 8.2.6 we see that this integral is bounded by

$$C \int_0^t s^{\frac{\gamma-1}{2}} \frac{\left(\frac{|y_l^2 - y_l^1|}{\sqrt{s}}\right)}{1 + \left(\frac{|y_l^2 - y_l^1|}{\sqrt{s}}\right)} ds. \tag{9.59}$$

An elementary calculation shows that if $\rho_e(\boldsymbol{y}^1, \boldsymbol{y}^2) < 1$, then this integral is bounded by a constant times $t^{\frac{\gamma}{2}} \rho_e(\boldsymbol{y}^1, \boldsymbol{y}^2)^\gamma$. If $\boldsymbol{y}^1$ and $\boldsymbol{y}^2$ differ in a coordinate other than the lth, then Lemmas 8.2.2 and 8.2.5 show that the estimate for $|\partial_{y_l} u(\boldsymbol{x}, \boldsymbol{y}^2, t) - \partial_{y_l} u(\boldsymbol{x}, \boldsymbol{y}^1, t)|$ reduces again to the integral in (9.59). We are therefore left to consider the off-diagonal cases: $|\partial_{x_j} u(\boldsymbol{x}, \boldsymbol{y}^2, t) - \partial_{x_j} u(\boldsymbol{x}, \boldsymbol{y}^1, t)|$ and $|\partial_{y_l} u(\boldsymbol{x}^2, \boldsymbol{y}, t) - \partial_{y_l} u(\boldsymbol{x}^1, \boldsymbol{y}, t)|$.

We can again assume that $\boldsymbol{x}^2 - \boldsymbol{x}^1$ and $\boldsymbol{y}^2 - \boldsymbol{y}^1$ each have exactly one non-zero entry, which we can assume is the first. We first consider

$$|\partial_{x_j} u(\boldsymbol{x}, \boldsymbol{y}^2, t) - \partial_{x_j} u(\boldsymbol{x}, \boldsymbol{y}^1, t)| \leq 2\|g\|_{WF,0,\gamma} \int_0^t \int_0^\infty \int_{-\infty}^\infty |\partial_{x_j} k_s^{b_j}(x_j, w_j)| \times \\ |\sqrt{x_j} - \sqrt{w_j}|^\gamma |k_s^e(y_1^2, z_1) - k_s^e(y_1^1, z_1)| dw_j dz_1 ds. \tag{9.60}$$

Applying Lemmas 8.2.2 and 6.1.10 shows that this is bounded by

$$|\partial_{x_j}u(\boldsymbol{x},\boldsymbol{y}^2,t)-\partial_{x_j}u(\boldsymbol{x},\boldsymbol{y}^1,t)|\leq C\|g\|_{WF,0,\gamma}\int\limits_0^t s^{\frac{\gamma}{2}-1}\left(\frac{\frac{|y_1^2-y_1^1|}{\sqrt{s}}}{1+\frac{|y_1^2-y_1^1|}{\sqrt{s}}}\right)ds. \tag{9.61}$$

An elementary estimate and Lemma 6.1.1 shows that therefore

$$|\partial_{x_j}u(\boldsymbol{x},\boldsymbol{y}^2,t)-\partial_{x_j}u(\boldsymbol{x},\boldsymbol{y}^1,t)|\leq C\|g\|_{WF,0,\gamma}\rho_e(\boldsymbol{y}^1,\boldsymbol{y}^2)^{\gamma}. \tag{9.62}$$

To estimate $|\partial_{y_m}u(\boldsymbol{x}^2,\boldsymbol{y},t)-\partial_{y_m}u(\boldsymbol{x}^1,\boldsymbol{y},t)|$, first bound $|\partial_{x_q}\partial_{y_l}u(\boldsymbol{x},\boldsymbol{y},t)|$. By relabeling, it suffices to consider $q=1$, for which we use the expression

$$\partial_{x_1}\partial_{y_m}u(\boldsymbol{x},\boldsymbol{y},t)=\int\limits_0^t\int\limits_0^\infty\cdots\int\limits_0^\infty\int\limits_{-\infty}^\infty\cdots\int\limits_{-\infty}^\infty\prod_{j\neq 1}k_s^{b_j}(x_j,w_j)\prod_{l\neq m}k_s^e(y_l,z_l)\times$$
$$\partial_{x_1}k_s^{b_1}(x_1,w_1)\partial_{y_m}k_s^e(y_m,z_m)[g(\boldsymbol{w},\boldsymbol{z},t-s)-g(x_1,\boldsymbol{w}_0'',\boldsymbol{z},t-s)]d\boldsymbol{w}d\boldsymbol{z}ds. \tag{9.63}$$

Putting in absolute values and using the standard estimate for the difference

$$g(\boldsymbol{w},\boldsymbol{z},t-s)-g(x_1,\boldsymbol{w}_0'',\boldsymbol{z},t-s),$$

gives the bound:

$$|\partial_{x_1}\partial_{y_m}u(\boldsymbol{x},\boldsymbol{y},t)|\leq$$
$$2\|g\|_{WF,0,\gamma}\int\limits_0^t\int\limits_{-\infty}^\infty\int\limits_0^\infty|\partial_{x_1}k_s^{b_1}(x_1,w_1)\partial_{y_m}k_s^e(y_m,z_m)||\sqrt{x_1}-\sqrt{w_1}|^{\gamma}dw_1dz_mds. \tag{9.64}$$

Applying Lemmas 6.1.10 and 8.2.5 we see that

$$\begin{aligned}|\partial_{x_1}\partial_{y_m}u(\boldsymbol{x},\boldsymbol{y},t)|&\leq C\|g\|_{WF,0,\gamma}\int\limits_0^t\frac{s^{\frac{\gamma}{2}-1}ds}{\sqrt{x_1}+\sqrt{s}}\\&\leq C\|g\|_{WF,0,\gamma}\frac{t^{\frac{\gamma}{2}}}{\sqrt{x_1}}.\end{aligned} \tag{9.65}$$

By integrating the last expression we see that

$$|\partial_{y_m}u(x_1^2,\boldsymbol{x}_0'',\boldsymbol{y},t)-\partial_{y_m}u(x_1^1,\boldsymbol{x}_0'',\boldsymbol{y},t)|\leq C\|g\|_{WF,0,\gamma}t^{\frac{\gamma}{2}}\left|\sqrt{x_1^2}-\sqrt{x_x^1}\right|. \tag{9.66}$$

This estimate implies that, for $\boldsymbol{x}^1,\boldsymbol{x}^2$ with $\rho_s(\boldsymbol{x}^1,\boldsymbol{x}^2)\leq 1$, we have

$$|\partial_{y_m}u(x_1^2,\boldsymbol{x}_0'',\boldsymbol{y},t)-\partial_{y_m}u(x_1^1,\boldsymbol{x}_0'',\boldsymbol{y},t)|\leq C\|g\|_{WF,0,\gamma}t^{\frac{\gamma}{2}}\rho_s(\boldsymbol{x}^1,\boldsymbol{x}^2)^{\gamma}, \tag{9.67}$$

completing the proof of the spatial part of (9.45).

To complete the estimates of the first derivatives we need to bound the difference $|\nabla u(\boldsymbol{x}, \boldsymbol{y}, t_2) - \nabla u(\boldsymbol{x}, \boldsymbol{y}, t_1)|$. From the estimates in (9.51), we see that for $t_1, t_2 \leq T$, and any $0 < c < 1$, there is a C_T so that if $t_1 < ct_2$, then

$$\begin{aligned}|\nabla_x u(\boldsymbol{x}, \boldsymbol{y}, t_2) - \nabla_x u(\boldsymbol{x}, \boldsymbol{y}, t_1)| &\leq C_T \|g\|_{WF,0,\gamma} |t_2 - t_1|^{\frac{\gamma}{2}} \\ |\nabla_y u(\boldsymbol{x}, \boldsymbol{y}, t_2) - \nabla_y u(\boldsymbol{x}, \boldsymbol{y}, t_1)| &\leq C_T \|g\|_{WF,0,\gamma} |t_2 - t_1|^{\frac{\gamma+1}{2}}.\end{aligned} \tag{9.68}$$

As usual, this reduces us to consideration of the case that $ct_2 < t_1 < t_2$. For this argument we fix a $\frac{1}{2} < c < 1$, and use a slightly different argument depending upon whether we are estimating an x-derivative or a y-derivative. The x-derivatives are done very much like the estimates in Chapter 7 beginning with (7.128). For example, to estimate the x_n-derivative we use the representation

$$\begin{aligned}&\partial_{x_n} u(\boldsymbol{x}, \boldsymbol{y}, t_2) - \partial_{x_n} u(\boldsymbol{x}, \boldsymbol{y}, t_1) = \\ &\int\limits_0^{t_2-t_1}\int\limits_{-\infty}^{\infty}\cdots\int\limits_{-\infty}^{\infty}\int\limits_0^{\infty}\cdots\int\limits_0^{\infty} \prod_{j=1}^{m} k_s^e(y_j, z_j) \prod_{j=1}^{n-1} k_s^{b_j}(x_j, w_j) \partial_{x_n} k_s^{b_n}(x_n, w_n) \times \\ &\qquad [g(\boldsymbol{w}_n', w_n, \boldsymbol{z}, t_2 - s) - g(\boldsymbol{w}_n', w_n, \boldsymbol{z}, t_1 - s)] dw_n d\boldsymbol{w}_n' d\boldsymbol{z} ds + \\ &\int\limits_0^{2t_1-t_2}\int\limits_{-\infty}^{\infty}\cdots\int\limits_{-\infty}^{\infty}\int\limits_0^{\infty}\cdots\int\limits_0^{\infty} \Bigg[\prod_{j=1}^{m} k_{t_2-s}^e(y_j, z_j) \prod_{j=1}^{n-1} k_{t_2-s}^{b_j}(x_j, w_j) \partial_{x_n} k_{t_2-s}^{b_n}(x_n, w_n) - \\ &\qquad \prod_{j=1}^{m} k_{t_1-s}^e(y_j, z_j) \prod_{j=1}^{n-1} k_{t_1-s}^{b_j}(x_j, w_j) \partial_{x_n} k_{t_1-s}^{b_n}(x_n, w_n)\Bigg] \times \\ &\qquad [g(\boldsymbol{w}_n', w_n, \boldsymbol{z}, s) - g(\boldsymbol{w}_n', x_n, \boldsymbol{z}, s)] dw_n d\boldsymbol{w}_n' d\boldsymbol{z} ds + \\ &\int\limits_{2t_1-t_2}^{t_1}\int\limits_{-\infty}^{\infty}\cdots\int\limits_{-\infty}^{\infty}\int\limits_0^{\infty}\cdots\int\limits_0^{\infty} \prod_{j=1}^{m} k_{t_2-s}^e(y_j, z_j) \prod_{j=1}^{n-1} k_{t_2-s}^{b_j}(x_j, w_j) \partial_{x_n} k_{t_2-s}^{b_n}(x_n, w_n) \times \\ &\qquad [g(\boldsymbol{w}_n', w_n, \boldsymbol{z}, s) - g(\boldsymbol{w}_n', x_n, \boldsymbol{z}, s)] dw_n d\boldsymbol{w}_n' d\boldsymbol{z} ds.\end{aligned} \tag{9.69}$$

The first and last terms are estimated exactly as before. To estimate the second integral we use the analogue of the expression in (7.129), first observing that

$$\begin{aligned}&\partial_{x_n}\left\{\prod_{j=1}^{m} k_{t_2-s}^e(y_j, z_j) \prod_{l=1}^{n} k_{t_2-s}^{b_l}(x_l, w_l) - \prod_{j=1}^{m} k_{t_1-s}^e(y_j, z_j) \prod_{l=1}^{n} k_{t_1-s}^{b_l}(x_l, w_l)\right\} = \\ &\quad \partial_{x_n}\left\{\prod_{j=1}^{m} k_{t_2-s}^e(y_j, z_j) \left[\prod_{l=1}^{n} k_{t_2-s}^{b_l}(x_l, w_l) - \prod_{l=1}^{n} k_{t_1-s}^{b_l}(x_l, w_l)\right] + \right. \\ &\qquad \left. \left[\prod_{j=1}^{m} k_{t_2-s}^e(y_j, z_j) - \prod_{j=1}^{m} k_{t_1-s}^e(y_j, z_j)\right] \prod_{l=1}^{n} k_{t_1-s}^{b_l}(x_l, w_l)\right\}.\end{aligned} \tag{9.70}$$

We use the expansion in (7.34) to replace the differences of products on the right-hand side of (9.70) with terms containing a single term of the form

$$\begin{aligned}&[k^{b_l}_{t_2-s}(x_l, w_l) - k^{b_l}_{t_1-s}(x_l, w_l)] \text{ or}\\&[k^{e}_{t_2-s}(y_j, z_j) - k^{e}_{t_1-s}(y_j, z_j)].\end{aligned} \tag{9.71}$$

If we always use the estimate

$$|g(\boldsymbol{w}'_n, w_n, \boldsymbol{z}, s) - g(\boldsymbol{w}'_n, x_n, \boldsymbol{z}, s)| \leq 2\|g\|_{WF,0,\gamma}|\sqrt{w_n} - \sqrt{x_n}|^{\gamma}, \tag{9.72}$$

then we see that there are three types of terms that must be bounded:

I.

$$\begin{aligned}\int\limits_0^{2t_1-t_2}\int\limits_0^{\infty}\int\limits_0^{\infty} &|k^{b_l}_{t_2-s}(x_l, w_l) - k^{b_l}_{t_1-s}(x_l, w_l)||\partial_{x_n}k^{b_n}_{t_2-s}(x_n, w_n)|\\&|\sqrt{x_n} - \sqrt{w_n}|^{\gamma}\,dw_l dw_n ds,\end{aligned} \tag{9.73}$$

II.

$$\int\limits_0^{2t_1-t_2}\int\limits_0^{\infty} |\partial_{x_n}[k^{b_n}_{t_2-s}(x_n, w_n) - k^{b_n}_{t_1-s}(x_n, w_n)]||\sqrt{x_n} - \sqrt{w_n}|^{\gamma}\,dw_n ds, \tag{9.74}$$

III.

$$\begin{aligned}\int\limits_0^{2t_1-t_2}\int\limits_{-\infty}^{\infty}\int\limits_0^{\infty} &|k^{e}_{t_2-s}(y_j, z_j) - k^{e}_{t_1-s}(y_j, z_j)||\partial_{x_n}k^{b_n}_{t_1-s}(x_n, w_n)|\\&|\sqrt{x_n} - \sqrt{w_n}|^{\gamma}\,dw_l dz_j ds.\end{aligned} \tag{9.75}$$

Terms of types I and II were shown, in the proof of (7.133), to be bounded by a constant times $|t_2-t_1|^{\frac{\gamma}{2}}$, leaving just the term of type III. Using Lemma 6.1.10 and Lemma 8.2.4 we see that these terms are bounded by

$$\int\limits_{t_2-t_1}^{t_1} (t_2 - t_1)\sigma^{\frac{\gamma}{2}-2}d\sigma \leq C|t_2 - t_1|^{\frac{\gamma}{2}}. \tag{9.76}$$

The argument for estimating the differences

$$|\partial_{y_j}u(\boldsymbol{x}, \boldsymbol{y}, t_2) - \partial_{y_j}u(\boldsymbol{x}, \boldsymbol{y}, t_1)| \tag{9.77}$$

is essentially identical, though the results are a bit different. We are free to assume that $j = m$ and use the analogue of (9.69) with ∂_{x_n} replaced with ∂_{y_m} and the difference $g(\boldsymbol{w}'_n, w_n, \boldsymbol{z}, s) - g(\boldsymbol{w}'_n, x_n, \boldsymbol{z}, s)$ replaced by

$$g(\boldsymbol{w}, \boldsymbol{z}'_m, z_m, s) - g(\boldsymbol{w}, \boldsymbol{z}'_m, y_m, s). \tag{9.78}$$

The contributions of the first and third integrals are then bounded by

$$2\|g\|_{WF,0,\gamma}\int_0^{2(t_2-t_1)}\int_{-\infty}^{\infty}|\partial_{y_m}k^e_s(y_m,z_m)||y_m-z_m|^{\gamma}\,dz_m ds. \tag{9.79}$$

Applying Lemma 8.2.5 we see that this integral is bounded by

$$C\int_0^{2(t_2-t_1)} s^{\frac{\gamma-1}{2}}ds=\frac{2}{1+\gamma}[2|t_2-t_1|]^{\frac{\gamma+1}{2}}, \tag{9.80}$$

which suffices to prove the desired estimate.

This leaves the analogue of the second integral in (9.69), which we expand using the analogue of (9.70) and (7.34), replacing ∂_{x_n} with ∂_{y_m}. We need to estimate three types of terms:

I.

$$\int_0^{2t_1-t_2}\int_0^{\infty}\int_{-\infty}^{\infty}|k^{b_l}_{t_2-s}(x_l,w_l)-k^{b_l}_{t_1-s}(x_l,w_l)||\partial_{y_m}k^e_{t_2-s}(y_m,z_m)||y_m-z_m|^{\gamma}\,dz_m dw_l ds, \tag{9.81}$$

II.

$$\int_0^{2t_1-t_2}\int_{-\infty}^{\infty}|\partial_{y_m}[k^e_{t_2-s}(y_m,z_m)-k^e_{t_1-s}(y_m,z_m)]||y_m-z_m|^{\gamma}\,dz_m ds, \tag{9.82}$$

III.

$$\int_0^{2t_1-t_2}\int_0^{\infty}\int_0^{\infty}|k^e_{t_2-s}(y_j,z_j)-k^e_{t_1-s}(y_j,z_j)||\partial_{y_m}k^e_{t_1-s}(y_m,z_m)||y_m-z_m|^{\gamma}\,dz_m dz_j ds. \tag{9.83}$$

Lemma 6.1.9 and Lemma 8.2.5 show that terms of type I are bounded by

$$\int_0^{2t_1-t_2}\frac{(t_2-t_1)(t_1-s)^{\frac{\gamma-1}{2}}ds}{(t_2-s)}\le\int_0^{2t_1-t_2}(t_2-t_1)(t_1-s)^{\frac{\gamma-3}{2}}ds\le C|t_2-t_1|^{\frac{\gamma+1}{2}}. \tag{9.84}$$

Using Lemma 8.2.7 we see that the terms of type II are bounded by

$$C|t_2-t_1|^{\frac{\gamma+1}{2}}. \tag{9.85}$$

The second estimate in Lemma 8.2.4 and Lemma 8.2.5 show that terms of type III are also bounded by

$$C\int_{t_2-t_1}^{t_1}\frac{s^{\frac{\gamma-1}{2}}(t_2-t_1)ds}{t_2-t_1+s}\leq C|t_2-t_1|^{\frac{\gamma+1}{2}}. \tag{9.86}$$

Thus we see that there is a constant C so that we have:

$$|\nabla_{\boldsymbol{y}}u(\boldsymbol{x},\boldsymbol{y},t_2)-\nabla_{\boldsymbol{y}}u(\boldsymbol{x},\boldsymbol{y},t_1)|\leq C\|g\|_{WF,0,\gamma}|t_2-t_1|^{\frac{\gamma+1}{2}}. \tag{9.87}$$

If $t_1 < t_2$, then (9.86) also gives the estimate

$$|\nabla_{\boldsymbol{y}}u(\boldsymbol{x},\boldsymbol{y},t_2)-\nabla_{\boldsymbol{y}}u(\boldsymbol{x},\boldsymbol{y},t_1)|\leq Ct_2^{\frac{1}{2}}\|g\|_{WF,0,\gamma}|t_2-t_1|^{\frac{\gamma}{2}}, \tag{9.88}$$

completing the proof of (9.45) as well as the proof that, if $t_1, t_2 < T$, then there is a constant C so that the first derivatives satisfy

$$\begin{aligned}|\nabla_{\boldsymbol{x},\boldsymbol{y}}u(\boldsymbol{x}^2,\boldsymbol{y}^2,t_2)-\nabla_{\boldsymbol{x},\boldsymbol{y}}u(\boldsymbol{x}^1,\boldsymbol{y}^1,t_1)|\leq \\ C(1+T^{\frac{\gamma}{2}})\|g\|_{WF,0,\gamma}[\rho_s(\boldsymbol{x}^1,\boldsymbol{x}^2)+\rho_e(\boldsymbol{y}^1,\boldsymbol{y}^2)+\sqrt{|t_2-t_1|}]^{\gamma}.\end{aligned} \tag{9.89}$$

9.2.2 Second Derivative Estimates

This brings us to the second derivatives. As it is essentially the same as the 1-dimensional case, Lemma 6.1.14 suffices to prove the bounds

$$|x_l\partial_{x_l}^2u(\boldsymbol{x},\boldsymbol{y},t)|\leq C\|g\|_{WF,0,\gamma}\min\{x_l^{\frac{\gamma}{2}},t^{\frac{\gamma}{2}}\}, \tag{9.90}$$

for $l = 1,\dots,n$. The calculations between (7.158) and (7.160) suffice to prove that for $1\leq l,k\leq n$, we have the estimates

$$|\sqrt{x_lx_k}\partial_{x_l}\partial_{x_k}u(\boldsymbol{x},\boldsymbol{y},t)|\leq C\|g\|_{WF,0,\gamma}\min\{x_l^{\frac{\gamma}{2}},x_k^{\frac{\gamma}{2}},t^{\frac{\gamma}{2}}\}. \tag{9.91}$$

Using Lemmas 8.2.5 and 8.2.8, we easily derive the estimates

$$|\partial_{y_j}\partial_{y_k}u(\boldsymbol{x},\boldsymbol{y},t)|\leq C\|g\|_{WF,0,\gamma}t^{\frac{\gamma}{2}}, \tag{9.92}$$

where $1\leq j,k\leq m$. Using Lemmas 6.1.10 and 8.2.5 we can also show that

$$|\sqrt{x_l}\partial_{x_l}\partial_{y_j}u(\boldsymbol{x},\boldsymbol{y},t)|\leq C\|g\|_{WF,0,\gamma}\min\{x_l^{\frac{\gamma}{2}},t^{\frac{\gamma}{2}}\}, \tag{9.93}$$

for $1\leq l\leq n$ and $1\leq j\leq m$.

To complete the spatial part of the estimate, we need to show that the second derivatives are Hölder continuous. As before, the earlier arguments suffice to show that

$$\begin{aligned}\left|\sqrt{x_l^2x_k^2}\partial_{x_l}\partial_{x_k}u(\boldsymbol{x}^2,\boldsymbol{y},t)-\sqrt{x_l^1x_k^1}\partial_{x_l}\partial_{x_k}u(\boldsymbol{x}^1,\boldsymbol{y},t)\right|\leq \\ C\|g\|_{WF,0,\gamma}\rho_s(\boldsymbol{x}^1,\boldsymbol{x}^2)^{\gamma}\text{ and} \\ |\partial_{y_j}\partial_{y_k}u(\boldsymbol{x},\boldsymbol{y}^2,t)-\partial_{y_j}\partial_{y_k}u(\boldsymbol{x},\boldsymbol{y}^1,t)|\leq C\|g\|_{WF,0,\gamma}\rho_e(\boldsymbol{y}^1,\boldsymbol{y}^2)^{\gamma}.\end{aligned} \tag{9.94}$$

Thus we are left to estimate

$$|\sqrt{x_l x_k}\partial_{x_l}\partial_{x_k}u(\boldsymbol{x},\boldsymbol{y}^2,t)-\sqrt{x_l x_k}\partial_{x_l}\partial_{x_k}u(\boldsymbol{x},\boldsymbol{y}^1,t)| \text{ and } \\ |\partial_{y_j}\partial_{y_k}u(\boldsymbol{x}^2,\boldsymbol{y},t)-\partial_{y_j}\partial_{y_k}u(\boldsymbol{x}^1,\boldsymbol{y},t)|, \tag{9.95}$$

and the mixed derivatives $\sqrt{x_l}\partial_{x_l}\partial_{y_j}u$.

We begin with the quantities in (9.95), by considering

$$|\sqrt{x_l x_k}\partial_{x_l}\partial_{x_k}u(\boldsymbol{x},\boldsymbol{y}^2,t)-\sqrt{x_l x_k}\partial_{x_l}\partial_{x_k}u(\boldsymbol{x},\boldsymbol{y}^1,t)|,$$

with $k\neq l$. Without loss of generality we can assume $k=1, l=2$ and therefore $\boldsymbol{y}^2-\boldsymbol{y}^1=(y_1^2-y_1^1,0,\dots,0)$. With these assumptions, using the observation that

$$\int_0^\infty \partial_{x_1}k_s^{b_1}(x_1,w_1)g(x_1,\boldsymbol{w}_0'',\boldsymbol{z},t-s)dw_1=0, \tag{9.96}$$

for all values of $(\boldsymbol{w}_0'',\boldsymbol{z},t-s)$, we get the estimate

$$|\sqrt{x_1x_2}\partial_{x_1}\partial_{x_2}u(\boldsymbol{x},\boldsymbol{y}^2,t)-\sqrt{x_l x_k}\partial_{x_l}\partial_{x_k}u(\boldsymbol{x},\boldsymbol{y}^1,t)|\leq \\ 2\|g\|_{WF,0,\gamma}\int_0^t\int_0^\infty\int_0^\infty\int_{-\infty}^\infty |\sqrt{x_1}\partial_{x_1}k_s^{b_1}(x_1,w_1)\sqrt{x_2}\partial_{x_2}k_s^{b_2}(x_2,w_2)||\sqrt{x_1}-\sqrt{w_1}|^\gamma\times \\ |k_s^e(y_1^2,z_1)-k_s^e(y_1^1,z_1)|dw_1dw_2dz_1ds. \tag{9.97}$$

Applying Lemmas 6.1.10 and 8.2.2, we see that the integral is estimated by

$$C\int_0^t\frac{\sqrt{x_1}s^{\frac{\gamma}{2}-1}}{1+\sqrt{x_1/s}}\frac{\sqrt{x_2}s^{-1}}{1+\sqrt{x_2/s}}\left(\frac{\frac{|y_1^2-y_1^1|}{\sqrt{s}}}{1+\frac{|y_1^2-y_1^1|}{\sqrt{s}}}\right)ds\leq C\int_0^t s^{\frac{\gamma}{2}-1}\left(\frac{\frac{|y_1^2-y_1^1|}{\sqrt{s}}}{1+\frac{|y_1^2-y_1^1|}{\sqrt{s}}}\right)ds, \tag{9.98}$$

and that

$$\int_0^t s^{\frac{\gamma}{2}-1}\left(\frac{\frac{|y_1^2-y_1^1|}{\sqrt{s}}}{1+\frac{|y_1^2-y_1^1|}{\sqrt{s}}}\right)ds\leq\int_0^{|y_1^2-y_1^1|^2}s^{\frac{\gamma}{2}-1}ds+\int_{|y_1^2-y_1^1|^2}^t|y_1^2-y_1^1|s^{\frac{\gamma-3}{2}}ds \\ \leq\frac{2}{\gamma(1-\gamma)}|y_1^2-y_1^1|^\gamma, \tag{9.99}$$

which completes this case.

We next consider this situation with $k=l$; we can take $k=l=1$, with $\boldsymbol{y}^2-\boldsymbol{y}^1$ as before. Using the fact that

$$\int_0^\infty \partial_{x_1}^2k_s^{b_1}(x_1,w_1)g(x_1,\boldsymbol{w}_0'',\boldsymbol{z},t-s)dw_1=0, \tag{9.100}$$

for all values of $(\boldsymbol{w}_0'', z, t-s)$, we get the estimate

$$|x_1\partial_{x_1}^2 u(\boldsymbol{x}, \boldsymbol{y}^2, t) - x_1\partial_{x_1}^2 u(\boldsymbol{x}, \boldsymbol{y}^1, t)| \leq$$
$$2\|g\|_{WF,0,\gamma}\int_0^t\int_0^\infty\int_{-\infty}^\infty |x_1\partial_{x_1}^2 k_s^{b_1}(x_1, w_1)||\sqrt{x_1}-\sqrt{w_1}|^\gamma \times$$
$$|k_s^e(y_1^2, z_1) - k_s^e(y_1^1, z_1)|dw_1dz_1ds. \quad (9.101)$$

We now apply Lemma 6.1.15 and Lemma 8.2.2 to see that this integral is bounded by

$$C\int_0^t \frac{\sqrt{x_1}s^{\frac{\gamma}{2}-1}}{\sqrt{x_1}+\sqrt{s}}\left(\frac{\frac{|y_1^2-y_1^1|}{\sqrt{s}}}{1+\frac{|y_1^2-y_1^1|}{\sqrt{s}}}\right)ds$$
$$\leq C\left[\int_0^{|y_1^2-y_1^1|^2} s^{\frac{\gamma}{2}-1}ds + \int_{|y_1^2-y_1^1|^2}^t |y_1^2-y_1^1|s^{\frac{\gamma-3}{2}}ds\right]$$
$$\leq \frac{2C}{\gamma(1-\gamma)}|y_1^2-y_1^1|^\gamma. \quad (9.102)$$

We have implicitly assumed that $t > |y_1^2 - y_1^1|^2$; if this is not the case, then one gets a single term in (9.99) and (9.102). Otherwise the argument is identical. This completes the spatial part of the Hölder estimate for the second x-derivatives.

We now turn to $|\partial_{y_j}\partial_{y_k}u(\boldsymbol{x}^2, \boldsymbol{y}, t) - \partial_{y_j}\partial_{y_k}u(\boldsymbol{x}^1, \boldsymbol{y}, t)|$ by considering $j \neq k$. We can assume that $j = 1$, $k = 2$, and $\boldsymbol{x}^2 - \boldsymbol{x}^1 = (x_1^2 - x_1^1, 0, \ldots, 0)$. We first need to consider the case where $x_1^1 = 0$. We use the fact that

$$\int_{-\infty}^\infty \partial_{y_1}k_s^e(y_1, z_1)[g(\boldsymbol{w}, z_1, z_0'', t-s) - g(\boldsymbol{w}, y_1, z_0'', t-s)]dz_1 = 0 \quad (9.103)$$

to see that

$$|\partial_{y_j}\partial_{y_k}u(\boldsymbol{x}^2, \boldsymbol{y}, t) - \partial_{y_j}\partial_{y_k}u(\boldsymbol{x}^1, \boldsymbol{y}, t)| \leq$$
$$2\|g\|_{WF,0,\gamma}\int_0^t\int_{-\infty}^\infty\int_{-\infty}^\infty\int_0^\infty |\partial_{y_1}k_s^e(y_1, z_1)\partial_{y_2}k_s^e(y_2, z_2)||y_1-z_1|^\gamma \times$$
$$|k_s^{b_1}(x_1^2, w_1) - k_s^{b_1}(0, w_1)|dw_1dz_1dz_2ds. \quad (9.104)$$

Applying Lemma 8.2.5 and the first estimate in Lemma 6.1.5 we see that the integral is estimated by

$$\int_0^t \frac{s^{\frac{\gamma}{2}-1}x_1^2 ds}{x_1^2+s} \leq \frac{4}{\gamma(2-\gamma)}(x_1^2)^{\frac{\gamma}{2}}. \quad (9.105)$$

Applying (6.72), this estimate implies that for any $0 < c < 1$, there is a constant C so that if $x_1^1 < cx_1^2$, then

$$|\partial_{y_j}\partial_{y_k}u(\boldsymbol{x}^2, \boldsymbol{y}, t) - \partial_{y_j}\partial_{y_k}u(\boldsymbol{x}^1, \boldsymbol{y}, t)| \leq C\|g\|_{WF,0,\gamma}\left|\sqrt{x_1^2} - \sqrt{x_1^1}\right|^{\gamma}. \tag{9.106}$$

We are therefore reduced to the case $cx_1^2 < x_1^1 < x_1^2$. In this case we have

$$|\partial_{y_j}\partial_{y_k}u(\boldsymbol{x}^2, \boldsymbol{y}, t) - \partial_{y_j}\partial_{y_k}u(\boldsymbol{x}^1, \boldsymbol{y}, t)| \leq$$
$$2\|g\|_{WF,0,\gamma}\int\limits_0^t\int\limits_{-\infty}^{\infty}\int\limits_{-\infty}^{\infty}\int\limits_0^{\infty}|\partial_{y_1}k_s^e(y_1, z_1)\partial_{y_2}k_s^e(y_2, z_2)||y_1 - z_1|^{\gamma}\times$$
$$|k_s^{b_1}(x_1^2, w_1) - k_s^{b_1}(x_1^1, w_1)|dw_1dz_1dz_2ds, \tag{9.107}$$

which can be estimated using Lemma 8.2.5 and the second estimate in Lemma 6.1.5. These lemmas show that the integral is bounded by

$$C\int\limits_0^t s^{\frac{\gamma}{2}-1}\left(\frac{\frac{\sqrt{x_1^2}-\sqrt{x_1^1}}{\sqrt{s}}}{1+\frac{\sqrt{x_1^2}-\sqrt{x_1^1}}{\sqrt{s}}}\right)ds \leq$$
$$C\left[\int\limits_0^{(\sqrt{x_1^2}-\sqrt{x_1^1})^2} s^{\frac{\gamma}{2}-1}ds + \int\limits_{(\sqrt{x_1^2}-\sqrt{x_1^1})^2}^{t} s^{\frac{\gamma-3}{2}}\left|\sqrt{x_1^2} - \sqrt{x_1^1}\right|ds\right]$$
$$\leq \frac{2}{\gamma(1-\gamma)}\left|\sqrt{x_1^2} - \sqrt{x_1^1}\right|^{\gamma}. \tag{9.108}$$

The case $j = k = 1$ follows exactly the same pattern. We use the fact that

$$\int\limits_{-\infty}^{\infty}\partial_{y_1}^2 k_s^e(y_1, z_1)g(\boldsymbol{w}, y_1, z_0'', t - s)dz_1 = 0 \tag{9.109}$$

for any values of $(\boldsymbol{w}, z_0'', t - s)$, and the estimate (8.28) to see that

$$\int\limits_{-\infty}^{\infty}|\partial_{y_1}^2 k_s^e(y_1, z_1)||y_1 - z_1|^{\gamma} \leq Cs^{\frac{\gamma}{2}-1}. \tag{9.110}$$

From this point the argument used for the case $j \neq k$ can be followed verbatim. We have again implicitly assumed that $t > \left(\sqrt{x_1^2} - \sqrt{x_1^1}\right)^2$. If this is not the case, then we get only the first term in the second line of (9.108); otherwise the argument is unchanged.

To complete the spatial estimates we need only show that the mixed partial derivatives $\sqrt{x_l}\partial_{x_l}\partial_{y_j}u(\boldsymbol{x}, \boldsymbol{y}, t)$ are Hölder continuous. Without loss of generality we can take $j = l = 1$. As usual we can assume that the points of evaluation differ in a single coordinate. We start by considering variations in the x-variables. There are two cases to consider: (1) The x-variable differs in the first slot, (2) The x-variable differs in another slot.

For case 1, we first need to take $x_1^1 = 0$. In this case we see that the second estimate in (9.93) and (6.72) imply that, if $0 < c < 1$, then there is a C so that, for $x_1^1 < cx_1^2$, we have the estimate

$$\left|\sqrt{x_1^2}\partial_{x_1}\partial_{y_1}u(\boldsymbol{x}^2, \boldsymbol{y}, t) - \sqrt{x_1^1}\partial_{x_1}\partial_{y_1}u(\boldsymbol{x}^1, \boldsymbol{y}, t)\right| \leq C\|g\|_{WF,0,\gamma}\left|\sqrt{x_1^2} - \sqrt{x_1^1}\right|^{\gamma}. \tag{9.111}$$

We are therefore left to consider $cx_1^2 < x_1^1 < x_1^2$. For this case we see that

$$\left|\sqrt{x_1^2}\partial_{x_1}\partial_{y_1}u(\boldsymbol{x}^2, \boldsymbol{y}, t) - \sqrt{x_1^1}\partial_{x_1}\partial_{y_1}u(\boldsymbol{x}^1, \boldsymbol{y}, t)\right| \leq$$

$$C\|g\|_{WF,0,\gamma}\int_0^t\int_0^{\infty}\int_{-\infty}^{\infty}\left|\sqrt{x_1^2}\partial_{x_1}k_s^{b_1}(x_1^2, w_1) - \sqrt{x_1^1}\partial_{x_1}k_s^{b_1}(x_1^1, w_1)\right| \times$$
$$\left|\sqrt{x_1^2} - \sqrt{w_1}\right|^{\gamma} |\partial_{y_1}k_s^e(y_1, z_1)|dw_1dz_1ds. \tag{9.112}$$

Applying Lemmas 8.2.5 and 6.1.11 we see that this integral is bounded by

$$C\int_0^t s^{\frac{\gamma}{2}-1}\frac{\left(\frac{\left|\sqrt{x_1^2}-\sqrt{x_1^1}\right|}{\sqrt{s}}\right)}{1+\left(\frac{\left|\sqrt{x_1^2}-\sqrt{x_1^1}\right|}{\sqrt{s}}\right)}ds. \tag{9.113}$$

As before we easily establish that, when $\left|\sqrt{x_1^2} - \sqrt{x_1^1}\right| < 1$, this is bounded by a constant times $\left|\sqrt{x_1^2} - \sqrt{x_1^1}\right|^{\gamma}$.

We now turn to the case that $\boldsymbol{x}^2 - \boldsymbol{x}^1$ is non-zero in the jth entry where $j > 1$. As in the previous case, we need to first consider $x_j^1 = 0$. In this case the difference is estimated by

$$\left|\sqrt{x_1^1}\partial_{x_1}\partial_{y_1}u(\boldsymbol{x}^2,\boldsymbol{y},t)-\sqrt{x_1^1}\partial_{x_1}\partial_{y_1}u(\boldsymbol{x}^1,\boldsymbol{y},t)\right|\leq$$

$$2\|g\|_{WF,0,\gamma}\int_0^t\int_0^\infty\int_0^\infty\int_{-\infty}^\infty\left|\sqrt{x_1^1}\partial_{x_1}k_s^{b_1}(x_1^1,w_1)\right|\left|\sqrt{x_1^1}-\sqrt{w_1}\right|^\gamma\times$$

$$|k_s^{b_j}(x_j^2,w_j)-k_s^{b_j}(0,w_j)||\partial_{y_1}k_s^e(y_1,z_1)|dz_1dw_jdw_1dt. \quad (9.114)$$

Applying Lemmas 6.1.10, 8.2.5 and the first estimate in Lemma 6.1.5 we see that this integral is estimated by

$$\int_0^t s^{\frac{\gamma}{2}-1}\frac{x_j^2ds}{s+x_j^2}\leq C(x_j^2)^{\frac{\gamma}{2}}. \quad (9.115)$$

Applying (6.72) we are reduced to consideration of the case $cx_j^2<x_j^1<x_j^2$, for a $0<c<1$. In this case we use the second estimate in Lemma 6.1.5 to see that the replacement for (9.115) is

$$C\int_0^t s^{\frac{\gamma}{2}-1}\frac{\left(\frac{|\sqrt{x_j^2}-\sqrt{x_j^1}|}{\sqrt{s}}\right)}{1+\left(\frac{|\sqrt{x_j^2}-\sqrt{x_j^1}|}{\sqrt{s}}\right)}ds\leq C\left|\sqrt{x_j^2}-\sqrt{x_j^1}\right|^\gamma. \quad (9.116)$$

This completes the proof that

$$\left|\sqrt{x_l^2}\partial_{x_l}\partial_{y_j}u(\boldsymbol{x}^2,\boldsymbol{y},t)-\sqrt{x_l^1}\partial_{x_l}\partial_{y_j}u(\boldsymbol{x}^1,\boldsymbol{y},t)\right|\leq C\|g\|_{WF,0,\gamma}\rho_s(\boldsymbol{x}^1,\boldsymbol{x}^2)^\gamma. \quad (9.117)$$

We are left to consider $\left|\sqrt{x_1}\partial_{x_1}\partial_{y_1}u(\boldsymbol{x},\boldsymbol{y}^2,t)-\sqrt{x_1}\partial_{x_1}\partial_{y_1}u(\boldsymbol{x},\boldsymbol{y}^2,t)\right|$, where, as before, we need to distinguish between the case that the y-variables differ in the first coordinate and in other coordinates. If $\boldsymbol{y}^2-\boldsymbol{y}^1=(y_1^2-y_1^1,0,\dots,0)$, then this difference is estimated by

$$2\|g\|_{WF,0,\gamma}\int_0^t\int_0^\infty\int_{-\infty}^\infty|\sqrt{x_1}k_s^{b_1}(x_1,w_1)|\times$$

$$|\partial_{y_1}k_s^e(y_1^2,z_1)-\partial_{y_1}k_s^e(y_1^1,z_1)||y_1^1-z_1|^\gamma dz_1dw_1ds. \quad (9.118)$$

Lemmas 6.1.10 and 8.2.6 show that this integral is estimated by

$$\int_0^t s^{\frac{\gamma}{2}-1}\frac{\left(\frac{|y_1^2-y_1^1|}{\sqrt{s}}\right)}{1+\left(\frac{|y_1^2-y_1^1|}{\sqrt{s}}\right)}ds\leq C\left|y_1^2-y_1^1\right|^\gamma. \quad (9.119)$$

If the y-variables differ in another coordinate, then the difference of second derivatives is estimated by

$$2\|g\|_{WF,0,\gamma}\int_0^t\int_0^\infty\int_{-\infty}^\infty\int_{-\infty}^\infty|\sqrt{x_1}k_s^{b_1}(x_1,w_1)|\times$$
$$|\partial_{y_1}k_s^e(y_1^1,z_1)||y_1^1-z_1|^\gamma\,|k_s^e(y_j^2,z_j)-k_s^e(y_j^1,z_j)|dz_jdz_1dw_1ds. \quad (9.120)$$

We now apply Lemmas 6.1.10, 8.2.5, and 8.2.2 to see that the integral is bounded by

$$\int_0^t s^{\frac{\gamma}{2}-1}\frac{\left(\frac{|y_j^2-y_j^1|}{\sqrt{s}}\right)}{1+\left(\frac{|y_j^2-y_j^1|}{\sqrt{s}}\right)}ds\le C\left|y_j^2-y_j^1\right|^\gamma. \quad (9.121)$$

This completes the proof that

$$\left|\sqrt{x_l^1}\partial_{x_l}\partial_{y_j}u(\boldsymbol{x},\boldsymbol{y}^2,t)-\sqrt{x_l^1}\partial_{x_l}\partial_{y_j}u(\boldsymbol{x},\boldsymbol{y}^1,t)\right|\le C\|g\|_{WF,0,\gamma}\,\rho_e(\boldsymbol{y}^1,\boldsymbol{y}^2)^\gamma. \quad (9.122)$$

To finish the proof of the proposition we need to establish the Hölder continuity, in the time-variable, of the second spatial-derivatives of u. Using (6.72) along with the estimates in (9.91), (9.92), and (9.93), we see that for any $0<c<1$, there is a constant C so that, if $t_1<ct_2$, then

$$\begin{aligned}
&|\sqrt{x_lx_k}\partial_{x_l}\partial_{x_k}u(\boldsymbol{x},\boldsymbol{y},t_2)-\sqrt{x_lx_k}\partial_{x_l}\partial_{x_k}u(\boldsymbol{x},\boldsymbol{y},t_1)|\le C\|g\|_{WF,0,\gamma}|t_2-t_1|^{\frac{\gamma}{2}}\\
&|\sqrt{x_l}\partial_{x_l}\partial_{y_j}u(\boldsymbol{x},\boldsymbol{y},t_2)-\sqrt{x_l}\partial_{x_l}\partial_{y_j}u(\boldsymbol{x},\boldsymbol{y},t_1)|\le C\|g\|_{WF,0,\gamma}|t_2-t_1|^{\frac{\gamma}{2}} \qquad (9.123)\\
&|\partial_{y_l}\partial_{y_j}u(\boldsymbol{x},\boldsymbol{y},t_2)-\partial_{y_l}\partial_{y_j}u(\boldsymbol{x},\boldsymbol{y},t_1)|\le C\|g\|_{WF,0,\gamma}|t_2-t_1|^{\frac{\gamma}{2}}.
\end{aligned}$$

We are left to consider these differences for $ct_2<t_1<t_2$, where we assume that $\frac{1}{2}<c<1$. For all these cases we use an expansion like that in (9.69), with the operator ∂_{x_n} replaced by the appropriate second order operator.

We first treat the pure x-derivatives, $\sqrt{x_lx_k}\partial_{x_l}\partial_{x_k}u$, where $l\neq k$. Without loss of generality we can assume that $k=n$. By replacing the difference

$$g(\boldsymbol{w}_n',w_n,\boldsymbol{z},t_2-s)-g(\boldsymbol{w}_n',w_n,\boldsymbol{z},t_1-s)$$

with

$$g(\boldsymbol{w}_n',w_n,\boldsymbol{z},t_2-s)-g(\boldsymbol{w}_n',x_n,\boldsymbol{z},t_2-s)+g(\boldsymbol{w}_n',x_n,\boldsymbol{z},t_1-s)-g(\boldsymbol{w}_n',w_n,\boldsymbol{z},t_1-s) \quad (9.124)$$

in the first integral in the analogue of (7.128), we see that it is estimated by

$$C\|g\|_{WF,0,\gamma}\int_0^{t_2-t_1}\int_0^\infty\int_0^\infty\sqrt{x_nx_l}|\partial_{x_n}k_s^{b_n}(x_n,w_n)\partial_{x_l}k_s^{b_l}(x_l,w_l)|\times$$
$$|\sqrt{w_n}-\sqrt{x_n}|^\gamma dw_ldw_nds. \quad (9.125)$$

Using Lemma 6.1.10 the integral in this term is estimated by

$$C\int_{0}^{t_2-t_1}\frac{\sqrt{x_lx_n}s^{\frac{\gamma}{2}-1}ds}{(\sqrt{x_l}+\sqrt{s})(\sqrt{x_n}+\sqrt{s})}\leq C|t_2-t_1|^{\gamma}. \tag{9.126}$$

The last integral in the analogue of (7.128) is easily seen to be bounded by

$$C\int_{t_2-t_1}^{2(t_2-t_1)}\frac{\sqrt{x_lx_n}s^{\frac{\gamma}{2}-1}ds}{(\sqrt{x_l}+\sqrt{s})(\sqrt{x_n}+\sqrt{s})}\leq C|t_2-t_1|^{\gamma}. \tag{9.127}$$

This leaves only the second integral in the analogue of (7.128), which we replace by a sum of terms using the analogue of (9.70) and (7.34). All the possible terms that arise from the analogue of the first term on the right-hand side of (9.70) are enumerated in (7.170)–(7.174), and shown to be bounded by $C\|g\|_{WF,0,\gamma}|t_2-t_1|^{\frac{\gamma}{2}}$. The second term on the right-hand side of (9.70) produces an additional type of term:

$$\int_{0}^{2t_1-t_2}\int_{-\infty}^{\infty}\int_{0}^{\infty}\int_{0}^{\infty}|k^e_{t_2-s}(y_j,z_j)-k^e_{t_1-s}(y_j,z_j)|\times$$
$$\sqrt{x_nx_l}|\partial_{x_n}k^{b_n}_{t_1-s}(x_n,w_n)\partial_{x_l}k^{b_l}_{t_1-s}(x_l,w_l)||\sqrt{w_n}-\sqrt{x_n}|^{\gamma}\,dw_ldw_ndz_jds. \tag{9.128}$$

Lemma 6.1.10 and the second estimate in Lemma 8.2.4 show that this integral is bounded by

$$\int_{t_2-t_1}^{t_1}\frac{\sqrt{x_lx_n}(t_2-t_1)s^{\frac{\gamma}{2}-1}ds}{(\sqrt{x_l}+\sqrt{s})(\sqrt{x_s}+\sqrt{s})(t_2-t_1+s)}\leq\int_{t_2-t_1}^{t_1}|t_2-t_1|s^{\frac{\gamma}{2}-2}ds\leq C|t_2-t_1|^{\frac{\gamma}{2}}. \tag{9.129}$$

Now we need to consider the case $k=l=n$. For these cases we are free to replace $x_n\partial^2_{x_n}$ with L_{b_n,x_n}. Most of the terms that arise in this case have been treated in the proof of Proposition 7.2.1. The only new type of term arises from expanding the second term on right-hand side of the analogue of (9.70) in the second integral. These are of the form

$$\int_{0}^{2t_1-t_2}\int_{-\infty}^{\infty}\int_{0}^{\infty}|k^e_{t_2-s}(y_j,z_j)-k^e_{t_1-s}(y_j,z_j)|\times$$
$$|x_n\partial^2_{x_n}k^{b_n}_{t_1-s}(x_n,w_n)||\sqrt{w_n}-\sqrt{x_n}|^{\gamma}\,dw_ndz_jds. \tag{9.130}$$

Lemma 8.2.4 and Lemma 6.1.15 show that this term is bounded by

$$C\int_{t_2-t_1}^{t_1}\frac{x_n(t_2-t_1)s^{\frac{\gamma}{2}-1}ds}{(s+x_n)(t_2-t_1+s)}\leq C|t_2-t_1|^{\frac{\gamma}{2}}. \tag{9.131}$$

This completes the proof that

$$|\sqrt{x_l x_k}\partial_{x_l}\partial_{x_k}[u(\boldsymbol{x},\boldsymbol{y},t_2)-u(\boldsymbol{x},\boldsymbol{y},t_1)]|\leq C\|g\|_{WF,0,\gamma}|t_2-t_1|^{\frac{\gamma}{2}}. \tag{9.132}$$

The verification that $\partial_{y_j}\partial_{y_k}u(\boldsymbol{x},\boldsymbol{y},t)$ satisfies the same estimate is essentially identical, simply interchanging estimates for k_s^b with estimates for k_s^e and vice versa. We leave the details to the interested reader.

To conclude the proof of Proposition 9.2.1 in the $k=0$ case, we verify that

$$|\sqrt{x_l}\partial_{x_l}\partial_{y_j}[u(\boldsymbol{x},\boldsymbol{y},t_2)-u(\boldsymbol{x},\boldsymbol{y},t_1)]|\leq C\|g\|_{WF,0,\gamma}|t_2-t_1|^{\frac{\gamma}{2}}. \tag{9.133}$$

To prove this estimate we use the expression in (7.128) with ∂_{x_n} replaced by $\sqrt{x_n}\partial_{x_n}\partial_{y_m}$. The first and third integrals are estimated by

$$\int_0^{2(t_2-t_1)}\int_{-\infty}^{\infty}\int_0^{\infty}|\partial_{y_m}k_s^e(y_m,z_m)\sqrt{x_n}\partial_{x_n}k_s^{b_n}(x_n,w_n)|\times \\ |\sqrt{x_n}-\sqrt{w_n}|^{\gamma}dz_m dw_n ds. \tag{9.134}$$

We use Lemma 6.1.10 and 8.2.5 to see that this integral is bounded by

$$C\int_0^{2(t_2-t_1)}\frac{\sqrt{x_n}s^{\frac{\gamma}{2}-1}ds}{\sqrt{x_n}+\sqrt{s}}\leq C|t_2-t_1|^{\frac{\gamma}{2}}. \tag{9.135}$$

Two cases arise in the estimation of the contribution of first term on the right-hand side of (9.70). In the first case we get terms of the form:

$$\int_0^{2t_1-t_2}\int_{-\infty}^{\infty}\int_0^{\infty}|\partial_{y_m}k_{t_2-s}^e(y_m,z_m)|\times \\ |\sqrt{x_n}\partial_{x_n}[k_{t_2-s}^{b_n}(x_n,w_n)-k_{t_1-s}^{b_n}(x_n,w_n)]||\sqrt{x_n}-\sqrt{w_n}|^{\gamma}dz_m dw_n ds. \tag{9.136}$$

Applying Lemmas 8.2.5 and 6.1.13 we see that this term is bounded by

$$\int_{t_2-t_1}^{t_1}\frac{\sqrt{x_n}(t_2-t_1)s^{\frac{\gamma}{2}-1}ds}{(\sqrt{x_n}+\sqrt{s})(t_2-t_1+s)}\leq\int_{t_2-t_1}^{t_1}(t_2-t_1)s^{\frac{\gamma}{2}-2}ds\leq C|t_2-t_1|^{\frac{\gamma}{2}}. \tag{9.137}$$

For the second case we have terms of the form

$$\int_0^{2t_1-t_2}\int_{-\infty}^{\infty}\int_0^{\infty}\int_0^{\infty}|\partial_{y_m}k_{t_2-s}^e(y_m,z_m)||\sqrt{x_n}\partial_{x_n}k_{t_1-s}^{b_n}(x_n,w_n)|\times \\ |k_{t_2-s}^{b_l}(x_l,w_l)-k_{t_1-s}^{b_l}(x_l,w_l)||\sqrt{x_n}-\sqrt{w_n}|^{\gamma}dz_m dw_l dw_n ds \tag{9.138}$$

and

$$\int_0^{2t_1-t_2}\int_{-\infty}^{\infty}\int_0^{\infty}\int_0^{\infty}|\partial_{y_m}k^e_{t_2-s}(y_m,z_m)||\sqrt{x_n}\partial_{x_n}k^{b_n}_{t_2-s}(x_n,w_n)|\times$$
$$|k^{b_l}_{t_2-s}(x_l,w_l)-k^{b_l}_{t_1-s}(x_l,w_l)||\sqrt{x_n}-\sqrt{w_n}|^{\gamma}\,dz_m dw_l dw_n ds. \quad (9.139)$$

Applying Lemmas 6.1.10, 8.2.5 and 6.1.9 shows that these terms are estimated by

$$C\int_{t_2-t_1}^{t_1}\frac{\sqrt{x_n}s^{\frac{\gamma-1}{2}}(t_2-t_1+s)^{-\frac{1}{2}}(t_2-t_1)ds}{(\sqrt{x_n}+\sqrt{s})(t_2-t_1+s)}\leq$$
$$C\int_{t_2-t_1}^{t_1}(t_2-t_1)s^{\frac{\gamma}{2}-2}ds\leq C|t_2-t_1|^{\frac{\gamma}{2}}. \quad (9.140)$$

Two cases also arise in the estimation of the contribution of second term on the right-hand side of (9.70). For the first case we get

$$\int_0^{2t_1-t_2}\int_{-\infty}^{\infty}\int_0^{\infty}|\partial_{y_m}k^e_{t_2-s}(y_m,z_m)-\partial_{y_m}k^e_{t_1-s}(y_m,z_m)|\times$$
$$|\sqrt{x_n}\partial_{x_n}k^{b_n}_{t_1-s}(x_n,w_n)||\sqrt{x_n}-\sqrt{w_n}|^{\gamma}\,dz_m dw_n ds. \quad (9.141)$$

Lemma 6.1.10 and Lemma 8.2.7 show that this integral is estimated by the expression in (9.137). For the second case we get

$$\int_0^{2t_1-t_2}\int_{-\infty}^{\infty}\int_0^{\infty}\int_0^{\infty}|\partial_{y_m}k^e_{t_2-s}(y_m,z_m)||\sqrt{x_n}\partial_{x_n}k^{b_n}_{t_1-s}(x_n,w_n)|\times$$
$$|k^e_{t_2-s}(y_k,z_k)-k^e_{t_1-s}(y_k,z_k)||\sqrt{x_n}-\sqrt{w_n}|^{\gamma}\,dz_m dz_k dw_n ds \quad (9.142)$$

and

$$\int_0^{2t_1-t_2}\int_{-\infty}^{\infty}\int_0^{\infty}\int_0^{\infty}|\partial_{y_m}k^e_{t_1-s}(y_m,z_m)||\sqrt{x_n}\partial_{x_n}k^{b_n}_{t_1-s}(x_n,w_n)|\times$$
$$|k^e_{t_2-s}(y_k,z_k)-k^e_{t_1-s}(y_k,z_k)||\sqrt{x_n}-\sqrt{w_n}|^{\gamma}\,dz_m dz_k dw_n ds. \quad (9.143)$$

Using Lemmas 6.1.10, 8.2.5 and 8.2.4 we see that these terms are estimated by

$$C\int_{t_2-t_1}^{t_1}\frac{\sqrt{x_n}s^{\frac{\gamma-1}{2}}s^{-\frac{1}{2}}(t_2-t_1)ds}{(\sqrt{x_n}+\sqrt{s})(t_2-t_1+s)}\leq C\int_{t_2-t_1}^{t_1}(t_2-t_1)s^{\frac{\gamma}{2}-2}ds\leq C|t_2-t_1|^{\frac{\gamma}{2}}. \quad (9.144)$$

This completes the proof that the second derivatives satisfy the appropriate Hölder estimates: there is a constant C uniformly bounded for $\mathbf{0} < \boldsymbol{b} < \boldsymbol{B}$ so that

$$\max_{1\le k,l\le n;\, 1\le i,j\le m} \Bigg\{ \left| \sqrt{x_l^2 x_k^2}\partial_{x_l}\partial_{x_k} u(\boldsymbol{x}^2, \boldsymbol{y}^2, t_2) - \sqrt{x_l^1 x_k^1}\partial_{x_l}\partial_{x_k} u(\boldsymbol{x}^1, \boldsymbol{y}^1, t_1) \right|, \\ \left| \sqrt{x_l^2}\partial_{x_l}\partial_{y_j} u(\boldsymbol{x}^2, \boldsymbol{y}^2, t_2) - \sqrt{x_l^1}\partial_{x_l}\partial_{y_j} u(\boldsymbol{x}^1, \boldsymbol{y}^1, t_1) \right|, \\ |\partial_{y_j}\partial_{y_i} u(\boldsymbol{x}^2, \boldsymbol{y}^2, t_2) - \partial_{y_j}\partial_{y_i} u(\boldsymbol{x}^1, \boldsymbol{y}^1, t_1)| \Bigg\} \\ \le C\|g\|_{WF,0,\gamma} \left[\rho_s(\boldsymbol{x}^1, \boldsymbol{x}^2) + \rho_e(\boldsymbol{y}^1, \boldsymbol{y}^2) + \sqrt{|t_2 - t_1|} \right]^{\gamma}. \tag{9.145}$$

To prove the Hölder continuity of $\partial_t u(\boldsymbol{x}, \boldsymbol{y}, t)$ we simply use the equation

$$\partial_t u = \sum_{l=1}^{n} L_{b_j, x_j} u + \sum_{j=1}^{m} \partial_{y_j}^2 u + g, \tag{9.146}$$

and the Hölder continuity of the expression appearing on the right-hand side of this relation. Arguing as before we can use Proposition 5.2.8 to allow components of $\boldsymbol{b}$ to tend to zero, and thereby extend these estimates to the case that $\mathbf{0} \le \boldsymbol{b}$. Using the estimates for scaled second derivatives (9.90), (9.91), and (9.93), along with Proposition 4.2.5, we argue as before to apply Lemma 5.2.7 and show that $u \in \mathscr{C}^{0,2+\gamma}_{WF}(\mathbb{R}_+^n \times \mathbb{R}^m \times [0, T])$. This completes the proof of the proposition in the $k = 0$ case.

For the $k > 0$ cases we assume that g is supported in $B_R^+(\mathbf{0}) \times \mathbb{R}^m \times [0, T]$. Using Propositions 4.2.4 and 5.2.9, as before, we easily obtain the desired conclusions for all $k \in \mathbb{N}$, and thereby complete the proof of the Proposition 9.2.1. □

9.3 OFF-DIAGONAL AND LONG-TIME BEHAVIOR

We next consider a general result describing the off-diagonal behavior of the solution kernel for (9.5). This result is important in the perturbation theory that follows in the next chapter.

PROPOSITION 9.3.1. *Let $\varphi, \psi \in \mathscr{C}_c^\infty(\mathbb{R}_+^n \times \mathbb{R}^m)$, and assume that*

$$\operatorname{dist}_{WF}(\operatorname{supp}\varphi, \operatorname{supp}\psi) = \eta > 0. \tag{9.147}$$

Let $0 < B$ and $0 \le b_j \le B$. For any $k \in \mathbb{N}_0$, the map

$$\mathcal{K}^{\boldsymbol{b}}_{\varphi,\psi,t} : g \mapsto \psi(w) K_t^{\boldsymbol{b}}[\varphi(z) g(z, \cdot)]$$

defines a bounded operator

$$\mathcal{K}^{\boldsymbol{b}}_{\varphi,\psi,t} : \dot{\mathscr{C}}^0(\mathbb{R}_+^n \times \mathbb{R}^m \times [0, T]) \to \dot{\mathscr{C}}^k(\mathbb{R}_+^n \times \mathbb{R}^m \times [0, T]). \tag{9.148}$$

There are positive constants c_η, C, where C depends on k, η and B so that the operator norm of this map is bounded, as $T \to 0^+$, by $Ce^{-\frac{c_\eta}{T}}$.

This proposition is a consequence of estimates on the 1-dimensional kernels. For the degenerate models we have

LEMMA 9.3.2. *Let $\eta > 0$ and for $x \in \mathbb{R}_+$ define the set*

$$J_{x,\eta} = \{y \in \mathbb{R}_+ : |\sqrt{x} - \sqrt{y}| \geq \eta\}. \tag{9.149}$$

For $0 < b \leq B$, $0 < \phi \leq \frac{\pi}{2}$, and $j \in \mathbb{N}_0$ there is a constant $C_{\eta,j,B,\phi}$ so that if $t = |t|e^{i\theta}$, where $|\theta| \leq \frac{\pi}{2} - \phi$, then

$$\int\limits_{J_{x,\eta}} |\partial_x^j k_t^b(x,y)|dy \leq C_{\eta,j,B,\phi} \frac{e^{-\cos\theta \frac{\eta^2}{2|t|}}}{|t|^j}. \tag{9.150}$$

For the Euclidean models we have

LEMMA 9.3.3. *Let $\eta > 0$ and for $x \in \mathbb{R}$ define the set*

$$J_{x,\eta} = \{y \in \mathbb{R} : |x - y| \geq \eta\}. \tag{9.151}$$

For $j \in \mathbb{N}_0$, $0 < \phi \leq \frac{\pi}{2}$, there is a constant $C_{\eta,j,\phi}$ so that if $t = |t|e^{i\theta}$, with the angle $|\theta| \leq \frac{\pi}{2} - \phi$, then

$$\int\limits_{J_{x,\eta}} |\partial_x^j k_t^e(x,y)|dy \leq C_{\eta,j,\phi} \frac{e^{-\cos\theta \frac{\eta^2}{8t}}}{|t|^{\frac{j}{2}}}. \tag{9.152}$$

The lemmas are proved in the Appendix.

PROOF OF PROPOSITION 9.3.1. We need to consider integrals of the form

$$I((\boldsymbol{x},\boldsymbol{y}),t) = \int\limits_{\mathbb{R}_+^n \times \mathbb{R}^m} |\partial_{\boldsymbol{y}}^{\boldsymbol{\beta}} \partial_{\boldsymbol{x}}^{\boldsymbol{\alpha}} \prod_{i=1}^n k_t^{b_i}(x_i, w_i) \prod_{l=1}^m k_t^e(y_l, z_l) \varphi(\boldsymbol{w},\boldsymbol{z}) f(\boldsymbol{w},\boldsymbol{z})| d\boldsymbol{w} d\boldsymbol{z}, \tag{9.153}$$

for $(\boldsymbol{x},\boldsymbol{y}) \in \operatorname{supp}\psi$, with $|\boldsymbol{\alpha}| = q$, $|\boldsymbol{\beta}| = p$. For such $(\boldsymbol{x},\boldsymbol{z})$ we let

$$\begin{aligned} U_{(\boldsymbol{x},\boldsymbol{y}),j} &= \{(\boldsymbol{w},\boldsymbol{z}) \in \operatorname{supp}\varphi : |\sqrt{x_j} - \sqrt{w_j}| \geq \frac{\eta}{n+m}\} \text{ for } 1 \leq j \leq n, \\ U_{(\boldsymbol{x},\boldsymbol{y}),j} &= \{(\boldsymbol{w},\boldsymbol{z}) \in \operatorname{supp}\varphi : |y_{j-n} - z_{j-n}| \geq \frac{\eta}{n+m}\} \text{ for } n+1 \leq j \leq n+m. \end{aligned} \tag{9.154}$$

Since

$$\operatorname{dist}_{WF}((\boldsymbol{x},\boldsymbol{y}),(\boldsymbol{w},\boldsymbol{z})) = \sum_{j=1}^n |\sqrt{x_j} - \sqrt{w_j}| + \sum_{l=1}^m |y_l - z_l|, \tag{9.155}$$

and $\text{dist}_{WF}(\text{supp}\,\psi, \text{supp}\,\varphi) \geq \eta$, it follows that

$$\text{supp}\,\varphi \subset \bigcup_{j=1}^{m+n} U_{(\boldsymbol{x},\boldsymbol{y}),j} \tag{9.156}$$

and that these sets are measurable. Thus we have the estimate

$$I((\boldsymbol{x},\boldsymbol{y}),t) \leq \|\varphi f\|_{L^\infty} \sum_{j=1}^{m+n} \int_{U_{(\boldsymbol{x},\boldsymbol{y}),j}} |\partial_{\boldsymbol{y}}^{\boldsymbol{\beta}} \partial_{\boldsymbol{x}}^{\boldsymbol{\alpha}} \prod_{i=1}^{n} k_t^{b_i}(x_i, w_i) \prod_{l=1}^{m} k_t^e(y_l, z_l)| d\boldsymbol{w} d\boldsymbol{z}. \tag{9.157}$$

Let $I_j((\boldsymbol{x},\boldsymbol{y}),t)$ denote the integral in this sum over $U_{(\boldsymbol{x},\boldsymbol{y}),j}$. We observe that

$$\begin{aligned} U_{(\boldsymbol{x},\boldsymbol{y}),j} &\subset \mathbb{R}_+^{j-1} \times J_{x_j,\frac{\eta}{n+m}} \times \mathbb{R}_+^{n-j} \times \mathbb{R}^m \text{ for } 1 \leq j \leq n, \\ U_{(\boldsymbol{x},\boldsymbol{y}),j} &\subset \mathbb{R}_+^{n} \times \mathbb{R}^{j-n-1} \times J_{y_{j-n},\frac{\eta}{n+m}} \times \mathbb{R}^{m+n-j} \text{ for } n+1 \leq j \leq n+m. \end{aligned} \tag{9.158}$$

Applying the 1-dimensional estimates we see that, if $1 \leq j \leq n$, then

$$\begin{aligned} &I_j((\boldsymbol{x},\boldsymbol{y}),t) \leq \\ &\|\varphi f\|_{L^\infty} \int_0^\infty \cdots \int_0^\infty \int_{\mathbb{R}^m} \int_{J_{x_j,\frac{\eta}{n+m}}} |\partial_{\boldsymbol{x}}^{\boldsymbol{\alpha}} \prod_{i=1}^{n} k_t^{b_i}(x_i, w_i) \partial_{\boldsymbol{y}}^{\boldsymbol{\beta}} \prod_{l=1}^{m} k_t^e(y_l, z_l)| dw_j \widehat{dw_j} dz, \end{aligned} \tag{9.159}$$

where, as usual, $\widehat{dw_j}$ is the volume form in $\mathbb{R}_+^n$ with dw_j omitted. Lemmas 9.3.2, 9.3.3, 6.1.21, and 8.2.15 show that

$$I_j((\boldsymbol{x},\boldsymbol{y}),t) \leq C\|\varphi f\|_{L^\infty} \frac{e^{-\cos\theta \frac{\eta^2}{2(n+m)^2 t}}}{t^{q+\frac{p}{2}}}. \tag{9.160}$$

A similar estimate applies for $n+1 \leq j \leq n+m$, which, upon summing, shows that

$$I((\boldsymbol{x},\boldsymbol{y}),t) \leq C\|\varphi f\|_{L^\infty} \frac{e^{-\cos\theta \frac{\eta^2}{8(n+m)^2 t}}}{t^{q+\frac{p}{2}}}. \tag{9.161}$$

The estimate on the right-hand side is independent of $(\boldsymbol{x},\boldsymbol{y}) \in \text{supp}\,\psi$, so we can integrate it to obtain

$$\int_0^T I((\boldsymbol{x},\boldsymbol{y}),t)dt \leq C\|\varphi f\|_{L^\infty} e^{-\cos\theta \frac{\eta^2}{16(n+m)^2 T}}. \tag{9.162}$$

Coupling this with the Leibniz formula, the proposition follows easily from these estimates. □

For each $t > 0$, we have defined the map $f \mapsto K_t^{\boldsymbol{b},m} f$, where

$$K_t^{\boldsymbol{b},m} f(\boldsymbol{x}, \boldsymbol{y}) = \int\limits_{\mathbb{R}_+^n \times \mathbb{R}^m} \prod_{i=1}^{n} k_t^{b_i}(x_i, w_i) \prod_{l=1}^{m} k_t^e(y_l, z_l) f(\boldsymbol{w}, z) dz d\boldsymbol{w}. \tag{9.163}$$

For any $t > 0$ and $f \in L^\infty(\mathbb{R}_+^n \times \mathbb{R}^m)$, it is clear that $K_t^{\boldsymbol{b},m} f \in \mathscr{C}^\infty(\mathbb{R}_+^n \times \mathbb{R}^m)$. The 1-dimensional estimates (6.52) and (8.36) imply the following result:

PROPOSITION 9.3.4. *For multi-indices $\boldsymbol{\alpha} \in \mathbb{N}_0^n$ and $\boldsymbol{\beta} \in \mathbb{N}_0^m$, there are constants $C_{\boldsymbol{\alpha},\boldsymbol{\beta}}$ so that*

$$|\partial_{\boldsymbol{x}}^{\boldsymbol{\alpha}} \partial_{\boldsymbol{y}}^{\boldsymbol{\beta}} K_t^{\boldsymbol{b},m} f(\boldsymbol{x}, \boldsymbol{y})| \leq C_{\boldsymbol{\alpha},\boldsymbol{\beta}} \frac{\|f\|_{L^\infty}}{t^{|\boldsymbol{\alpha}|+\frac{|\boldsymbol{\beta}|}{2}}}. \tag{9.164}$$

If we let $(\sqrt{\boldsymbol{x}}) = (\sqrt{x_1}, \ldots, \sqrt{x_n})$, then we also have

$$|(\sqrt{\boldsymbol{x}})^{\boldsymbol{\alpha}} \partial_{\boldsymbol{x}}^{\boldsymbol{\alpha}} \partial_{\boldsymbol{y}}^{\boldsymbol{\beta}} K_t^{\boldsymbol{b},m} f(\boldsymbol{x}, \boldsymbol{y})| \leq C_{\boldsymbol{\alpha},\boldsymbol{\beta}} \frac{\|f\|_{L^\infty}}{t^{\frac{|\boldsymbol{\alpha}|+|\boldsymbol{\beta}|}{2}}}. \tag{9.165}$$

This proposition follows easily from (6.52), (6.53) and (8.36). We leave the details to the interested reader.

9.4 THE RESOLVENT OPERATOR

We close this section by stating a proposition summarizing the properties of the resolvent operator, $R(\mu)$, as an operator on the Hölder spaces $\mathscr{C}_{WF}^{k,\gamma}(\mathbb{R}_+^n \times \mathbb{R}^m)$. As contrasted with the case of the heat equation, we do not need to assume that the data has compact support in the x-variables to prove estimates when $k > 0$. As before we use Proposition 4.2.4 to commute differential operators of the form $\partial_{\boldsymbol{x}}^{\boldsymbol{\alpha}} \partial_{\boldsymbol{y}}^{\boldsymbol{\beta}}$ past the heat kernel. Since we are only proving spatial estimates we do not need to commute ∂_t^j past the kernel, and hence do not encounter the need for weighted estimates on the data.

PROPOSITION 9.4.1. *The resolvent operator $R(\mu)$ is analytic in the complement of $(-\infty, 0]$, and is given by the integral in* (9.13) *provided that* $\operatorname{Re}(\mu e^{i\theta}) > 0$. *For $\alpha \in (0, \pi]$, there are constants $C_{\boldsymbol{b},\alpha}$ so that if*

$$\alpha - \pi \leq \arg \mu \leq \pi - \alpha, \tag{9.166}$$

then for $f \in \dot{\mathscr{C}}^0(\mathbb{R}_+^n \times \mathbb{R}^m)$ we have

$$\|R(\mu) f\|_{L^\infty} \leq \frac{C_{\boldsymbol{b},\alpha}}{|\mu|} \|f\|_{L^\infty}, \tag{9.167}$$

with $C_{\boldsymbol{b},\pi} = 1$. Moreover, for $0 < \gamma < 1$, and $k \in \mathbb{N}_0$, there are constants $C_{k,\boldsymbol{b},\alpha,\gamma}$ so that if $f \in \mathscr{C}_{WF}^{k,\gamma}(\mathbb{R}_+^n \times \mathbb{R}^m)$, then

$$\|R(\mu) f\|_{WF,k,\gamma} \leq \frac{C_{k,\boldsymbol{b},\alpha,\gamma}}{|\mu|} \|f\|_{WF,k,\gamma}. \tag{9.168}$$

We also have the estimates

$$\|\nabla_{\boldsymbol{y}} R(\mu) f\|_{WF,k,\gamma} \leq C_{k,\alpha}\left[\frac{1}{|\mu|^{\frac{\gamma}{2}}}+\frac{1}{|\mu|^{\frac{\gamma+1}{2}}}\right]\|f\|_{WF,k,\gamma}, \tag{9.169}$$

$$\begin{aligned}\|\sqrt{\boldsymbol{x}}\cdot\nabla_{\boldsymbol{x}} R(\mu) f\|_{WF,0,\gamma} &\leq C_{k,\alpha}\left[\frac{1}{|\mu|^{\frac{\gamma}{2}}}+\frac{1}{|\mu|^{\frac{\gamma+1}{2}}}\right]\|f\|_{WF,0,\gamma},\\ \|\psi(\boldsymbol{x})\boldsymbol{x}\cdot\nabla_{\boldsymbol{x}} R(\mu) f\|_{WF,k,\gamma} &\leq \sqrt{X} C_{k,\alpha}\left[\frac{1}{|\mu|^{\frac{\gamma}{2}}}+\frac{1}{|\mu|}\right]\|f\|_{WF,k,\gamma}.\end{aligned} \tag{9.170}$$

Here $\psi(\boldsymbol{x})$ *is a smooth function with* $|\nabla\psi(\boldsymbol{x})| \leq 1$ *and* $\operatorname{supp}\psi \subset [0, X]^n$.

If for a $k \in \mathbb{N}_0$, *and* $0 < \gamma < 1$, $f \in \mathscr{C}^{k,\gamma}_{WF}(\mathbb{R}^n_+ \times \mathbb{R}^m)$, *then*

$$R(\mu) f \in \mathscr{C}^{k,2+\gamma}_{WF}(\mathbb{R}^n_+ \times \mathbb{R}^m),$$

and we have

$$(\mu - L_{\boldsymbol{b},m}) R(\mu) f = f. \tag{9.171}$$

If $f \in \mathscr{C}^{0,2+\gamma}_{WF}(\mathbb{R}^n_+ \times \mathbb{R}^m)$, *then*

$$R(\mu)(\mu - L_{\boldsymbol{b},m}) f = f. \tag{9.172}$$

There are constants $C_{\boldsymbol{b},k,\alpha}$ *so that, for* μ *satisfying* (7.185), *we have*

$$\|R(\mu) f\|_{WF,k,2+\gamma} \leq C_{\boldsymbol{b},k,\alpha}\left[1+\frac{1}{|\mu|}\right]\|f\|_{WF,k,\gamma}. \tag{9.173}$$

For any $B > 0$, *these constants are uniformly bounded for* $0 \leq \boldsymbol{b} < B\mathbf{1}$.

PROOF. The proof of this proposition, with $k = 0$, is almost immediate from the proof of Proposition 9.2.1. If $\kappa^{\boldsymbol{b},m}_t(\boldsymbol{x}, z; \boldsymbol{y}, \boldsymbol{w})$ denotes the heat kernel for $L_{\boldsymbol{b},m}$, then this proof estimated the integrals

$$\int_0^t \int_{\mathbb{R}^n_+} \int_{\mathbb{R}^m} \kappa^{\boldsymbol{b},m}_s(\boldsymbol{x}, z; \boldsymbol{y}, \boldsymbol{w}) g(z; \boldsymbol{w}, t-s)\, dz\, d\boldsymbol{w}\, ds. \tag{9.174}$$

The only estimate on g that is used in these arguments is

$$|g(\boldsymbol{x}^1, \boldsymbol{y}^1, t) - g(\boldsymbol{x}^2, \boldsymbol{y}^2, t)| \leq \|g\|_{WF,0,\gamma}\, \rho((\boldsymbol{x}^1, \boldsymbol{y}^1), (\boldsymbol{x}^2, \boldsymbol{y}^2))^{\gamma}. \tag{9.175}$$

To prove the present theorem we consider integrals of the form

$$\int_0^\infty \int_{\mathbb{R}^n_+} \int_{\mathbb{R}^m} \kappa^{\boldsymbol{b},m}_{se^{i\theta}}(\boldsymbol{x}, z; \boldsymbol{y}, \boldsymbol{w}) f(z; \boldsymbol{w}) e^{-se^{i\phi}\mu}\, dz\, d\boldsymbol{w}\, e^{i\phi} ds, \tag{9.176}$$

where $\mu = |\mu|e^{-i\psi}$, with $|\psi| \leq \pi - \alpha$, for an $0 < \alpha \leq \pi$. We can choose $|\phi| < \frac{\pi-\alpha}{2}$ so that

$$-\frac{\pi}{2} + \frac{\alpha}{2} < \phi - \psi < \frac{\pi}{2} - \frac{\alpha}{2}, \tag{9.177}$$

leading to an absolutely convergent integral. All the arguments used in the proof of Proposition 9.2.1 apply with the modification that the time integrals now extend from 0 to ∞ and include a factor of $e^{-\cos\theta s}$, where $\theta = \phi - \alpha$. In light of this we only give a detailed outline for the proof of the current proposition, with references to formulæ in the previous argument.

As in the proofs of the previous results it suffices to establish these results for the $k = 0$ case, and arbitrary $\mathbf{0} < \boldsymbol{b}$. The case where some components of $\boldsymbol{b}$ vanish and arbitrary $k \in \mathbb{N}$ are then obtained using Proposition 5.2.8 and Lemma 4.1.1, respectively. We fix a $0 < \alpha \leq \pi$.

We begin by showing that if $f \in \mathscr{C}^{0,\gamma}_{WF}(\mathbb{R}^n_+ \times \mathbb{R}^m)$, then $R(\mu)f \in \mathscr{C}^{2}_{WF}(\mathbb{R}^n_+ \times \mathbb{R}^m)$. First we see that Lemma 6.1.3 implies that

$$\|R(\mu)f\|_{L^\infty} \leq C_{\boldsymbol{b},\alpha} \frac{\|f\|_{L^\infty}}{|\mu|}. \tag{9.178}$$

To prove the estimate on $\|R(\mu)f\|_{WF,k,\gamma}$, for $k = 0$, we observe that the argument in the proof of Proposition 9.1.1 showing that $v(\boldsymbol{x}, \boldsymbol{y}, t)$, with initial data $f \in \mathscr{C}^{0,\gamma}_{WF}(\mathbb{R}^n_+ \times \mathbb{R}^m)$, satisfies Hölder estimates and applies equally well to complex times $t \in S_\phi$. Thus we know that there is a constant $C_{\boldsymbol{b},\phi}$ so that, for $t \in S_\phi$, we have

$$|v(\boldsymbol{x}^1, \boldsymbol{y}^1, t) - v(\boldsymbol{x}^2, \boldsymbol{y}^2, t)| \leq C_{\boldsymbol{b},\phi}\|f\|_{WF,0,\gamma}[\rho((\boldsymbol{x}^1, \boldsymbol{y}^1), (\boldsymbol{x}^2, \boldsymbol{y}^2))]^\gamma. \tag{9.179}$$

Integrating the estimate that this implies for $[v(\cdot,\cdot,t)]_{WF,0,\gamma}$, shows that there is a constant $C_{\boldsymbol{b},\alpha,\gamma}$ for which

$$\|R(\mu)f\|_{WF,0,\gamma} \leq C_{\boldsymbol{b},\alpha,\gamma} \frac{\|f\|_{WF,0,\gamma}}{|\mu|}. \tag{9.180}$$

We obtain this estimate for $k > 0$ by using the formulæ in Proposition 4.2.2 to commute the derivatives $\partial_{\boldsymbol{x}}^{\boldsymbol{\alpha}} \partial_{\boldsymbol{y}}^{\boldsymbol{\beta}}$ through the heat kernel and onto the data, f. As noted above, in this context there is no need for time derivatives, hence we do not need to assume that f has compact support in the x-variables.

Next observe that we can use the single variable estimates in formulæ analogous to those in (7.105) to show that

$$|\partial_{x_j} R(\mu)f(\boldsymbol{x}, \boldsymbol{y})| \leq C_{\boldsymbol{b},\alpha} \frac{\|f\|_{WF,0,\gamma}}{|\mu|^{\frac{\gamma}{2}}} \text{ and } |\sqrt{x_j}\partial_{x_j} R(\mu)f(\boldsymbol{x}, \boldsymbol{y})| \leq C_{\boldsymbol{b},\alpha} \frac{\|f\|_{WF,0,\gamma}}{|\mu|^{\frac{\gamma+1}{2}}}, \tag{9.181}$$

and

$$|\partial_{y_l} R(\mu)f(\boldsymbol{x}, \boldsymbol{y})| \leq C_{\boldsymbol{b},\alpha} \frac{\|f\|_{WF,0,\gamma}}{|\mu|^{\frac{\gamma+1}{2}}}. \tag{9.182}$$

The simple 1-dimensional estimates also suffice to prove that

$$\begin{aligned}|x_j\partial_{x_j}^2 R(\mu)f(\boldsymbol{x},\boldsymbol{y})| &\leq C_{\boldsymbol{b},\alpha}\frac{\|f\|_{WF,0,\gamma}}{|\mu|^{\frac{\gamma}{2}}}\\ |x_j\partial_{x_j}^2 R(\mu)f(\boldsymbol{x},\boldsymbol{y})| &\leq C_{\boldsymbol{b},\alpha}\|f\|_{WF,0,\gamma}x_j^{\frac{\gamma}{2}},\end{aligned}\tag{9.183}$$

and

$$|\partial_{y_l}^2 R(\mu)f(\boldsymbol{x},\boldsymbol{y})| \leq C_{\boldsymbol{b},\alpha}\frac{\|f\|_{WF,0,\gamma}}{|\mu|^{\frac{\gamma}{2}}}.\tag{9.184}$$

Using a formula like that in (7.158) we can show that

$$\begin{aligned}|\sqrt{x_i x_j}\partial_{x_i}\partial_{x_j} R(\mu)f(\boldsymbol{x},\boldsymbol{y})| &\leq C_{\boldsymbol{b},\alpha}\frac{\|f\|_{WF,0,\gamma}}{|\mu|^{\frac{\gamma}{2}}}\\ |\partial_{y_l}\partial_{y_m} R(\mu)f(\boldsymbol{x},\boldsymbol{y})| &\leq C_{\boldsymbol{b},\alpha}\frac{\|f\|_{WF,0,\gamma}}{|\mu|^{\frac{\gamma}{2}}}.\end{aligned}\tag{9.185}$$

Finally, using an expression like that in (9.64), we can show that

$$|\sqrt{x_i}\partial_{x_i}\partial_{y_l} R(\mu)f(\boldsymbol{x},\boldsymbol{y})| \leq C_{\boldsymbol{b},\alpha}\frac{\|f\|_{WF,0,\gamma}}{|\mu|^{\frac{\gamma}{2}}}.\tag{9.186}$$

This establishes that $R(\mu)f \in \mathscr{C}^2_{WF}(\mathbb{R}_+^n \times \mathbb{R}^m)$.

We can now use a standard integration by parts argument, see (4.52)–(4.53), to show that

$$(\mu - L_{\boldsymbol{b},m})R(\mu)f = f.\tag{9.187}$$

As in the 1-d case, we demonstrate below that, if $f \in \mathscr{C}^{0,\gamma}_{WF}(\mathbb{R}_+^n \times \mathbb{R}^m)$, then

$$R(\mu)f \in \mathscr{C}^{0,2+\gamma}_{WF}(\mathbb{R}_+^n \times \mathbb{R}^m),$$

and therefore, by the open mapping theorem, to show that $R(\mu)$ is also a left inverse it suffices to show that the null-space $(\mu - L_{\boldsymbol{b},m})$ is trivial for $\mu \in S_0$. If there were a non-trivial eigenfunction f, for such a μ, then $e^{\mu t}f$ would solve the Cauchy problem, and grow exponentially with t. As this contradicts (9.14), it follows that this null-space is empty. We can therefore conclude that if $f \in \mathscr{C}^{0,2+\gamma}_{WF}(\mathbb{R}_+^n \times \mathbb{R}^m)$, and $\mu \in S_0$, then

$$R(\mu)(\mu - L_{\boldsymbol{b},m})f = f.\tag{9.188}$$

As the left-hand side is analytic in $\mu \in \mathbb{C} \setminus (-\infty, 0]$, it follows that this relation also holds in the complement of the negative real axis.

It remains to establish the Hölder continuity of the first and second derivatives of $R(\mu)f$. Equation (9.184) implies that if $\boldsymbol{y}$ and $\boldsymbol{y}'$ differ only in the lth coordinate then

$$|\partial_{y_l} R(\mu)f(\boldsymbol{x},\boldsymbol{y}) - \partial_{y_l} R(\mu)f(\boldsymbol{x},\boldsymbol{y}')| \leq C_{\boldsymbol{b},\alpha}\frac{\|f\|_{WF,0,\gamma}}{|\mu|^{\frac{\gamma}{2}}}|y_l - y_l'|.\tag{9.189}$$

As observed earlier, if $\boldsymbol{y} - \boldsymbol{y}'$ is supported in the mth place and $m \neq l$, then Lemmas 8.2.2 and 8.2.5 show that we have the bound:

$$\begin{aligned}
&|\partial_{y_l} R(\mu) f(\boldsymbol{x}, \boldsymbol{y}) - \partial_{y_l} R(\mu) f(\boldsymbol{x}, \boldsymbol{y}')| \\
&\qquad \leq C_{\boldsymbol{b},\alpha} \|f\|_{WF,0,\gamma} \int_0^\infty s^{\frac{\gamma-1}{2}} \frac{\left(\frac{|y_m - y'_m|}{\sqrt{s}}\right)}{1 + \left(\frac{|y_m - y'_m|}{\sqrt{s}}\right)} e^{-\cos\theta|\mu|s} ds \\
&\qquad \leq C_{\boldsymbol{b},\alpha} \|f\|_{WF,0,\gamma} \int_0^\infty s^{\frac{\gamma}{2}-1} |y_m - y'_m| e^{-\cos\theta|\mu|s} ds,
\end{aligned} \tag{9.190}$$

from which it follows easily that, for $\rho_e(\boldsymbol{y}, \boldsymbol{y}') < 1$, we have:

$$|\partial_{y_l} R(\mu) f(\boldsymbol{x}, \boldsymbol{y}) - \partial_{y_l} R(\mu) f(\boldsymbol{x}, \boldsymbol{y}')| \leq C_{\boldsymbol{b},\alpha} \frac{\|f\|_{WF,0,\gamma} [\rho_e(\boldsymbol{y}, \boldsymbol{y}')]^\gamma}{|\mu|^{\frac{\gamma}{2}}}. \tag{9.191}$$

To complete the estimate of the first $\boldsymbol{y}$-derivatives, we need to bound the difference $|\partial_{y_l} R(\mu) f(\boldsymbol{x}, \boldsymbol{y}) - \partial_{y_l} R(\mu) f(\boldsymbol{x}', \boldsymbol{y})|$. We can assume that $\boldsymbol{x} - \boldsymbol{x}'$ is supported in the first slot. The derivation of (9.65) implies that

$$|\partial_{x_1} \partial_{y_l} R(\mu) f(\boldsymbol{x}, \boldsymbol{y})| \leq C_{\boldsymbol{b},\alpha} \|f\|_{WF,0,\gamma} \int_0^\infty \frac{s^{\frac{\gamma}{2}-1} e^{-\cos\theta|\mu|s} ds}{\sqrt{x_1} + \sqrt{s}}. \tag{9.192}$$

Splitting the integral into the part from 0 to x_1 and the rest we see that

$$|\partial_{x_1} \partial_{y_l} R(\mu) f(\boldsymbol{x}, \boldsymbol{y})| \leq C_{\boldsymbol{b},\alpha} \frac{\|f\|_{WF,0,\gamma}}{\sqrt{x_1} |\mu|^{\frac{\gamma}{2}}}, \tag{9.193}$$

which upon integration implies that

$$|\partial_{y_l} R(\mu) f(\boldsymbol{x}, \boldsymbol{y}) - \partial_{y_l} R(\mu) f(\boldsymbol{x}', \boldsymbol{y})| \leq C_{\boldsymbol{b},\alpha} \frac{\|f\|_{WF,0,\gamma} \rho_s(\boldsymbol{x}, \boldsymbol{x}')}{|\mu|^{\frac{\gamma}{2}}}. \tag{9.194}$$

As we only require an estimate when $\rho_s(\boldsymbol{x}, \boldsymbol{x}') < 1$, this shows that

$$|\partial_{y_l} R(\mu) f(\boldsymbol{x}, \boldsymbol{y}) - \partial_{y_l} R(\mu) f(\boldsymbol{x}', \boldsymbol{y})| \leq C_{\boldsymbol{b},\alpha} \frac{\|f\|_{WF,0,\gamma} \rho_s(\boldsymbol{x}, \boldsymbol{x}')^\gamma}{|\mu|^{\frac{\gamma}{2}}}. \tag{9.195}$$

By commuting *spatial* derivatives past the kernel using Proposition 4.2.2, we obtain (9.169) for all $k \in \mathbb{N}_0$.

To obtain an estimate for $\|\sqrt{x_j} \partial_{x_j} R(\mu) f\|_{WF,0,\gamma}$, we integrate the estimate of $|\sqrt{x_j} \partial_{x_j} v(\cdot, t)|$ afforded by Lemma 6.1.10 to conclude that

$$|\sqrt{x_j} \partial_{x_j} R(\mu) f(\boldsymbol{x}, \boldsymbol{y})| \leq C_\alpha \frac{x_j^{\frac{\gamma}{2}}}{|\mu|^{\frac{\gamma}{2}}}. \tag{9.196}$$

As usual this implies that for $0 < c < 1$, there is a C so that if $x'_j < cx_j$ and $\boldsymbol{x} - \boldsymbol{x}'$ is supported in the jth place, then

$$\left|\sqrt{x_j}\partial_{x_j} R(\mu) f(\boldsymbol{x}, \boldsymbol{y}) - \sqrt{x'_j}\partial_{x_j} R(\mu) f(\boldsymbol{x}', \boldsymbol{y})\right| \leq C \frac{\left|\sqrt{x_j} - \sqrt{x'_j}\right|^{\gamma}}{|\mu|^{\frac{\gamma}{2}}}. \tag{9.197}$$

To obtain a similar estimate when $cx_j < x'_j < x_j$, we use Lemma A.142. Integrating the estimates in (9.185) and (9.186) we easily complete the proof that, for $\rho((\boldsymbol{x}, \boldsymbol{y}), (\boldsymbol{x}', \boldsymbol{y}')) < 1$, we have

$$|\sqrt{x_j}\partial_{x_j} R(\mu) f(\boldsymbol{x}, \boldsymbol{y}) - \sqrt{x'_j}\partial_{x_j} R(\mu) f(\boldsymbol{x}', \boldsymbol{y}')| \leq C \frac{|\rho((\boldsymbol{x}, \boldsymbol{y}), (\boldsymbol{x}', \boldsymbol{y}'))|^{\gamma}}{|\mu|^{\frac{\gamma}{2}}}, \tag{9.198}$$

finishing the proof of the first estimate in (9.170). The second estimate is proved by using Proposition 4.2.2 to commute derivatives past the heat kernel, and applying the first estimate in (9.170) and the Leibniz formula, (5.62), to terms of the form:

$$\psi(\boldsymbol{x})\sqrt{x_j}\sqrt{x_j}\partial_{x_j} R(\mu)\partial_{\boldsymbol{x}}^{\boldsymbol{\alpha}}\partial_{\boldsymbol{y}}^{\boldsymbol{\beta}} f.$$

This explains the appearance of the $\sqrt{X}$. All other terms are of lower order and easily estimated. This completes the proof of the estimates in (9.170).

We still need to establish the Hölder continuity of the unscaled first derivatives in the $\boldsymbol{x}$-variables. By integrating the second estimate in (9.183) we can show that if $\boldsymbol{x}$ and $\boldsymbol{x}'$ differ only in the jth coordinate, then

$$|\partial_{x_j} R(\mu) f(\boldsymbol{x}, \boldsymbol{y}) - \partial_{x_j} R(\mu) f(\boldsymbol{x}', \boldsymbol{y})| \leq C_{\boldsymbol{b},\alpha} \|f\|_{WF,0,\gamma} |\sqrt{x_j} - \sqrt{x'_j}|^{\gamma}. \tag{9.199}$$

To do the off-diagonal cases, we assume that $\boldsymbol{x} - \boldsymbol{x}'$ is supported in the mth slot, with $j \neq m$. If $x'_m = 0$, then by arguing as in (7.124) we see that

$$|\partial_{x_j} R(\mu) f(\boldsymbol{x}, \boldsymbol{y}) - \partial_{x_j} R(\mu) f(\boldsymbol{x}', \boldsymbol{y})| C_{\boldsymbol{b},\alpha} \|f\|_{WF,0,\gamma} \times \int_0^{\infty} s^{\frac{\gamma}{2}-1} \frac{x_m/s}{1 + x_m/s} e^{-\cos\theta|\mu|s} ds, \tag{9.200}$$

which is easily seen to be bounded by $C_{\boldsymbol{b},\alpha}\|f\|_{WF,0,\gamma} x_m^{\frac{\gamma}{2}}$. Applying (6.72) we see that, if $0 < c < 1$, then there is a constant $C_{\boldsymbol{b},\alpha}$ so that for $cx_m > x'_m$, we have

$$|\partial_{x_j} R(\mu) f(\boldsymbol{x}, \boldsymbol{y}) - \partial_{x_j} R(\mu) f(\boldsymbol{x}', \boldsymbol{y})| \leq C_{\boldsymbol{b},\alpha} \|f\|_{WF,0,\gamma} |\sqrt{x_m} - \sqrt{x'_m}|^{\gamma}. \tag{9.201}$$

We are therefore reduced to considering $cx_m < x'_m < x_m$, for a $c < 1$. If we use (6.33) it follows that

$$|\partial_{x_j} R(\mu) f(\boldsymbol{x}, \boldsymbol{y}) - \partial_{x_j} R(\mu) f(\boldsymbol{x}', \boldsymbol{y})| \leq C_{\boldsymbol{b},\alpha} \|f\|_{WF,0,\gamma} \times \int_0^{\infty} s^{\frac{\gamma}{2}-1} \left(\frac{\frac{\sqrt{x_m}-\sqrt{x'_m}}{\sqrt{s}}}{1 + \frac{\sqrt{x_m}-\sqrt{x'_m}}{\sqrt{s}}} \right) e^{-\cos\theta|\mu|s} ds. \tag{9.202}$$

We split this into an integral from 0 to $(\sqrt{x_m} - \sqrt{x'_m})^2$ and the rest, to obtain:

$$|\partial_{x_j} R(\mu) f(\boldsymbol{x}, \boldsymbol{y}) - \partial_{x_j} R(\mu) f(\boldsymbol{x}', \boldsymbol{y})| \leq C_{\boldsymbol{b},\alpha} \|f\|_{WF,0,\gamma} \times$$
$$\left[\int\limits_0^{(\sqrt{x_m}-\sqrt{x'_m})^2} s^{\frac{\gamma}{2}-1} + \int\limits_{(\sqrt{x_m}-\sqrt{x'_m})^2}^{\infty} s^{\frac{\gamma}{2}-1} \left(\frac{\sqrt{x_m} - \sqrt{x'_m}}{\sqrt{s}} \right) e^{-\cos\theta|\mu|s} ds \right]. \quad (9.203)$$

Performing these integrals shows that this term is also estimated by

$$|\partial_{x_j} R(\mu) f(\boldsymbol{x}, \boldsymbol{y}) - \partial_{x_j} R(\mu) f(\boldsymbol{x}', \boldsymbol{y})| \leq C_{\boldsymbol{b},\alpha} \|f\|_{WF,0,\gamma} |\sqrt{x_m} - \sqrt{x'_m}|^{\gamma}. \quad (9.204)$$

We now estimate $|\partial_{x_j} R(\mu) f(\boldsymbol{x}, \boldsymbol{y}) - \partial_{x_j} R(\mu) f(\boldsymbol{x}, \boldsymbol{y}')|$, with $\boldsymbol{y} - \boldsymbol{y}'$ supported at a single index, which we label 1. Arguing exactly as in the derivation of (9.61), we see that

$$|\partial_{x_j} R(\mu) f(\boldsymbol{x}, \boldsymbol{y}) - \partial_{x_j} R(\mu) f(\boldsymbol{x}, \boldsymbol{y}')|$$
$$\leq C_{\boldsymbol{b},\alpha} \|f\|_{WF,0,\gamma} \int\limits_0^{\infty} s^{\frac{\gamma}{2}-1} \left(\frac{\frac{|y_1^2 - y_1^1|}{\sqrt{s}}}{1 + \frac{|y_1^2 - y_1^1|}{\sqrt{s}}} \right) e^{-\cos\theta|\mu|s} ds. \quad (9.205)$$

The same argument used to prove (9.204) can be employed to show that

$$|\partial_{x_j} R(\mu) f(\boldsymbol{x}, \boldsymbol{y}) - \partial_{x_j} R(\mu) f(\boldsymbol{x}, \boldsymbol{y}')| \leq C_{\boldsymbol{b},\alpha} \|f\|_{WF,0,\gamma} \rho_e(\boldsymbol{y}, \boldsymbol{y}')^{\gamma}. \quad (9.206)$$

All that remains is to prove the Hölder continuity of the second derivatives. Using the second estimate in (9.183) and the argument used in the proof of (6.162) we can show that

$$|x_j \partial_{x_j}^2 R(\mu) f(\boldsymbol{x}'_j, x_j, \boldsymbol{x}''_j, \boldsymbol{y}) - y_j \partial_{x_j}^2 R(\mu) f(\boldsymbol{x}'_j, y_j, \boldsymbol{x}''_j, \boldsymbol{y})| \leq$$
$$C_{b,\alpha} \|f\|_{WF,0,\gamma} |\sqrt{x_j} - \sqrt{y_j}|^{\gamma}. \quad (9.207)$$

Arguing as in the derivation of (7.142), we see that

$$|x_j \partial_{x_j}^2 R(\mu) f(\boldsymbol{x}'_m, x_m, \boldsymbol{x}''_m, \boldsymbol{y}) - x_j \partial_{x_j}^2 R(\mu)(\boldsymbol{x}'_m, 0, \boldsymbol{x}''_m, \boldsymbol{y})| \leq$$
$$C_{b,\alpha} \|f\|_{WF,0,\gamma} \int\limits_0^{\infty} s^{\frac{\gamma}{2}-1} e^{-\cos\theta|\mu|s} \frac{x_m/s}{1 + x_m/s} ds. \quad (9.208)$$

As before the integral is bounded by a constant times $x_m^{\frac{\gamma}{2}}$, which suffices to prove the Hölder estimate for $x'_m < c x_m$, for a fixed $0 < c < 1$. If we fix such a c, then for $c x_m < x'_m < x_m$, the argument leading to (7.143) gives

$$|x_j \partial_{x_j}^2 R(\mu) f(\boldsymbol{x}'_m, x_m, \boldsymbol{x}''_m, \boldsymbol{y}) - x_j \partial_{x_j}^2 R(\mu) f(\boldsymbol{x}'_m, x'_m, \boldsymbol{x}''_m, \boldsymbol{y})| \leq$$
$$C_{b,\alpha} \|f\|_{WF,0,\gamma} \int\limits_0^{\infty} s^{\frac{\gamma}{2}-1} \frac{\frac{\sqrt{x_m} - \sqrt{x'_m}}{\sqrt{s}}}{1 + \frac{\sqrt{x_m} - \sqrt{x'_m}}{\sqrt{s}}} e^{-\cos\theta|\mu|s} ds. \quad (9.209)$$

The argument used to estimate the integral in (9.202) applies to show that

$$|x_j\partial^2_{x_j}R(\mu)f(\boldsymbol{x}'_m, x_m, \boldsymbol{x}''_m, \boldsymbol{y}) - x_j\partial^2_{x_j}R(\mu)f(\boldsymbol{x}'_m, x'_m, \boldsymbol{x}''_m, \boldsymbol{y})| \leq C_{b,\alpha}\|f\|_{WF,0,\gamma}|\sqrt{x_m} - \sqrt{x'_m}|^\gamma. \quad (9.210)$$

To estimate $|x_j\partial^2_{x_j}R(\mu)f(\boldsymbol{x}, \boldsymbol{y}) - x_j\partial^2_{x_j}R(\mu)f(\boldsymbol{x}, \boldsymbol{y}')|$, we argue as in the derivation of (9.102) to see that, if $\boldsymbol{y} - \boldsymbol{y}'$ is supported in the first argument, then

$$|x_j\partial^2_{x_j}R(\mu)f(\boldsymbol{x}, \boldsymbol{y}) - x_j\partial^2_{x_j}R(\mu)f(\boldsymbol{x}, \boldsymbol{y}')| \leq C_{b,\alpha}\|f\|_{WF,0,\gamma}\int_0^\infty \frac{\sqrt{x_1}s^{\frac{\gamma}{2}-1}}{\sqrt{x_1}+\sqrt{s}}\left(\frac{\frac{|y_1^2-y_1^1|}{\sqrt{s}}}{1+\frac{|y_1^2-y_1^1|}{\sqrt{s}}}\right)e^{-\cos\theta|\mu|s}ds. \quad (9.211)$$

As before, this integral is estimated by a constant times $|y_1^2 - y_1^1|^\gamma$, completing the proof that

$$|x_j\partial^2_{x_j}R(\mu)f(\boldsymbol{x}, \boldsymbol{y}) - x_j\partial^2_{x_j}R(\mu)f(\boldsymbol{x}', \boldsymbol{y}')| \leq C_{b,\alpha}\|f\|_{WF,0,\gamma}[\rho((\boldsymbol{x}, \boldsymbol{y}), (\boldsymbol{x}', \boldsymbol{y}'))]^\gamma. \quad (9.212)$$

The argument between (7.155) and (7.164) applies with small modifications (largely by replacing the upper limit in the s-integrations with infinity and the measure ds with $e^{-\cos\theta|\mu|s}ds$) to show that, with $j \neq k$, we have:

$$|\sqrt{x_jx_k}\partial_{x_j}\partial_{x_k}R(\mu)f(\boldsymbol{x}, \boldsymbol{y}) - \sqrt{x'_jx'_k}\partial_{x_j}\partial_{x_k}R(\mu)f(\boldsymbol{x}', \boldsymbol{y})| \leq C_{b,\alpha}\|f\|_{WF,0,\gamma}[\rho_s(\boldsymbol{x}, \boldsymbol{x}')]^\gamma. \quad (9.213)$$

Similarly, the derivation of the estimate in (9.97)–(9.98) shows that, if $\boldsymbol{y} - \boldsymbol{y}'$ is supported in the first slot, then

$$|\sqrt{x_jx_k}\partial_{x_j}\partial_{x_k}R(\mu)f(\boldsymbol{x}, \boldsymbol{y}) - \sqrt{x_jx_k}\partial_{x_j}\partial_{x_k}R(\mu)f(\boldsymbol{x}, \boldsymbol{y}')| \leq C_{b,\alpha}\|f\|_{WF,0,\gamma}\int_0^\infty s^{\frac{\gamma}{2}-1}\left(\frac{\frac{|y_1^2-y_1^1|}{\sqrt{s}}}{1+\frac{|y_1^2-y_1^1|}{\sqrt{s}}}\right)e^{-\cos\theta|\mu|s}ds, \quad (9.214)$$

which completes the proof that

$$|\sqrt{x_jx_k}\partial_{x_j}\partial_{x_k}R(\mu)f(\boldsymbol{x}, \boldsymbol{y}) - \sqrt{x'_jx'_k}\partial_{x_j}\partial_{x_k}R(\mu)f(\boldsymbol{x}', \boldsymbol{y}')| \leq C_{b,\alpha}\|f\|_{WF,0,\gamma}[\rho((\boldsymbol{x}, \boldsymbol{y}), (\boldsymbol{x}', \boldsymbol{y}'))]^\gamma. \quad (9.215)$$

As before we can use the analogous estimates for the Euclidean kernel to show that

$$|\partial_{y_j}\partial_{y_k}R(\mu)f(\boldsymbol{x}, \boldsymbol{y}) - \partial_{y_j}\partial_{y_k}R(\mu)f(\boldsymbol{x}, \boldsymbol{y}')| \leq C_{b,\alpha}\|f\|_{WF,0,\gamma}[\rho_e(\boldsymbol{y}, \boldsymbol{y}')]^\gamma. \quad (9.216)$$

The argument between (9.104) and (9.110) applies with the usual small changes to show that

$$|\partial_{y_j}\partial_{y_k}R(\mu)f(\boldsymbol{x},\boldsymbol{y})-\partial_{y_j}\partial_{y_k}R(\mu)f(\boldsymbol{x}',\boldsymbol{y})| \leq C_{b,\alpha}\|f\|_{WF,0,\gamma}[\rho_s(\boldsymbol{x},\boldsymbol{x}')]^{\gamma}. \quad (9.217)$$

To estimate the mixed derivatives $\sqrt{x_j}\partial_{x_j}\partial_{y_k}R(\mu)f(\boldsymbol{x},\boldsymbol{y})$, we slightly modify the argument between (9.111) and (9.122). In each case we are reduced to estimating an integral of one of the following two forms:

$$\int_0^\infty s^{\frac{\gamma}{2}-1}\frac{De^{-\cos\theta|\mu|s}ds}{\sqrt{s}+D} \text{ or } \int_0^\infty s^{\frac{\gamma}{2}-1}\frac{De^{-\cos\theta|\mu|s}ds}{s+D}. \quad (9.218)$$

The first integral is estimated by $C_\gamma D^\gamma$ and the second by $C_\gamma D^{\frac{\gamma}{2}}$. Using these estimates we complete the proof that

$$|\sqrt{x_j}\partial_{x_j}\partial_{y_k}R(\mu)f(\boldsymbol{x},\boldsymbol{y})-\sqrt{x'_j}\partial_{x_j}\partial_{y_k}R(\mu)f(\boldsymbol{x}',\boldsymbol{y}')| \leq C_{b,\alpha}\|f\|_{WF,0,\gamma}[\rho((\boldsymbol{x},\boldsymbol{y}),(\boldsymbol{x}',\boldsymbol{y}'))]^{\gamma}. \quad (9.219)$$

This completes the $k=0$ case for $\boldsymbol{b}>\boldsymbol{0}$. The constants are again uniformly bounded for $\boldsymbol{0}<\boldsymbol{b}\leq B\boldsymbol{1}$, and so we can apply Proposition 5.2.8 to extend these results to $\boldsymbol{0}\leq\boldsymbol{b}$.

Finally, to treat $k>0$, we use Proposition 4.2.2 to commute the spatial derivatives past the heat kernel and follow the argument above to establish this theorem for arbitrary $k\in\mathbb{N}$. The only terms that require additional consideration are contributions to the left-hand side of (9.170) from terms of the form:

$$\|\partial_{\boldsymbol{x}}^{\boldsymbol{\alpha}}\partial_{\boldsymbol{y}}^{\boldsymbol{\beta}}x_j\partial_{x_j}R(\mu)f\|_{WF,0,\gamma}, \text{ for } f\in\mathscr{C}^{k,\gamma}_{WF}, \quad (9.220)$$

where $\alpha_j\neq 0$. In all other cases

$$[\partial_{\boldsymbol{x}}^{\boldsymbol{\alpha}}\partial_{\boldsymbol{y}}^{\boldsymbol{\beta}},x_j\partial_{x_j}]=0$$

and the estimate follows easily using Proposition 4.2.2. If $\alpha_j\neq 0$, then

$$\partial_{\boldsymbol{x}}^{\boldsymbol{\alpha}}\partial_{\boldsymbol{y}}^{\boldsymbol{\beta}}x_j\partial_{x_j}=x_j\partial_{x_j}\partial_{\boldsymbol{x}}^{\boldsymbol{\alpha}}\partial_{\boldsymbol{y}}^{\boldsymbol{\beta}}+\alpha_j\partial_{\boldsymbol{x}}^{\boldsymbol{\alpha}}\partial_{\boldsymbol{y}}^{\boldsymbol{\beta}}, \quad (9.221)$$

which shows that

$$\|\partial_{\boldsymbol{x}}^{\boldsymbol{\alpha}}\partial_{\boldsymbol{y}}^{\boldsymbol{\beta}}x_j\partial_{x_j}R(\mu)f\|_{WF,0,\gamma}\leq\|x_j\partial_{x_j}\partial_{\boldsymbol{x}}^{\boldsymbol{\alpha}}\partial_{\boldsymbol{y}}^{\boldsymbol{\beta}}R(\mu)f\|_{WF,0,\gamma}+\|R(\mu)f\|_{WF,k,\gamma}. \quad (9.222)$$

It now follows from Proposition 4.2.2 and the $k=0$ case that this term is bounded by

$$\leq C_\alpha\left[\frac{1}{|\mu|^{\frac{\gamma}{2}}}+\frac{1}{|\mu|}\right]\|f\|_{WF,k,\gamma}. \quad (9.223)$$

This completes the proof of the proposition. □

Part III

Analysis of Generalized Kimura Diffusions

Chapter Ten

Existence of Solutions

We now return to the principal goal of this monograph, the analysis of a generalized Kimura diffusion operator, L, defined on a manifold with corners, P. The estimates proved in the previous chapters for the solutions to model problems, along with the adapted local coordinates introduced in Chapter 2.2, allow the use of the Schauder method to prove existence of solutions to the inhomogeneous problem

$$(\partial_t - L)w = g \text{ in } P \times (0, T) \text{ with } w(x, 0) = f. \tag{10.1}$$

Ultimately we will show that if

$$f \in \mathscr{C}_{WF}^{k,2+\gamma}(P) \text{ and } g \in \mathscr{C}_{WF}^{k,\gamma}(P \times [0, T]), \tag{10.2}$$

then the unique solution $w \in \mathscr{C}_{WF}^{k,2+\gamma}(P \times [0, T])$. In this chapter we prove the basic existence result:

THEOREM 10.0.2. *For $k \in \mathbb{N}_0$ and $0 < \gamma < 1$, if the data f, g satisfy* (10.2), *then equation* (10.1) *has a unique solution $w \in \mathscr{C}_{WF}^{k,2+\gamma}(P \times [0, T])$. There are constants $C_{k,\gamma}$, $M_{k,\gamma}$ so that*

$$\|w\|_{WF,k,2+\gamma,T} \leq M_{k,\gamma} \exp(C_{k,\gamma} T)[\|g\|_{WF,k,\gamma,T} + \|f\|_{WF,k,2+\gamma}]. \tag{10.3}$$

Remark. The hypothesis that $f \in \mathscr{C}_{WF}^{k,2+\gamma}(P)$ is not what one should expect: the result suggested by the non-degenerate case would be that for $f \in \mathscr{C}_{WF}^{k,\gamma}(P)$, there is a solution in $\mathscr{C}_{WF}^{k,2+\gamma}(P \times (0, \infty)) \cap \mathscr{C}_{WF}^{k,\gamma}(P \times [0, \infty))$. This is true for the model problems. For basic applications to probability theory, Theorem 10.0.2 suffices. We resolve this issue, according to expectations, in Chapter 11.

As we have done before, we write the solution $w = v + u$, where v solves the homogeneous Cauchy problem with $v(x, 0) = f(x)$ and u solves the inhomogeneous problem with $u(x, 0) = 0$. Each part is estimated separately. In the early sections of this chapter we treat the $k = 0$ case, returning to the problem of higher regularity at the end. The issues with the support of the data that arose in the analysis of higher regularity for the model problems does not arise in the present context. This is because whenever a model solution operator appears as part of a parametrix it is always multiplied on the right by a smooth compactly supported function. Hence it can be regarded as acting on data with fixed compact support.

With $k = 0$, we begin by proving the existence of u for $t \in [0, T_0]$, where $T_0 > 0$ is independent of u. A similar argument establishes the existence of v. Using these arguments together, we obtain existence up to time T, and the estimate given in the theorem in the $k = 0$ case. Before delving into the details of the argument, we first give definitions for the WF-Hölder spaces on a general compact manifold with corners, and then a brief account of the steps involved in the existence proof.

10.1 WF-HÖLDER SPACES ON A MANIFOLD WITH CORNERS

We now give precise definitions for various function spaces, $\mathscr{C}^{k,\gamma}_{WF}(P)$, $\mathscr{C}^{k,\gamma}_{WF}(P\times[0,T])$, etc. which we need to use. For the $(0,\gamma)$-case we could use an intrinsic definition, using the singular, incomplete metric, g_{WF}, determined by the principal symbol of L, to define a distance function, $d_{WF}(\boldsymbol{x}, \boldsymbol{y})$. We could then define the global $(0,\gamma)$-WF semi-norm by setting

$$[f]_{WF,0,\gamma} = \overset{1}{\sup_{\boldsymbol{x}\neq\boldsymbol{y}}} \frac{|f(p_1) - f(p_2)|}{d_{WF}(p_1, p_2)^{\gamma}}, \tag{10.4}$$

and a norm on $\mathscr{C}^{0,\gamma}_{WF}(P)$ by letting

$$\|f\|_{WF,0,\gamma} = \|f\|_{L^\infty(P)} + [f]_{WF,0,\gamma}\,. \tag{10.5}$$

For computations it is easier to build the global norms out of locally defined norms.

By Proposition 2.2.3, there are coordinate charts covering a neighborhood of bP in which the operator L assumes a simple normal form. At a point $q \in bP$ of codimension n this coordinate system $(x_1,\dots,x_n; y_1,\dots,y_m)$ is parametrized by a subset of the form

$$C^{n,m}(l) = [0, l^2)^n \times \left(-\frac{l}{2}, \frac{l}{2}\right)^m, \tag{10.6}$$

where $q \leftrightarrow (\boldsymbol{0}_n; \boldsymbol{0}_m)$. Let U denote the open set centered at q covered by this coordinate patch and $\phi : C^{n,m}(l) \to U$, the coordinate map. We call this a *normal cubic coordinate* or NCC patch centered at q. The parameter domain, $C^{n,m}(l)$, is called a "positive cube" of side length l in $\mathbb{R}^n_+ \times \mathbb{R}^m$. In these coordinates the operator L takes the form

$$L = \sum_{i=1}^{n} x_i\partial_{x_i}^2 + \sum_{1\le k,l\le m} c'_{kl}(\boldsymbol{x},\boldsymbol{y})\partial_{y_k}\partial_{y_l} + \sum_{i=1}^{n} b_i(\boldsymbol{x},\boldsymbol{y})\partial_{x_i} +$$

$$\sum_{1\le i\neq j\le n} x_ix_ja'_{ij}(\boldsymbol{x},\boldsymbol{y})\partial_{x_i}\partial_{x_j} + \sum_{i=1}^{n}\sum_{l=1}^{m} x_ib'_{il}(\boldsymbol{x},\boldsymbol{y})\partial_{x_i}\partial_{y_l} + \sum_{l=1}^{m} d_l(\boldsymbol{x},\boldsymbol{y})\partial_{y_l}. \tag{10.7}$$

The principal part of L at q is given by

$$L_q^p = \sum_{i=1}^{n} x_i\partial_{x_i}^2 + \sum_{1\le k,l\le m} c'_{kl}(\boldsymbol{0}_n,\boldsymbol{0}_m)\partial_{y_k}\partial_{y_l} + \sum_{i=1}^{n} b_i(\boldsymbol{0}_n,\boldsymbol{0}_m)\partial_{x_i}. \tag{10.8}$$

The matrix $c'_{kl}(\mathbf{0}_n, \mathbf{0}_m)$ is positive definite and the coefficients $\{b_i(\mathbf{0}_n, \mathbf{0}_m)\}$ are non-negative. The estimates in the previous chapter show that $L - L_q^p$ is, in a precise sense, a residual term.

If $\psi \in \mathscr{C}_c^\infty(U)$, and f is defined in U, then we can use the local definitions of the various WF norms to define the local WF norms:

$$\begin{aligned}\|\psi f\|^U_{WF,k,\gamma} &= \|(\psi f)\circ\phi\|_{WF,k,\gamma}\\ \|\psi f\|^U_{WF,k,2+\gamma} &= \|(\psi f)\circ\phi\|_{WF,k,2+\gamma}.\end{aligned} \tag{10.9}$$

If g is defined in $U \times [0, T]$ then we similarly define the local (in space and time) norm:

$$\begin{aligned}\|\psi g\|^U_{WF,k,\gamma,T} &= \|(\psi g)(\phi,\cdot)\|_{WF,k,\gamma,T}\\ \|\psi g\|^U_{WF,k,2+\gamma,T} &= \|(\psi g)(\phi,\cdot)\|_{WF,k,2+\gamma,T}.\end{aligned} \tag{10.10}$$

DEFINITION 10.1.1. *Let $\mathfrak{W} = \{(W_j, \phi_j) : j = 1, \dots, K\}$ be a cover of bP by NCC charts, $W_0 \subset\subset \operatorname{int} P$, covering $P \setminus \cup_{j=1}^K W_j$, and let $\{\varphi_j : j = 0, \dots, K\}$ be a partition of unity subordinate to this cover. A function $f \in \mathscr{C}^{k,\gamma}_{WF}(P)$ provided $(\varphi_j f)\circ\phi_j \in \mathscr{C}^{k,\gamma}_{WF}(W_j)$, for each j. We define a global norm on $\mathscr{C}^{k,\gamma}_{WF}(P)$ by setting*

$$\|f\|_{WF,k,\gamma} = \sum_{j=0}^{K} \|(\varphi_j f)\circ\phi_j\|^{W_j}_{WF,k,\gamma}. \tag{10.11}$$

There are analogous definitions for the spaces $\mathscr{C}^{k,2+\gamma}_{WF}(P)$, $\mathscr{C}^{k,\gamma}_{WF}(P\times[0,T])$, and $\mathscr{C}^{k,2+\gamma}_{WF}(P\times[0,T])$. The corresponding norms are denoted by

$$\|f\|_{WF,k,2+\gamma},\ \|g\|_{WF,k,\gamma,T},\ \|g\|_{WF,k,2+\gamma,T}.$$

It is straightforward to show that different NCC covers define equivalent norms and therefore, in all cases, the topological vector spaces do not depend on the choice of NCC cover. Once we have fixed such a cover, then the definitions of the norm on $\mathscr{C}^{0,\gamma}_{WF}(P)$ in (10.5) and (10.11) are also equivalent. In fact, if U is an NCC patch of codimension n, with local coordinates $(x_1, \dots, x_n; y_1, \dots, y_m)$, then there is a constant C so that, for $(\boldsymbol{x}^1, \boldsymbol{y}^1), (\boldsymbol{x}^2, \boldsymbol{y}^2) \in U$, we have

$$\begin{aligned}C d_{WF}((\boldsymbol{x}^1, \boldsymbol{y}^1), (\boldsymbol{x}^2, \boldsymbol{y}^2)) \le [\rho_s(\boldsymbol{x}^1, \boldsymbol{x}^2) + \rho_e(\boldsymbol{y}^1, \boldsymbol{y}^2)] \le\\ C^{-1} d_{WF}((\boldsymbol{x}^1, \boldsymbol{y}^1), (\boldsymbol{x}^2, \boldsymbol{y}^2)).\end{aligned} \tag{10.12}$$

In the remainder of this chapter we fix the cover $\mathfrak{W}$.

10.1.1 Properties of WF-Hölder Spaces

The details of the construction of the parametrix rely on some general results about the local function spaces $\mathscr{C}^{0,\gamma}_{WF}(\mathbb{R}^n_+ \times \mathbb{R}^m)$, $\mathscr{C}^{0,\gamma}_{WF}(\mathbb{R}^n_+ \times \mathbb{R}^m \times [0, T])$, for which it is useful to recall the local semi-norms

$$[f]_{WF,0,\gamma} = \sup_{(x^1,y^1)\neq(x^2,y^2)}^{1} \frac{|f(x^1, y^1) - f(x^2, y^2)|}{[\rho_s(x^1, x^2) + \rho_e(y^1, y^2)]^\gamma}, \tag{10.13}$$

$$[g]_{WF,0,\gamma} = \sup_{(x^1,y^1,t^1)\neq(x^2,y^2,t^2)}^{1} \frac{|g(x^1, y^1, t^1) - g(x^2, y^2, t^2)|}{[\rho_s(x^1, x^2) + \rho_e(y^1, y^2) + \sqrt{|t^2 - t^1|}]^\gamma}, \tag{10.14}$$

and the Leibniz formula:

LEMMA 10.1.2. *Suppose that $f, g \in \mathscr{C}^{0,\gamma}_{WF}(\mathbb{R}^n_+ \times \mathbb{R}^m)$ or $\mathscr{C}^{0,\gamma}_{WF}(\mathbb{R}^n_+ \times \mathbb{R}^m \times [0, T])$. The semi-norm of the product fg satisfies the estimate:*

$$[fg]_{WF,0,\gamma} \leq \|f\|_{L^\infty} [g]_{WF,0,\gamma} + \|g\|_{L^\infty} [f]_{WF,0,\gamma} \,. \tag{10.15}$$

PROOF. These estimates follow easily from the observation that, with $\boldsymbol{w}_j = (\boldsymbol{x}_j, \boldsymbol{y}_j)$, or $\boldsymbol{w}_j = (\boldsymbol{x}_j, \boldsymbol{y}_j, t_j)$, $j = 1, 2$, we have

$$\frac{|f(\boldsymbol{w}_1)g(\boldsymbol{w}_1) - f(\boldsymbol{w}_2)g(\boldsymbol{w}_2)|}{[\rho(\boldsymbol{w}_1, \boldsymbol{w}_2)]^\gamma} \leq \frac{|[f(\boldsymbol{w}_1) - f(\boldsymbol{w}_2)]g(\boldsymbol{w}_1)| + |f(\boldsymbol{w}_2)[g(\boldsymbol{w}_1) - g(\boldsymbol{w}_2)]|}{[\rho(\boldsymbol{w}_1, \boldsymbol{w}_2)]^\gamma}, \tag{10.16}$$

from which the assertions of the lemma are immediate. □

We also have a result about the behavior of WF norms under the scaling of cutoff functions.

LEMMA 10.1.3. *Suppose that $f \in \mathscr{C}^1_c(\mathbb{R}^n_+ \times \mathbb{R}^m)$ has support in the positive cube $[0, l^2]^n \times [-l, l]^m$. If $\epsilon > 0$ and we define*

$$f_\epsilon(\boldsymbol{x}, \boldsymbol{y}) = f\left(\frac{\boldsymbol{x}}{\epsilon^2}, \frac{\boldsymbol{y}}{\epsilon}\right), \tag{10.17}$$

then there is a constant C_l depending on the support of f so that

$$\|f_\epsilon\|_{WF,0,\gamma} \leq C_l[\epsilon^{-\gamma} + 1]\|f\|_{\mathscr{C}^1}. \tag{10.18}$$

PROOF. First observe that $\|f_\epsilon\|_{L^\infty} = \|f\|_{L^\infty}$; so we only need to estimate $[f_\epsilon]_{WF,0,\gamma}$. This estimate follows from the observation that

$$\epsilon\rho\left(\left(\frac{\boldsymbol{x}_1}{\epsilon^2}, \frac{\boldsymbol{y}_1}{\epsilon}\right), \left(\frac{\boldsymbol{x}_2}{\epsilon^2}, \frac{\boldsymbol{y}_2}{\epsilon}\right)\right) = \rho((\boldsymbol{x}_1, \boldsymbol{y}_1), (\boldsymbol{x}_2, \boldsymbol{y}_2)), \tag{10.19}$$

and therefore

$$\frac{|f_\epsilon(\boldsymbol{x}_2, \boldsymbol{y}_2) - f_\epsilon(\boldsymbol{x}_1, \boldsymbol{y}_1)|}{[\rho((\boldsymbol{x}_1, \boldsymbol{y}_1), (\boldsymbol{x}_2, \boldsymbol{y}_2))]^\gamma} = \epsilon^{-\gamma} \frac{\left|f\left(\frac{\boldsymbol{x}_2}{\epsilon^2}, \frac{\boldsymbol{y}_2}{\epsilon}\right) - f\left(\frac{\boldsymbol{x}_1}{\epsilon^2}, \frac{\boldsymbol{y}_1}{\epsilon}\right)\right|}{\left[\rho\left(\left(\frac{\boldsymbol{x}_1}{\epsilon^2}, \frac{\boldsymbol{y}_1}{\epsilon}\right), \left(\frac{\boldsymbol{x}_2}{\epsilon^2}, \frac{\boldsymbol{y}_2}{\epsilon}\right)\right)\right]^\gamma}. \tag{10.20}$$

Letting $\boldsymbol{w}_j = \boldsymbol{x}_j/\epsilon^2$, $z_j = \boldsymbol{y}_j/\epsilon$, for $j = 1, 2$, this becomes:

$$\begin{aligned}\frac{|f_\epsilon(\boldsymbol{x}_2, \boldsymbol{y}_2) - f_\epsilon(\boldsymbol{x}_1, \boldsymbol{y}_1)|}{[\rho((\boldsymbol{x}_1, \boldsymbol{y}_1), (\boldsymbol{x}_2, \boldsymbol{y}_2))]^\gamma} &= \epsilon^{-\gamma}\frac{|f(\boldsymbol{w}_2, z_2) - f(\boldsymbol{w}_1, z_1)|}{[\rho((\boldsymbol{w}_2, z_2), (\boldsymbol{w}_1, z_1))]^\gamma} \\ &\leq \epsilon^{-\gamma}\frac{\|\nabla f\|_{L^\infty}[|\boldsymbol{w}_2 - \boldsymbol{w}_1| + |z_2 - z_1|]}{[\rho((\boldsymbol{w}_2, z_2), (\boldsymbol{w}_1, z_1))]^\gamma}\end{aligned} \tag{10.21}$$

where we used the mean value theorem on the right-hand side of (10.21). The second line in (10.21) is estimated by

$$\epsilon^{-\gamma}\|\nabla f\|_{L^\infty}\left[|z_2 - z_1|^{1-\gamma} + \sum_{l=1}^{n}|\sqrt{w_{2l}} - \sqrt{w_{1l}}|^{1-\gamma}|\sqrt{w_{2l}} + \sqrt{w_{1l}}|\right]. \tag{10.22}$$

Taking the supremum of the quantity in the brackets in (10.22) for pairs $(\boldsymbol{w}_1, z_1)$, $(\boldsymbol{w}_2, z_2)$ lying in $[0, 4l^2]^n \times [-2l, 2l]^m$, shows that there is a constant C_l, so that for such pairs:

$$\frac{|f_\epsilon(\boldsymbol{x}_2, \boldsymbol{y}_2) - f_\epsilon(\boldsymbol{x}_1, \boldsymbol{y}_1)|}{[\rho((\boldsymbol{x}_1, \boldsymbol{y}_1), (\boldsymbol{x}_2, \boldsymbol{y}_2))]^\gamma} \leq C_l\epsilon^{-\gamma}\|\nabla f\|_{L^\infty}. \tag{10.23}$$

This covers the case where both $(\boldsymbol{x}_1, \boldsymbol{y}_1)$ and $(\boldsymbol{x}_2, \boldsymbol{y}_2)$ lie in certain neighborhood of the supp f_ϵ. If neither point lies in supp f, then the numerator is zero. Hence the only case remaining is when $(\boldsymbol{w}_1, z_1) \in [0, l^2]^n \times [-l, l]^m$, and $(\boldsymbol{w}_2, z_2) \notin [0, 4l^2]^n \times [-2l, 2l]^m$. In this case the denominator in the first line of (10.21) is bounded below by l^γ, and the numerator is bounded above by $2\|f\|_{L^\infty}$, which completes the proof of the lemma. □

LEMMA 10.1.4. *Suppose that $f \in \mathscr{C}^1(\mathbb{R}_+^n \times \mathbb{R}^m)$ and $a \in \mathscr{C}^1(\mathbb{R}_+^n \times \mathbb{R}^m)$ with support in a positive cube of side length l, and $a(0, 0) = 0$. There is a constant C, depending on l and the dimension, so that, if $m = 0$, then we have*

$$[af_\epsilon]_{WF,0,\gamma} \leq C\|f\|_{\mathscr{C}^1}\|a\|_{\mathscr{C}^1}\epsilon^{2-\gamma}. \tag{10.24}$$

If $m \geq 1$, then

$$[af_\epsilon]_{WF,0,\gamma} \leq C\|f\|_{\mathscr{C}^1}\|a\|_{\mathscr{C}^1}\epsilon^{1-\gamma}. \tag{10.25}$$

If a is a $\mathscr{C}^1$-function of the variables $(\sqrt{\boldsymbol{x}}, \boldsymbol{y}) = (\sqrt{x_1}, \dots, \sqrt{x_n}, y_1, \dots, y_m)$, that is

$$a(\boldsymbol{x}, \boldsymbol{y}) = A(\sqrt{\boldsymbol{x}}, \boldsymbol{y}), \text{ where } A \in \mathscr{C}^1(\mathbb{R}_+^n \times \mathbb{R}^m), \tag{10.26}$$

then the estimate in (10.25) holds for af_ϵ, with $\|a\|_{\mathscr{C}^1}$ replaced by $\|A\|_{\mathscr{C}^1}$.

PROOF. We begin with the case $m = 0$. The triangle inequality shows that

$$\begin{aligned}&\frac{|a(\boldsymbol{x}^1)f_\epsilon(\boldsymbol{x}^1) - a(\boldsymbol{x}^2)f_\epsilon(\boldsymbol{x}^2)|}{[\rho(\boldsymbol{x}^1, \boldsymbol{x}^2)]^\gamma} \leq \\ &\qquad \frac{|a(\boldsymbol{x}^1) - a(\boldsymbol{x}^2)||f_\epsilon(\boldsymbol{x}^1)|}{[\rho(\boldsymbol{x}^1, \boldsymbol{x}^2)]^\gamma} + \frac{|a(\boldsymbol{x}^2)||f_\epsilon(\boldsymbol{x}^1) - f_\epsilon(\boldsymbol{x}^2)|}{[\rho(\boldsymbol{x}^1, \boldsymbol{x}^2)]^\gamma}.\end{aligned} \tag{10.27}$$

We first assume that $\boldsymbol{x}^1, \boldsymbol{x}^2 \in \operatorname{supp} f_\epsilon$. In this case the second term on the right-hand side of (10.27) is bounded by

$$l\epsilon^2 \|\nabla a\|_{L^\infty} [f_\epsilon]_{WF,0,\gamma} \leq C_l \epsilon^{2-\gamma} \|a\|_{\mathscr{C}^1} \|f\|_{\mathscr{C}^1}, \tag{10.28}$$

where we use Lemma 10.1.2 and (10.23) to bound $[f_\epsilon]_{WF,0,\gamma}$, along with the fact that $a(\mathbf{0}) = 0$.

The first term is bounded by

$$\frac{\|f\|_{L^\infty} \|a\|_{\mathscr{C}^1} |\boldsymbol{x}^1 - \boldsymbol{x}^2|}{[\rho(\boldsymbol{x}^1, \boldsymbol{x}^2)]^\gamma} \leq C\epsilon^{2-\gamma} \|f\|_{L^\infty} \|a\|_{\mathscr{C}^1}. \tag{10.29}$$

This proves (10.24) when both $\boldsymbol{x}^1, \boldsymbol{x}^2 \in \operatorname{supp} f_\epsilon$. Essentially the same argument applies if $\boldsymbol{x}^2 \in \operatorname{supp} f_\epsilon$, and $\boldsymbol{x}^1/\epsilon^2 \in [0, 4l^2]^n \setminus \operatorname{supp} f_\epsilon$, though only the second term on the right-hand side of (10.27) is non-zero. The final case we need to consider is $\boldsymbol{x}^2 \in \operatorname{supp} f_\epsilon$, and $\boldsymbol{x}^1/\epsilon^2 \notin [0, 4l^2]^n$. For this case, the denominator in

$$\frac{|a(\boldsymbol{x}^2)||f_\epsilon(\boldsymbol{x}^1) - f_\epsilon(\boldsymbol{x}^2)|}{[\rho(\boldsymbol{x}^1, \boldsymbol{x}^2)]^\gamma} \tag{10.30}$$

is bounded below by $(l\epsilon)^{-\gamma}$, and the numerator is bounded above by

$$\epsilon^2 \|a\|_{\mathscr{C}^1} \|f\|_{L^\infty}, \tag{10.31}$$

thus completing the proof of (10.24) in case $m = 0$.

For the case $m \neq 0$, observe that

$$\frac{|a(\boldsymbol{x}^1, \boldsymbol{y}^1) f_\epsilon(\boldsymbol{x}^1, \boldsymbol{y}^1) - a(\boldsymbol{x}^2, \boldsymbol{y}^2) f_\epsilon(\boldsymbol{x}^2, \boldsymbol{y}^2)|}{[\rho((\boldsymbol{x}^1, \boldsymbol{y}^1), (\boldsymbol{x}^2, \boldsymbol{y}^2))]^\gamma} \leq \\ \frac{|a(\boldsymbol{x}^1, \boldsymbol{y}^1) - a(\boldsymbol{x}^2, \boldsymbol{y}^2)||f_\epsilon(\boldsymbol{x}^1, \boldsymbol{y}^1)| + |f_\epsilon(\boldsymbol{x}^1, \boldsymbol{y}^1) - f_\epsilon(\boldsymbol{x}^2, \boldsymbol{y}^2)||a(\boldsymbol{x}^2, \boldsymbol{y}^2)|}{[\rho((\boldsymbol{x}^1, \boldsymbol{y}^1), (\boldsymbol{x}^2, \boldsymbol{y}^2))]^\gamma}. \tag{10.32}$$

Note that $\operatorname{supp} f_\epsilon \subset [0, \epsilon^2 l^2]^n \times [-\epsilon l, \epsilon l]^m$. If both points again belong to the $\operatorname{supp} f_\epsilon$, then the quantity on the right-hand side of (10.32) is bounded by

$$(l\epsilon)^{1-\gamma} \|a\|_{\mathscr{C}^1} \|f\|_{L^\infty} + \epsilon\, [f_\epsilon]_{WF,0,\gamma} \leq C_l \epsilon^{1-\gamma} \|a\|_{\mathscr{C}^1} \|f\|_{\mathscr{C}^1}. \tag{10.33}$$

If now $(\boldsymbol{x}^2, \boldsymbol{y}^2) \in \operatorname{supp} f_\epsilon$, but $(\boldsymbol{x}^1, \boldsymbol{y}^1) \in [0, 4\epsilon^2 l^2]^n \times [-2\epsilon l, 2\epsilon l]^m \setminus \operatorname{supp} f_\epsilon$, then only the second term on the right-hand side of (10.32) is non-zero; it is estimated by

$$(l\epsilon)\|a\|_{\mathscr{C}^1} [f_\epsilon]_{WF,0,\gamma} \leq C_l \epsilon^{1-\gamma} \|a\|_{\mathscr{C}^1} \|f\|_{\mathscr{C}^1}. \tag{10.34}$$

Finally, if $(\boldsymbol{x}^2, \boldsymbol{y}^2) \in \operatorname{supp} f_\epsilon$, but $(\boldsymbol{x}^1, \boldsymbol{y}^1) \notin [0, 4\epsilon^2 l^2]^n \times [-2\epsilon l, 2\epsilon l]^m$, then the denominator is bounded below by $(l\epsilon)^\gamma$, and the numerator is bounded above by the product $\epsilon \|a\|_{\mathscr{C}^1} \|f\|_{L^\infty}$, which completes the proof in this case.

The argument if a is a $\mathscr{C}^1$-function of $(\sqrt{x}, y)$ is again quite similar.

$$\frac{|A(\sqrt{x^1}, y^1)f_\epsilon(x^1, y^1) - A(\sqrt{x^2}, y^2)f_\epsilon(x^2, y^2)|}{|[\rho((x^1, y^1), (x^2, y^2))]|^\gamma} \leq$$
$$\frac{|A(\sqrt{x^1}, y^1) - A(\sqrt{x^2}, y^2)||f_\epsilon(x^1, y^1)| + |A(\sqrt{x^2}, y^2)||f_\epsilon(x^1, y^1) - f_\epsilon(x^2, y^2)|}{|[\rho((x^1, y^1), (x^2, y^2))]|^\gamma}. \tag{10.35}$$

We observe that

$$\begin{aligned}|A(\sqrt{x^1}, y^1) - A(\sqrt{x^2}, y^2)| &\leq \|\nabla A\|_{L^\infty}\left[\sum_{j=1}^{n}|\sqrt{x_j^1} - \sqrt{x_j^2}| + |y^1 - y^2|\right] \\ &= \|\nabla A\|_{L^\infty}\rho((x^1, y^1), (x^2, y^2)).\end{aligned} \tag{10.36}$$

If both points are in supp f_ϵ, then the right-hand side of (10.35) is bounded by

$$\|\nabla A\|_{L^\infty}\left\{[\rho((x^1, y^1), (x^2, y^2))]^{1-\gamma}\|f\|_{L^\infty} + \epsilon\,[f_\epsilon]_{WF,0,\gamma}\right\}. \tag{10.37}$$

Since both points are in supp f_ϵ, Lemma 10.1.2 shows that this is bounded by

$$C\epsilon^{1-\gamma}\|A\|_{\mathscr{C}^1}\|f\|_{\mathscr{C}^2}. \tag{10.38}$$

The other cases follow similarly. □

10.2 OVERVIEW OF THE PROOF

The domain P is assumed to be a manifold with corners of dimension $N > 1$. The boundary of P is a stratified space with

$$bP = \bigcup_{j=1}^{M}\Sigma_j, \tag{10.39}$$

where Σ_j is the (open) stratum of codimension j boundary points. From the definition of manifold with corners it follows that

$$\overline{\Sigma}_k = \bigcup_{j=k}^{M}\Sigma_j. \tag{10.40}$$

To prove the existence of a solution to the equation (10.1), we use an induction over M the maximal codimension of a stratum of bP.

The argument begins by assuming that P is a manifold with boundary, i.e., $M = 1$. Using the estimates proved in the previous chapter we can easily show that there is a function $\varphi \in \mathscr{C}_c^\infty(P)$, equal to 1 in a neighborhood of bP and an operator

$$\widehat{Q}_b^t : \mathscr{C}_{WF}^{k,\gamma}(P \times [0, T]) \to \mathscr{C}_{WF}^{k,2+\gamma}(P \times [0, T]), \tag{10.41}$$

so that

$$(\partial_t - L)\widehat{Q}^t_b g = \varphi g + (E^{0,t}_b + E^{1,t}_b)g, \tag{10.42}$$

where

$$E^{0,t}_b, E^{1,t}_b : \mathscr{C}^{k,\gamma}_{WF}(P \times [0,T]) \to \mathscr{C}^{k,\gamma}_{WF}(P \times [0,T]) \tag{10.43}$$

are bounded and $E^{1,t}_b$ is a compact operator on this space, which tends to zero in norm as T tends to zero. If $k = 0$, then we can arrange for $E^{0,t}_b$ to have norm as small as we please.

Let U be a neighborhood of bP so that $bU \cap \operatorname{int} P$ is a smooth hypersurface in P, and $\overline{U} \subset\subset \varphi^{-1}(1)$. The subset $P_U = P \cap U^c$ is a smooth compact manifold with boundary, and $L \restriction_{P_U}$ is a non-degenerate elliptic operator. We can double P_U across its boundary to obtain $\widetilde{P}_U$, which is a manifold without boundary. The operator L can be extended to a classically elliptic operator $\widetilde{L}$ defined on all of $\widetilde{P}_U$. The classical theory of non-degenerate parabolic equations on compact manifolds, without boundary, applies to construct an *exact* solution operator $u_i = \widetilde{Q}^t[(1-\varphi)g]$ to the inhomogeneous equation:

$$\begin{aligned}&(\partial_t - \widetilde{L})u_i = (1-\varphi)\widetilde{g} \text{ in } \widetilde{P}_U \times [0,T]\\ &\text{with } u_i(p,0) = 0,\ p \in \widetilde{P}_U.\end{aligned} \tag{10.44}$$

This operator defines bounded maps from $\mathscr{C}^{k,\gamma}_{WF}(\widetilde{P}_U \times [0,T]) \to \mathscr{C}^{k,2+\gamma}_{WF}(\widetilde{P}_U \times [0,T])$, for any $0 < \gamma < 1$ and $k \in \mathbb{N}_0$. Of course, in $\widetilde{P}_U$ these spaces are equivalent to the classical heat Hölder spaces $\mathscr{C}^{k,\gamma}(\widetilde{P}_U \times [0,T])$ and $\mathscr{C}^{k+2,\gamma}(\widetilde{P}_U \times [0,T])$, respectively.

To complete the construction when $M = 1$, choose $\psi \in \mathscr{C}^\infty_c(P_U)$ so that $\psi \equiv 1$ on a neighborhood of the support of $(1-\varphi)$, and set

$$\widehat{Q}^t g = \widehat{Q}^t_b g + \widehat{Q}^t_i g \tag{10.45}$$

where interior parametrix is defined to be

$$\widehat{Q}^t_i = \psi \widetilde{Q}^t[(1-\varphi)g]. \tag{10.46}$$

Here it is understood that $(1-\varphi)g$ and $\psi\widetilde{Q}^t[(1-\varphi)g]$ are extended by zero to all of $\widetilde{P}_U$ and P, respectively. Applying the operator gives

$$(\partial_t - L)\widehat{Q}^t g = g + (E^{0,t}_b + E^{1,t}_b)g + E^{\infty,t}_i g, \tag{10.47}$$

where

$$E^{\infty,t}_i g = [\psi, L]\widehat{Q}^t_i[(1-\varphi)g]. \tag{10.48}$$

Since $\psi \equiv 1$ on a neighborhood of the support of $(1-\varphi)$ it follows from classical results that $E^{\infty,t}_i$ is a smoothing operator which tends to zero exponentially as $T \to 0^+$. More generally, assume by induction that $E^{\infty,t}_i$ is a compact operator tending to zero, as $T_0 \to 0$, in the operator norm defined by $\mathscr{C}^{k,\gamma}_{WF}(P \times [0,T_0])$. If T_0 is sufficiently small, then the operator

$$E^{0,t}_b + E^{1,t}_b + E^{\infty,t}_i = E^t : \mathscr{C}^{0,\gamma}_{WF}(P \times [0,T_0]) \longrightarrow \mathscr{C}^{0,\gamma}_{WF}(P \times [0,T_0]) \tag{10.49}$$

has norm strictly less than 1, and therefore $(\mathrm{Id}+E^t)$ is invertible. Thus the operator

$$\mathcal{Q}^t = \widehat{Q}^t(\mathrm{Id}+E^t)^{-1} \tag{10.50}$$

is a right inverse to $(\partial_t - L)$ up to time T_0 and is a bounded map

$$\mathcal{Q}^t : \mathcal{C}^{0,\gamma}_{WF}(P \times [0, T_0]) \to \mathcal{C}^{0,2+\gamma}_{WF}(P \times [0, T_0]). \tag{10.51}$$

At the end of this chapter we use a result from [14] to show that the Neumann series for $(\mathrm{Id}+E^t)^{-1}$ converges in the operator norm topology of $\mathcal{C}^{k,\gamma}_{WF}(P \times [0, T_0])$, for any $k \in \mathbb{N}$.

To handle the case of higher codimension boundaries we use the following induction hypotheses:

[Inhomogeneous Case:] Let P be any manifold with corners such that the maximal codimension of bP is less than or equal to M, and let L be a generalized Kimura diffusion on P. We assume that the solution operator $\mathcal{Q}^t$ of the initial value problem

$$(\partial_t - L)u = g \text{ with } u(x, 0) = 0 \tag{10.52}$$

exists and has the following properties:

1. For $k \in \mathbb{N}_0$, $0 < T$, and $0 < \gamma < 1$, the maps

$$\mathcal{Q}^t : \mathcal{C}^{k,\gamma}_{WF}(P \times [0, T]) \longrightarrow \mathcal{C}^{k,2+\gamma}_{WF}(P \times [0, T]) \tag{10.53}$$

are bounded. The maps

$$\mathcal{Q}^t : \mathcal{C}^{k,\gamma}_{WF}(P \times [0, T]) \longrightarrow \mathcal{C}^{k,\gamma}_{WF}(P \times [0, T]) \tag{10.54}$$

tend to zero in norm as $T \to 0^+$.

2. Let $\psi_1, \psi_2 \in \mathcal{C}^\infty(P)$ be such that $\mathrm{dist}(\mathrm{supp}\,\psi_1, \mathrm{supp}\,\psi_2) > 0$. Then the operator

$$\psi_1\mathcal{Q}^t\psi_2 : \mathcal{C}^{k,\gamma}_{WF}(P \times [0, T]) \longrightarrow \mathcal{C}^{k,2+\gamma}_{WF}(P \times [0, T]) \tag{10.55}$$

is compact, and its norm tends to zero as $T \to 0^+$. We call this the *small time localization property*.

[Homogeneous Case:] We also assume the existence of a solution operator $\mathcal{Q}^t_0$ for the homogeneous Cauchy problem:

$$(\partial_t - L)v = v \text{ with } v(x, 0) = f, \tag{10.56}$$

with the following properties:

1. For $k \in \mathbb{N}_0$, $0 < T$, and $0 < \gamma < 1$, the maps

$$\mathcal{Q}^t_0 : \mathcal{C}^{k,2+\gamma}_{WF}(P) \longrightarrow \mathcal{C}^{k,2+\gamma}_{WF}(P \times [0, T]) \tag{10.57}$$

are bounded.

2. As $t \to 0^+$, for $f \in \mathscr{C}^{k,2+\gamma}_{WF}(P)$, and $0 < \widetilde{\gamma} < \gamma$, we have that

$$\lim_{t\to 0^+} \|\mathfrak{Q}_0^t f - f\|_{WF,k,\widetilde{\gamma}} = 0. \tag{10.58}$$

3. If $\psi_1, \psi_2 \in \mathscr{C}^\infty(P)$ have $\operatorname{dist}(\operatorname{supp}\psi_1, \operatorname{supp}\psi_2) > 0$, then the operator

$$\psi_1\mathfrak{Q}_0^t\psi_2 : \mathscr{C}^{k,2+\gamma}_{WF}(P \times [0,T]) \longrightarrow \mathscr{C}^{k,2+\gamma}_{WF}(P \times [0,T]) \tag{10.59}$$

is compact and tends to zero in norm as $T \to 0^+$.

To carry out the induction step we require the following basic geometric result, which is related to Lemma 2.1.7:

THEOREM 10.2.1. *Let P be a compact manifold with corners with maximal codimension of bP equal to $M \geq 1$, and L a generalized Kimura diffusion operator on P. Suppose that*

$$bP = \Sigma_1 \cup \cdots \cup \Sigma_M, \tag{10.60}$$

where each Σ_j is the boundary component of bP of codimension j, and let $U \subset P$ be a neighborhood of Σ_M. There exists a compact manifold with corners $\widetilde{P}$ so that the maximal codimension of $b\widetilde{P}$ is $M-1$, with a generalized Kimura diffusion operator $\widetilde{L}$ defined on $\widetilde{P}$. The subset $P_U = P \cap U^c$ is diffeomorphic to a subset of $\widetilde{P}$ under a map Ψ which carries $L_U = L\restriction_{P\cap U^c}$ to $\widetilde{L}\restriction_{\Psi(P\cap U^c)}$.

Remark. Informally we say that (P_U, L_U) is embedded into $(\widetilde{P}, \widetilde{L})$. An example of the doubling construction is shown in Figure 10.1.

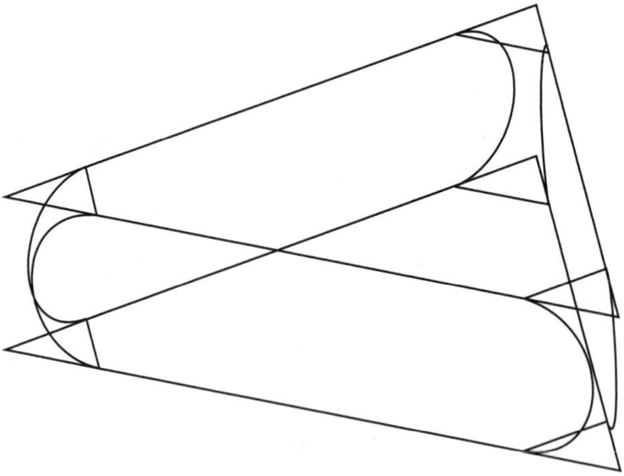

Figure 10.1: The doubling construction used to remove the maximal codimension component of bP when P is an equilateral triangle.

The proof of Theorem 10.2.1 is given later in this chapter. To carry out the induction step, we use Theorem 10.2.1 to embed (P_U, L_U) into $(\widetilde{P}, \widetilde{L})$, where $\widetilde{P}$ is a manifold with corners, of codimension at most $M - 1$. The induction hypothesis shows that there is an exact solution operator $\widetilde{Q}_i^t$ for the equation $(\partial_t - \widetilde{L})\widetilde{u} = \widetilde{g}$ on $\widetilde{P}$. In the sequel we refer to this as the *interior term*, which explain the i subscript. In the context of inductive arguments over the maximal codimension of bP, we use the adjective "interior" to refer to the things coming from parts of P disjoint from the maximal codimensional part of bP.

We use the codimension M model operators to build a boundary parametrix, $\widetilde{Q}_b^t$, in a neighborhood of Σ_M. Arguing much as in the codimension 1 case, we can glue $\widetilde{Q}_i^t$ to $\widetilde{Q}_b^t$ to obtain an operator

$$\widetilde{Q}^t : \mathscr{C}_{WF}^{k,\gamma}(P \times [0,T]) \to \mathscr{C}_{WF}^{k,2+\gamma}(P \times [0,T]),$$

so that

$$(\partial_t - L)\widetilde{Q}^t = \mathrm{Id} + E^t, \text{ with } E^t : \mathscr{C}_{WF}^{k,\gamma}(P \times [0,T]) \to \mathscr{C}_{WF}^{k,\gamma}(P \times [0,T]). \quad (10.61)$$

As before, if $k = 0$, then we can arrange to have the norm of the error term E^t bounded by any fixed $\delta < 1$, as $T \to 0$. Thus, for some $T_0 > 0$, we obtain the exact solution operator for (P, L) by setting $\mathfrak{Q}^t = \widetilde{Q}^t(\mathrm{Id} + E^t)^{-1}$; this operator defines a bounded map

$$\mathfrak{Q}^t : \mathscr{C}_{WF}^{0,\gamma}(P \times [0,T_0]) \to \mathscr{C}_{WF}^{0,2+\gamma}(P \times [0,T_0]). \quad (10.62)$$

In Section 10.4 we give the detailed construction of a boundary parametrix for the maximal codimension stratum of the boundary. Combining this with the estimates in Section 10.3 we verify the induction hypothesis in the base case that $M = 1$ and also the inductive step itself, which completes the proof for the $k = 0$ case. The estimates with $k > 0$ are left for the end of this chapter.

10.3 THE INDUCTION ARGUMENT

To complete the proof of the theorem we need only verify the induction hypothesis. Assume that P is a manifold with corners so that the maximal codimension of bP is $M+1$, and that L is a generalized Kimura diffusion operator on P. Using the estimates in the previous chapters we show in Section 10.4 that there is a function $\varphi \in \mathscr{C}_c^\infty(P)$, that equals 1 on a small neighborhood of Σ_{M+1}, and vanishes outside a slightly larger neighborhood, and an operator $\widehat{Q}_b^t$, with the mapping properties in (10.41), so that, for $g \in \mathscr{C}_{WF}^{0,\gamma}(P \times [0,T])$, we have:

$$(\partial_t - L)\widehat{Q}_b^t g = \varphi g + (E_b^{0,t} + E_b^{1,t})g. \quad (10.63)$$

Here $E_b^{0,t}$ and $E_b^{1,t}$ are bounded maps of $\mathscr{C}_{WF}^{k,\gamma}(P \times [0,T])$, for any $k \in \mathbb{N}_0$ and $0 < \gamma < 1$. Below we show that for any $\delta > 0$ we can construct $E_b^{0,t}$ so that its norm, acting on $\mathscr{C}_{WF}^{0,\gamma}(P \times [0,T])$, is less than δ and $E_b^{1,t}$ is a compact map of this space to itself, which tends to zero in norm as $T \to 0$. At the end of the chapter this is verified for $\mathscr{C}_{WF}^{k,\gamma}(P \times [0,T])$, with $k \in \mathbb{N}$.

Let U be a neighborhood of Σ_{M+1} so that $\overline{U}\subset\subset\operatorname{int}\varphi^{-1}(1)$; set

$$P_U = P\cap U^c \text{ and } L_U = L\restriction_{P_U}. \tag{10.64}$$

We now apply Theorem 10.2.1 to find a manifold with corners $\widetilde{P}$, of maximal codimension M, and a generalized Kimura diffusion operator $\widetilde{L}$ so that (P_U, L_U) is embedded into $(\widetilde{P}, \widetilde{L})$. The induction hypothesis implies that there is a solution operator $\widetilde{Q}^t$ to the equation $(\partial_t - \widetilde{L})\widetilde{u} = \widetilde{g}$ on $\widetilde{P}$ with the desired mapping properties with respect to the WF-Hölder spaces on $\widetilde{P}$. As before, we choose $\psi\in\mathcal{C}^\infty_c(P_U)$ so that $\psi\equiv 1$ on a neighborhood of the support of $(1-\varphi)$ and define

$$\widehat{Q}^t_i g = \psi\widetilde{Q}^t[(1-\varphi)g], \tag{10.65}$$

where it is understood that we extend $(1-\varphi)g$ by zero to $\widetilde{P}$, and $\psi\widetilde{Q}^t[(1-\varphi)g]$ by zero to P.

If we let $\widehat{Q}^t = \widehat{Q}^t_b + \widehat{Q}^t_i$, then

$$(\partial_t - L)\widehat{Q}^t g = g + (E^{0,t}_b + E^{1,t}_b + E^t_i)g, \tag{10.66}$$

where, as before, $E^t_i g = [\psi, L]\widetilde{Q}^t[(1-\varphi)g]$. The support of the kernel of E^t_i is a positive distance from the diagonal and therefore the induction hypothesis implies that this is again a compact operator, tending to zero, as $T\to 0^+$, in the operator norms defined by $\mathcal{C}^{k,\gamma}_{WF}(P\times[0,T_0])$. If we choose T_0 sufficiently small, then, with $E^t = (E^{0,t}_b + E^{1,t}_b + E^t_i)$, the operator $\mathrm{Id}+E^t$ is invertible as a map from $\mathcal{C}^{k,\gamma}_{WF}(P\times[0,T_0])$ to itself. We set

$$\mathcal{Q}^t = \widehat{Q}^t(\mathrm{Id}+E^t)^{-1} \tag{10.67}$$

to get a right inverse to $(\partial_t - L)$ on the time interval $[0, T_0]$, which clearly has the correct mapping properties with respect to the WF-Hölder spaces on P.

But for the construction of the boundary parametrix, which is done in Section 10.4, we can complete the proof of the induction step in this case by showing that $\mathcal{Q}^t$ has the small time localization property. That is, if φ', ψ' are smooth functions on P with

$$\operatorname{dist}(\operatorname{supp}\varphi', \operatorname{supp}\psi') > 0, \tag{10.68}$$

then the operator

$$\psi'\mathcal{Q}^t\varphi' : \mathcal{C}^{k,\gamma}_{WF}(P\times[0,T]) \longrightarrow \mathcal{C}^{k,2+\gamma}_{WF}(P\times[0,T]) \tag{10.69}$$

is a compact operator that tends to zero in norm, as $T\to 0^+$.

The operator $(\mathrm{Id}+E^t)^{-1}$ as a map from $\mathcal{C}^{k,\gamma}_{WF}(P\times[0,T])$ to itself is defined as a convergent Neumann series

$$(\mathrm{Id}+E^t)^{-1} = \sum_{j=0}^{\infty}(-E^t)^j. \tag{10.70}$$

Given $\eta > 0$, there is an N so that for any $T < T_0$, we have that

$$\left\| \sum_{j=N+1}^{\infty} (-E^t)^j \right\|_{WF,k,\gamma,T} \leq \eta. \tag{10.71}$$

The induction hypothesis and the properties of the solution operators to the model problems show that the operator $\psi' \widehat{Q}^t \varphi'$ has the small time localization property. Therefore the essential point is to see that this is true of a composition $A^t B^t$.

LEMMA 10.3.1. *Suppose that for $t \in [0, T]$, the maps*

$$\begin{aligned} A^t &: \mathcal{C}^{k,\gamma}_{WF}(P \times [0, T]) \longrightarrow \mathcal{C}^{k,\gamma}_{WF}(P \times [0, T]) \\ B^t &: \mathcal{C}^{k,\gamma}_{WF}(P \times [0, T]) \longrightarrow \mathcal{C}^{k,\gamma}_{WF}(P \times [0, T]) \end{aligned} \tag{10.72}$$

are bounded, so that if φ and ψ are smooth functions with disjoint supports, then $\varphi A^t \psi$ and $\varphi B^t \psi$ have the small time localization property, i.e.,, are compact and tend to zero in norm as $T \to 0^+$. Moreover, the composition

$$A^t B^t : \mathcal{C}^{k,\gamma}_{WF}(P \times [0, T]) \longrightarrow \mathcal{C}^{k,\gamma}_{WF}(P \times [0, T]) \tag{10.73}$$

has the same property.

PROOF. Let φ, ψ be as above. Choose $\theta \in \mathcal{C}^\infty(P)$ with the properties:

$$\operatorname{dist}(\operatorname{supp} \psi, \operatorname{supp} \theta) > 0, \text{ and } \operatorname{dist}(\operatorname{supp} \varphi, \operatorname{supp}(1 - \theta)) > 0, \tag{10.74}$$

so that $\theta \equiv 1$ on a neighborhood of $\operatorname{supp} \varphi$. We observe that

$$\psi A^t B^t \varphi = [\psi A^t \theta] B^t \varphi + \psi A^t [(1 - \theta) B^t \varphi]. \tag{10.75}$$

The operators $[\psi A^t \theta]$ and $[(1 - \theta) B^t \varphi]$ have the small time localization property. Hence $\psi A^t B^t \varphi$ is compact and converges in norm to zero as $T \to 0^+$. □

If we let

$$\operatorname{Id} + F^t_N = \sum_{j=0}^{N} (-E^t)^j \text{ and } \mathfrak{Q}^t_N = \widehat{Q}^t (\operatorname{Id} + F^t_N), \tag{10.76}$$

then this lemma shows that the operator $\psi' (\operatorname{Id} + F^t_N) \varphi'$ is compact as a map from $\mathcal{C}^{k,\gamma}_{WF}(P \times [0, T])$ to itself and tends to zero in norm, as $T \to 0^+$. Furthermore, the difference

$$\mathfrak{Q}^t - \mathfrak{Q}^t_N : \mathcal{C}^{k,\gamma}_{WF}(P \times [0, T]) \longrightarrow \mathcal{C}^{k,2+\gamma}_{WF}(P \times [0, T]) \tag{10.77}$$

tends to zero in the norm topology. With θ as above:

$$\psi' \mathfrak{Q}^t_N \varphi' = [\psi' \widehat{Q}^t \theta] (\operatorname{Id} + F^t_N) \varphi' + \psi' \widehat{Q}^t [(1 - \theta)(\operatorname{Id} + F^t_N) \varphi'], \tag{10.78}$$

which shows, as above, that $\psi'\mathcal{Q}^t_N\varphi'$ has the small time localization property, and therefore $\psi'\mathcal{Q}^t\varphi'$ is also compact. Finally for any $\eta > 0$, there is an N so that as a map from $\mathcal{C}^{k,\gamma}_{WF}(P \times [0, T])$ to $\mathcal{C}^{k,2+\gamma}_{WF}(P \times [0, T])$ for $T < T_0$ we have

$$\|\psi'(\mathcal{Q}^t - \mathcal{Q}^t_N)\varphi'\| \leq \eta. \tag{10.79}$$

This shows that the norm of

$$\psi'\mathcal{Q}^t\varphi' : \mathcal{C}^{k,\gamma}_{WF}(P \times [0, T]) \longrightarrow \mathcal{C}^{k,2+\gamma}_{WF}(P \times [0, T]) \tag{10.80}$$

tends to zero as $T \to 0^+$. This establishes that as an operator from $\mathcal{C}^{k,\gamma}_{WF}(P \times [0, T])$ to $\mathcal{C}^{k,2+\gamma}_{WF}(P \times [0, T])$, the solution operator $\mathcal{Q}^t$ has the small time localization property.

To complete this part of the argument, we need only show that for any $k \geq 0$ the Neumann series for $(\mathrm{Id} + E^t)^{-1}$ converges in operator norm topology defined by $\mathcal{C}^{k,\gamma}_{WF}(P \times [0, T])$, and that $E^t : \mathcal{C}^{k,\gamma}_{WF}(P \times [0, T]) \to \mathcal{C}^{k,\gamma}_{WF}(P \times [0, T])$ has the small time localization property. The induction hypothesis shows that the interior error term E^t_i has this property, so it only needs to be verified for the boundary contribution to E^t. The detailed construction of the boundary parametrix is done in the following section, for $k = 0$. The argument for $k > 0$ is presented at the end of the chapter.

10.4 THE BOUNDARY PARAMETRIX CONSTRUCTION

In this section we give the details of the argument that if P is a manifold with corners so that the maximal codimension of bP is $M+1$, and L is a generalized Kimura diffusion defined on P, then given $\delta > 0$, there is an operator $\widehat{Q}^t_b$ and a function $\varphi \in \mathcal{C}^\infty_c(P)$ so that

1. φ equals 1 in a neighborhood of Σ_{M+1}.
2. For any $0 < \gamma < 1$ and some $T_0 > 0$,

$$\widehat{Q}^t_b : \mathcal{C}^{k,\gamma}_{WF}(P \times [0, T_0]) \to \mathcal{C}^{k,2+\gamma}_{WF}(P \times [0, T_0])$$

   is a bounded operator. As a map from $\mathcal{C}^{k,\gamma}_{WF}(P \times [0, T_0])$ to itself, this operator tends, as $T_0 \to 0^+$, to zero in norm.
3. For $g \in \mathcal{C}^{k,\gamma}_{WF}(P \times [0, T_0])$,

$$(\partial_t - L)\widehat{Q}^t_b g = \varphi g + (E^{0,t}_b g + E^{1,t}_b g), \tag{10.81}$$

   where $E^{0,t}_b$ has norm at most δ as an operator on $\mathcal{C}^{k,\gamma}_{WF}(P \times [0, T_0])$ and $E^{1,t}_b$ is a compact operator on this space with norm tending to zero as $T_0 \to 0^+$.
4. The family of operators $E^{0,t}_b$ has the small time localization property.

In this section we verify claims 1–4 in the case that $k = 0$.

10.4.1 The Codimension N Case

The argument is a little simpler if $M + 1 = N = \dim P$, so that the stratum Σ_N consists of a finite number of isolated points. We begin the construction by choosing an $0 < \eta << 1$. The set Σ_N is finite and consists of p points, which we generically denote by q. For each $1 \leq j \leq p$ we let $\mathfrak{U}_N = \{(U_j, \varphi_j) : j = 1, \ldots, p\}$ be an NCC covering of a neighborhood of Σ_N. By shrinking these neighborhoods, if needed, we can assume that these sets are disjoint, each containing a single element of Σ_N. For consistency with later cases we let $F_j = \Sigma_N \cap U_j$. We use the sets in $\mathfrak{U}_N$ to define local norms, $\| \cdot \|^j_{WF,k,\gamma,T}$ on $\mathscr{C}^{k,\gamma}_{WF}$.

Let $(x_1, \ldots, x_N)$ denote normal cubic coordinates in one of these neighborhoods, U_j, centered at the point q. In these coordinates the operator L takes the form

$$L = \sum_{j=1}^{N} x_j \partial_{x_j}^2 + \sum_{j=1}^{N} (b_j + \widetilde{b}_j(\boldsymbol{x}))\partial_{x_j} + \sum_{1 \leq j \neq k \leq N} \sqrt{x_j x_k} a'_{jk}(\boldsymbol{x}) \sqrt{x_j x_k} \partial_{x_j} \partial_{x_k}. \tag{10.82}$$

Here $\widetilde{b}_j$ are smooth functions vanishing at $\boldsymbol{x} = 0$ and a'_{jk} are smooth functions, and we let $\boldsymbol{b} = (b_1, \ldots, b_N)$. We let $\chi \in \mathscr{C}^\infty_c(\mathbb{R}^N_+)$ be a non-negative function which equals 1 in the positive cube of side length 2, centered at $\boldsymbol{x} = 0$ and vanishes outside the positive cube of side length 3, and $\psi \in \mathscr{C}^\infty_c(\mathbb{R}^N_+)$ be a non-negative function, which equals 1 in the positive cube of side length 4 centered at $\boldsymbol{x} = 0$ and vanishes outside the positive cube of side length 5. We define

$$\chi_\epsilon(\boldsymbol{x}) = \chi\left(\frac{\boldsymbol{x}}{\epsilon^2}\right) \text{ and } \psi_\epsilon(\boldsymbol{x}) = \psi\left(\frac{\boldsymbol{x}}{\epsilon^2}\right). \tag{10.83}$$

Let $K^{\boldsymbol{b},t}_{i,q}$ denote the solution operator for the model problem

$$\left(\partial_t - L_{i,q}\right) u = g, \quad u(x, 0) = 0, \tag{10.84}$$

where

$$L_{i,q} = \sum_{j=1}^{N} [x_j \partial_{x_j}^2 + b_j \partial_{x_j}]. \tag{10.85}$$

In the calculations that follow we suppress the explicit changes of variable, but understand that they introduce bounded constants into the estimates that are independent of ϵ. We have that

$$(\partial_t - L)\psi_\epsilon K^{\boldsymbol{b},t}_{i,q}[\chi_\epsilon g] = \chi_\epsilon g + [\psi_\epsilon, L] K^{\boldsymbol{b},t}_{i,q}[\chi_\epsilon g] + \psi_\epsilon (L_{i,q} - L) K^{\boldsymbol{b},t}_{i,q}[\chi_\epsilon g]. \tag{10.86}$$

As $\psi_\epsilon = 1$ on the ϵ-neighborhood of the supp χ_ϵ, the support of the kernel function of the commutator term is contained in the complement of the ϵ-neighborhood of the diagonal. Hence, for any $\epsilon > 0$, Proposition 9.3.1 shows that this term converges exponentially to zero in the $\mathscr{C}^{k,\gamma}_{WF}$-operator norm for any $k \in \mathbb{N}_0$ and $\gamma \in (0,1)$. That is, for any k, γ, T there are positive constants $C(k,\gamma,T)$ and $\mu(k,\gamma)$, so that, with

$$E^{\infty,t}_{i,q} g = [\psi_\epsilon, L] K^{\boldsymbol{b},t}_{i,q}[\chi_\epsilon g], \tag{10.87}$$

we have

$$\| E^{\infty,t}_{i,q} g \|_{WF,k,\gamma,T} \leq C(T,k,\gamma)\epsilon^{-\mu(k,\gamma)} \|g\|_{WF,0,\gamma,T}. \tag{10.88}$$

The constant $C(T,k,\gamma)$ tends to zero as $T \to 0^+$.

This leaves only the last term:

$$E^{0,t}_{i,q} g = \psi_\epsilon (L_{i,q} - L) K^{\boldsymbol{b},t}_{i,q}[\chi_\epsilon g] = \\ -\psi_\epsilon \left[\sum_{i=1}^N \widetilde{b}_i(\boldsymbol{x})\partial_{x_i} + \sum_{1 \leq i \neq j \leq N} \sqrt{x_i x_j} a'_{ij}(\boldsymbol{x}) \sqrt{x_i x_j} \partial_{x_i}\partial_{x_j} \right] K^{\boldsymbol{b},t}_{i,q}[\chi_\epsilon g]. \tag{10.89}$$

We need to estimate the $\mathscr{C}^{0,\gamma}_{WF}$-norm of this term, which involves two parts, the sup-norm part

$$I = \left\| \psi_\epsilon \left[\sum_{i=1}^N \widetilde{b}_i(\boldsymbol{x})\partial_{x_i} + \sum_{1 \leq i \neq j \leq N} \sqrt{x_i x_j} a'_{ij}(\boldsymbol{x}) \sqrt{x_i x_j} \partial_{x_i}\partial_{x_j} \right] K^{\boldsymbol{b},t}_{i,q}[\chi_\epsilon g] \right\|_{L^\infty}, \tag{10.90}$$

and the $(0,\gamma)$-semi-norm part

$$II = \left[\!\left[\psi_\epsilon \left[\sum_{i=1}^N \widetilde{b}_i(\boldsymbol{x})\partial_{x_i} + \sum_{1 \leq i \neq j \leq N} \sqrt{x_i x_j} a'_{ij}(\boldsymbol{x}) \sqrt{x_i x_j} \partial_{x_i}\partial_{x_j} \right] K^{\boldsymbol{b},t}_{i,q}[\chi_\epsilon g] \right]\!\right]_{WF,0,\gamma,T}, \tag{10.91}$$

which we estimate using Lemma 10.1.2. Since the function ψ_ϵ is supported in the set where $x_i \leq 5\epsilon^2$, Proposition 7.2.1 implies that the first term is estimated by

$$C\epsilon^2 \|\chi_\epsilon g\|_{WF,0,\gamma,T}. \tag{10.92}$$

Applying Lemmas 10.1.2 and 10.1.3, we see that

$$I \leq C\epsilon^{2-\gamma} \|g\|_{WF,0,\gamma,T}, \tag{10.93}$$

where the constant C is independent of ϵ.

To estimate II, Lemma 10.1.2 shows that we need to consider terms of the forms

$$\|\psi_\epsilon \widetilde{b}_i(\boldsymbol{x})\|_{L^\infty} \left[\!\left[\partial_{x_i} K^{\boldsymbol{b},t}_{i,q}[\chi_\epsilon g] \right]\!\right]_{WF,0,\gamma,T}, \left[\!\left[\psi_\epsilon \widetilde{b}_i(\boldsymbol{x}) \right]\!\right]_{WF,0,\gamma,T} \|\partial_{x_i} K^{\boldsymbol{b},t}_{i,q}[\chi_\epsilon g]\|_{L^\infty} \tag{10.94}$$

and

$$\|\psi_\epsilon \sqrt{x_i x_j} a'_{ij}(\boldsymbol{x})\|_{L^\infty} \left[\!\left[\sqrt{x_i x_j} \partial_{x_i}\partial_{x_j} K^{\boldsymbol{b},t}_{i,q}[\chi_\epsilon g] \right]\!\right]_{WF,0,\gamma,T}, \\ \left[\!\left[\psi_\epsilon \sqrt{x_i x_j} a'_{ij}(\boldsymbol{x}) \right]\!\right]_{WF,0,\gamma,T} \|\sqrt{x_i x_j} \partial_{x_i}\partial_{x_j} K^{\boldsymbol{b},t}_{i,q}[\chi_\epsilon g]\|_{L^\infty}. \tag{10.95}$$

Lemma 10.1.3 and Proposition 7.2.1 show that the terms where the sup-norm is on the coefficients are estimated by $C\epsilon^{2-\gamma}$. Applying Lemma 10.1.4 we see that there is a C independent of ϵ so that

$$[\psi_\epsilon \widetilde{b}_i(\boldsymbol{x})]_{WF,0,\gamma,T} + \left[\!\left[\psi_\epsilon \sqrt{x_i x_j} a'_{ij}(\boldsymbol{x})\right]\!\right]_{WF,0,\gamma,T} \leq C\epsilon^{2-\gamma}. \tag{10.96}$$

We get an additional order of vanishing in the second term because the coefficients vanish to second order in the variables $\{\sqrt{x_i}\}$. We again use the estimate from Proposition 7.2.1 to see that

$$\|\partial_{x_i} K^{\boldsymbol{b},t}_{i,q}[\chi_\epsilon g]\|_{L^\infty} + \|\sqrt{x_i x_j}\partial_{x_i}\partial_{x_j} K^{\boldsymbol{b},t}_{i,q}[\chi_\epsilon g]\|_{L^\infty} \leq C\epsilon^{-\gamma}\|g\|_{WF,0,\gamma,T}, \tag{10.97}$$

showing that these products in (10.94) and (10.95) are bounded by a constant times $\epsilon^{2(1-\gamma)}\|g\|_{WF,0,\gamma,T}$.

Altogether the right-hand side of (10.89) contributes M_N terms of these types, which allows us to conclude that there is a C independent of ϵ, and $T \leq T_0$, so that

$$I + II \leq CM_N\epsilon^{2(1-\gamma)}\|g\|_{WF,0,\gamma,T}, \tag{10.98}$$

whence

$$\|E^{0,t}_{i,q} g\|_{WF,0,\gamma,T} \leq CM_N\epsilon^{2(1-\gamma)}\|g\|_{WF,0,\gamma,T}. \tag{10.99}$$

These calculations apply at each of the p points in Σ_N.

For each $\epsilon > 0$ we let $\chi_{i,q}$ ($\psi_{i,q}$, resp.) denote the function χ_ϵ (ψ_ϵ, resp.) in the ith-coordinate patch, with this choice of ϵ. The contribution of Σ_N to the boundary parametrix is given by

$$\widehat{Q}^t_b = \sum_{j=1}^{p}\sum_{q\in F_j} \psi_{i,q} K^{\boldsymbol{b},t}_{i,q}\chi_{i,q}. \tag{10.100}$$

We therefore have

$$\begin{aligned}(\partial_t - L)\widehat{Q}^t_b g &= \sum_{j=1}^{p}\sum_{q\in F_j}\left[\chi_{i,q} g + E^{0,t}_{i,q} g + E^{\infty,t}_{i,q} g\right] \\ &= \varphi_\epsilon g + E^{0,t}_\epsilon g + E^{\infty,t}_\epsilon g,\end{aligned} \tag{10.101}$$

where

$$\varphi_\epsilon = \sum_{i=1}^{p}\chi_{i,q}, \quad E^{0,t}_\epsilon = \sum_{j=1}^{p}\sum_{q\in F_j} E^{0,t}_{i,q}, \quad \text{and} \quad E^{\infty,t}_\epsilon = \sum_{j=1}^{p}\sum_{q\in F_j} E^{\infty,t}_{i,q}. \tag{10.102}$$

The local estimate (10.98) shows that there is a constant C so that for any $\epsilon > 0$ we have

$$\|E^{0,t}_\epsilon g\|_{WF,0,\gamma} \leq C\epsilon^{2(1-\gamma)}\|g\|_{WF,0,\gamma,T}. \tag{10.103}$$

We can therefore choose $\epsilon > 0$ so that

$$C\epsilon^{2(1-\gamma)} = \delta. \tag{10.104}$$

With this choice of ϵ we let

$$\varphi = \sum_{i=1}^{p} \chi_{i,q}. \tag{10.105}$$

For this fixed $\epsilon > 0$, the estimate in (10.88) shows that

$$\|E_{\epsilon}^{\infty,t} g\|_{WF,k,\gamma,T} \leq C(T,k,\gamma)\epsilon^{-\mu(k,\gamma)}\|g\|_{WF,0,\gamma,T}, \tag{10.106}$$

where $C(T,k,\gamma) \to 0$ as $T \to 0^+$. Thus with

$$E_{\epsilon}^{t} = E_{\epsilon}^{0,t} + E_{\epsilon}^{\infty,t}, \tag{10.107}$$

we have the norm estimate

$$\|E_{\epsilon}^{t}\|_{0,\gamma} \leq [\delta + C(T,0,\gamma)\epsilon^{-\mu(0,\gamma)}]. \tag{10.108}$$

The function φ equals 1 in a neighborhood of Σ_N, and we have estimate

$$\|(\partial_t - L)\widehat{Q}_b^t g - \varphi g\|_{WF,0,\gamma,T} \leq [\delta + C(T,0,\gamma)\epsilon^{-\mu(0,\gamma)}]\|g\|_{WF,0,\gamma,T}. \tag{10.109}$$

It only remains to verify the small time localization property for the error term. The operator E_{ϵ}^{t} is built from a finite combination of terms of the form $GK^t\theta$, where G is a differential operator, θ is a smooth function, and K^t is the heat kernel of a model operator. If φ and ψ are smooth functions with disjoint supports, then we can choose another smooth function φ' so that

$$\operatorname{supp}\varphi' \cap \operatorname{supp}\psi = \emptyset \text{ and } \varphi' = 1 \text{ on } \operatorname{supp}\varphi. \tag{10.110}$$

Since G is a differential operator, it is immediate that

$$\varphi G K^t \theta \psi = \varphi G \varphi' K^t \theta \psi. \tag{10.111}$$

As the supports of φ' and ψ are disjoint, it follows that $\varphi' K^t \theta \psi$ is a family of smoothing operators tending to zero as $T \to 0^+$ as a map from the space $\mathscr{C}^0(P \times [0,T])$ to $\mathscr{C}^k(P \times [0,T])$, for any $k \in \mathbb{N}$. This completes the construction of the boundary parametrix in this case.

10.4.2 Intermediate Codimension Case

Now assume that $n = M + 1 < \dim P$, and that Σ_{M+1} is the maximal codimensional stratum of bP. This includes the case that $n = 1$, which is the base case needed to start the induction.

We let $N^+\Sigma_{M+1}$ denote the inward pointing normal bundle of Σ_{M+1}. Since Σ_{M+1} is the maximal codimensional stratum, the tubular neighborhood theorem for manifolds with corners implies that there is a neighborhood W of Σ_{M+1} in P that is diffeomorphic to a neighborhood W_0, of the zero section in $N^+\Sigma_{M+1}$. Let $\Psi : W_0 \to W$ be such a diffeomorphism, which reduces to the inclusion map along the zero section. We let Φ denote a $\mathscr{C}^\infty$-function defined on $N^+\Sigma_{M+1}$ so that $\Phi = 1$ in a neighborhood W_1

of zero section and $\Phi = 0$ outside a somewhat larger neighborhood W_2. We define a family of functions $\{\Phi_\epsilon : \epsilon \in (0, 1]\}$ in $\mathscr{C}^\infty(W)$ by setting

$$\Phi_\epsilon(r) = \Phi\left(\epsilon^{-2} \cdot \Psi^{-1}(r)\right). \tag{10.112}$$

Here $\epsilon^{-2}\cdot$ denotes the usual action of $\mathbb{R}_+$ on the fiber of $N^+\Sigma_{M+1}$.

Let $\mathfrak{U} = \{U_j : j = 1, \dots, p\}$ denote a covering of a neighborhood of Σ_{M+1} by NCC charts. The fact that Σ_{M+1} is the maximum codimensional stratum implies that all of these charts have coordinates lying in $\mathbb{R}_+^n \times \mathbb{R}^m$. Let

$$(\boldsymbol{x}; \boldsymbol{y}) = (x_1, \dots, x_n; y_1, \dots, y_m)$$

be the normal cubic coordinates in a subset U_j, so that in these coordinates L is given by

$$L = \sum_{i=1}^{n} x_i \partial_{x_i}^2 + \sum_{1\le k,l\le m} c_{kl}(\boldsymbol{x}, \boldsymbol{y})\partial_{y_k}\partial_{y_l} + \sum_{i=1}^{n} b_i(\boldsymbol{x}, \boldsymbol{y})\partial_{x_i} +$$

$$\sum_{1\le i\ne j\le n} x_i x_j a'_{ij}(\boldsymbol{x}, \boldsymbol{y})\partial_{x_i}\partial_{x_j} + \sum_{i=1}^{n}\sum_{l=1}^{m} x_i b'_{il}(\boldsymbol{x}, \boldsymbol{y})\partial_{x_i}\partial_{y_l} + \sum_{l=1}^{m} d_l(\boldsymbol{x}, \boldsymbol{y})\partial_{y_l}. \tag{10.113}$$

We let L^p denote the sum on the first line; this is the principal part of L.

There is a positive constant K so that within the coordinate chart the coefficient matrix c_{kl} satisfies

$$K\,\mathrm{Id}_m \le c_{kl}(\boldsymbol{x}, \boldsymbol{y}) \le K^{-1}\,\mathrm{Id}_m. \tag{10.114}$$

For each point in $q \in \Sigma_{M+1,j} = \Sigma_{M+1} \cap U_j$ we could choose an affine change of coordinates in the $\boldsymbol{y}$-variables, which we denote by $(\boldsymbol{x}, \widetilde{\boldsymbol{y}})$, so that in these variables $q \leftrightarrow (\boldsymbol{0}; \boldsymbol{0})$ and

$$L^p \restriction_q = \sum_{i=1}^{n} x_i \partial_{x_i}^2 + \sum_{1\le k,l\le m} (\delta_{kl} + \widetilde{c}_{kl}(\boldsymbol{x}, \widetilde{\boldsymbol{y}}))\partial_{\widetilde{y}_k}\partial_{\widetilde{y}_l} + \sum_{i=1}^{n} (b_i + \widetilde{b}_i(\boldsymbol{x}, \widetilde{\boldsymbol{y}}))\partial_{x_i}, \tag{10.115}$$

where

$$\widetilde{c}_{kl}(\boldsymbol{0}_n, \boldsymbol{0}_m) = \widetilde{b}_i(\boldsymbol{0}_n, \boldsymbol{0}_m) = 0. \tag{10.116}$$

In light of the bounds (10.114) these affine changes of variable come from a compact subset of $GL_m \ltimes \mathbb{R}^m$ and therefore, under all these changes of variable, the coefficients $\widetilde{c}_{kl}, b_i, \widetilde{b}_i$ and a'_{ij}, b'_{il}, d_l remain uniformly bounded in the $\mathscr{C}^\infty$-topology.

In fact we do not use these changes of variables in our construction, but simply note that the constants in the estimates for the model operators at points $q = (\boldsymbol{0}; \boldsymbol{y}_q) \in U_i$, which we can take to be

$$L_{i,q} = \sum_{i=1}^{n} [x_i \partial_{x_i}^2 + b_i(\boldsymbol{0}, \boldsymbol{y}_q)\partial_{x_i}] + \sum_{k,l=1}^{m} c_{kl}(\boldsymbol{0}, \boldsymbol{y}_q)\partial_{y_l}\partial_{y_k}, \tag{10.117}$$

are uniformly bounded. We have

$$L = L_{i,q} + L^r_{i,q} + \sum_{l=1}^{m} d_l(\boldsymbol{x}, \boldsymbol{y})\partial_{y_l}, \tag{10.118}$$

where the residual "second order" part at q is

$$\begin{aligned} L^r_{i,q} = &\sum_{1\leq k,l\leq m} c'_{kl,q}(\boldsymbol{x}, \boldsymbol{y})\partial_{y_k}\partial_{y_l} + \sum_{i=1}^{n} b'_{i,q}(\boldsymbol{x}, \boldsymbol{y})\partial_{x_i} + \\ &\sum_{1\leq i\neq j\leq n} \sqrt{x_i x_j} a'_{ij}(\boldsymbol{x}, \boldsymbol{y})\sqrt{x_i x_j}\partial_{x_i}\partial_{x_j} + \sum_{i=1}^{n}\sum_{l=1}^{m} \sqrt{x_i} b'_{il}(\boldsymbol{x}, \boldsymbol{y})\sqrt{x_i}\partial_{x_i}\partial_{y_l}. \end{aligned} \tag{10.119}$$

The coefficients of $L^r_{i,q}$ are smooth functions of $(\boldsymbol{x}, \boldsymbol{y})$, and

$$c'_{kl,q}(\boldsymbol{x}, \boldsymbol{y}) = c_{kl}(\boldsymbol{x}, \boldsymbol{y}) - c_{kl}(\boldsymbol{0}, \boldsymbol{y}_q) \text{ and } b'_{i,q}(\boldsymbol{x}, \boldsymbol{y}) = b_i(\boldsymbol{x}, \boldsymbol{y}) - b_i(0, \boldsymbol{y}_q), \tag{10.120}$$

so that

$$c'_{kl,q}(\boldsymbol{0}, \boldsymbol{y}_q) = b'_{i,q}(0, \boldsymbol{y}_q) = 0. \tag{10.121}$$

We let $\chi(\boldsymbol{x}, \boldsymbol{y})$ and $\psi(\boldsymbol{x}, \boldsymbol{y})$ be functions in $\mathscr{C}^\infty_c(\mathbb{R}^n_+ \times \mathbb{R}^m)$ so that

$$\begin{aligned} &\chi \equiv 1 \text{ in } [0, 4]^m \times (-2, 2)^n \\ &\operatorname{supp}\chi \subset [0, 9]^m \times (-3, 3)^n, \end{aligned} \tag{10.122}$$

and

$$\begin{aligned} &\psi \equiv 1 \text{ in } [0, 16]^m \times (-4, 4)^n \\ &\operatorname{supp}\psi \subset [0, 25]^m \times (-5, 5)^n. \end{aligned} \tag{10.123}$$

With $(\boldsymbol{0}; \boldsymbol{y}_q)$ the coordinates of q, we define

$$\widetilde{\chi}_{i,q} = \chi\left(\frac{\boldsymbol{x}}{\epsilon^2}, \frac{\boldsymbol{y} - \boldsymbol{y}_q}{\epsilon}\right), \tag{10.124}$$

and

$$\psi_{i,q} = \psi\left(\frac{\boldsymbol{x}}{\epsilon^2}, \frac{\boldsymbol{y} - \boldsymbol{y}_q}{\epsilon}\right). \tag{10.125}$$

Of course these functions depend on the choice of ϵ, but to simplify the notation, we leave this dependence implicit. We let $F_{i,\epsilon}$ be the points in $U_i \cap \Sigma_{M+1}$ with coordinates $\{(0, \epsilon\boldsymbol{j}) : \boldsymbol{j} \in \mathbb{Z}^m\}$.

The following lemma is immediate from these definitions.

LEMMA 10.4.1. *Every point lies in the support of at most a fixed finite number of the functions* $\{\psi_{i,q} : q \in F_{i,\epsilon};\ i = 1, \ldots, p\}$, *independently of* ϵ.

From the definition of the sets $F_{i,\epsilon}$ it is clear that there is a constant S, independent of ϵ, so that for $r \in P$ we have the estimate

$$X_\epsilon(r) = \sum_{i=1}^{p} \sum_{q \in F_{i,\epsilon}} \widetilde{\chi}_{i,q}(r) \leq S. \tag{10.126}$$

It is also clear that $X_\epsilon(r) \geq 1$ for $r \in \Sigma_{M+1}$. By choosing the neighborhoods W_1, W_2 (independently of ϵ) used in the definition of Φ_ϵ (see (10.112)), we can arrange to have $\Phi_\epsilon = 1$ on the set where $X_\epsilon \geq \frac{1}{2}$, and

$$\operatorname{supp} \Phi_\epsilon \subset X_\epsilon^{-1}([\frac{1}{16}, S]). \tag{10.127}$$

To get a partition of unity of a neighborhood of Σ_{M+1}, we replace the functions $\{\widetilde{\chi}_{i,q}\}$ with

$$\chi_{i,q} = \Phi_\epsilon \left[\frac{\widetilde{\chi}_{i,q}}{\sum_{i=1}^{p} \sum_{q \in F_{i,\epsilon}} \widetilde{\chi}_{i,q}} \right]. \tag{10.128}$$

For any choice of $\epsilon > 0$, these functions are smooth and define a partition of unity in a neighborhood of Σ_{M+1}. By repeated application of Lemmas 10.1.2 and 10.1.3 and (10.127), it follows that there is a constant C independent of $\epsilon > 0$ so that

$$\|\psi_{i,q}\|_{WF,0,\gamma} + \|\chi_{i,q}\|_{WF,0,\gamma} \leq C\epsilon^{-\gamma}. \tag{10.129}$$

For each $\epsilon > 0$ we define a boundary parametrix by setting

$$\widehat{Q}_b^t = \sum_{i=1}^{p} \sum_{q \in F_{i,\epsilon}} \psi_{i,q} K_{i,q}^{\boldsymbol{b},t} \chi_{i,q}, \tag{10.130}$$

where $K_{i,q}^{\boldsymbol{b},t}$ denotes the solution operator constructed above for the model problem

$$(\partial_t - L_{i,q})u = g \quad u(r, 0) = 0,$$

with $L_{i,q}$ defined in (10.117), and with $\boldsymbol{b} = (b_1(\mathbf{0}, \boldsymbol{y}_q), \ldots, b_n(\mathbf{0}, \boldsymbol{y}_q))$. We now consider the typical term appearing in the parametrix. If g is a Hölder continuous function defined in a neighborhood of $\operatorname{supp} \chi_{i,q}$, then

$$u_{i,q} = \psi_{i,q} K_{i,q}^{\boldsymbol{b},t}[\chi_{i,q} g] \tag{10.131}$$

is well defined throughout U_i and can be extended, by zero, to all of P. We apply the operator to $u_{i,q}$ obtaining:

$$\begin{aligned}(\partial_t - L)u_i^q &= \psi_{i,q}(\partial_t - L_{i,q} + L_{i,q} - L)K_{i,q}^{\boldsymbol{b},t}[\chi_{i,q} g] + [\psi_{i,q}, L]K_{i,q}^{\boldsymbol{b},t}[\chi_{i,q} g] \\ &= \chi_{i,q} g + \psi_{i,q}(L_{i,q} - L)K_{i,q}^{\boldsymbol{b},t}[\chi_{i,q} g] + [\psi_{i,q}, L]K_{i,q}^{\boldsymbol{b},t}[\chi_i^q g].\end{aligned} \tag{10.132}$$

The estimates for the sizes of these errors will be in terms of $\|\chi_{i,q} g\|^{i}_{WF,0,\gamma,T}$. It follows from Lemmas 10.1.2 and 10.1.3 and (10.129) that there is a constant C, independent of ϵ, T so that

$$\|\chi_{i,q} g\|^{i}_{WF,0,\gamma,T} \leq C\epsilon^{-\gamma} \|g\|_{WF,0,\gamma,T}. \tag{10.133}$$

There are three types of error terms:

$$E^{\infty,t}_{i,q} g = [\psi_{i,q}, L] K^{\boldsymbol{b},t}_{i,q}[\chi_{i,q} g], \quad E^{1,t}_{i,q} g = \psi_{i,q} \left[\sum_{l=1}^{m} d_l(\boldsymbol{x}, \boldsymbol{y}) \partial_{y_l}\right] K^{\boldsymbol{b},t}_{i,q}[\chi_{i,q} g],$$
$$E^{0,t}_{i,q} = \psi_{i,q} L^{r}_{i,q} K^{\boldsymbol{b},t}_{i,q}[\chi_{i,q} g], \tag{10.134}$$

where $L^{r}_{i,q}$ is given by (10.119). For each $\epsilon > 0$ and $d \in \{0, 1, \infty\}$ we define

$$E^{d,t}_{\epsilon} = \sum_{i=1}^{p} \sum_{q \in F_{i,\epsilon}} E^{d,t}_{i,q}. \tag{10.135}$$

Observe that the support of the coefficients of $[\psi_{i,q}, L]$ is disjoint from that of $\chi_{i,q}$ and therefore Proposition 9.3.1 shows that $E^{\infty,t}_{i,q}$ is a smoothing operator tending exponentially to zero as $T \to 0^+$. As before there are positive constants $C(T, k, \gamma)$ and $\mu(k, \gamma)$ so that, for any $\epsilon > 0$, we have

$$\|E^{\infty,t}_{i,q} g\|_{WF,k,\gamma,T} \leq C(T, k, \gamma)\epsilon^{-\mu(k,\gamma)} \|g\|_{WF,k,\gamma,T}, \tag{10.136}$$

where $C(T, k, \gamma) = O(T)$ as $T \to 0^+$.

The error term $E^{1,t}_{i,q} g$ produced by the tangential first derivatives is of lower order, but more importantly, equation (9.45) shows that the norm of this term also tends to zero as $T \to 0^+$. Hence there is a positive constant C independent of ϵ so that

$$\|E^{1,t}_{i,q} g\|_{WF,0,\gamma,T} \leq C\epsilon^{-\gamma} T^{\frac{\gamma}{2}} \|g\|_{WF,0,\gamma,T}. \tag{10.137}$$

Recalling Lemma 10.4.1, each point will lie in the support of at most S of the functions $\psi_{i,q}$ and therefore there is a constant C independent of ϵ so that the sum of these terms satisfies an estimate of the form

$$\|[E^{1,t}_{\epsilon} + E^{\infty,t}_{\epsilon}] g\|_{WF,0,\gamma,T} \leq SC\epsilon^{-(\mu(0,\gamma)+\gamma)} T^{\frac{\gamma}{2}} \|g\|_{WF,0,\gamma,T}. \tag{10.138}$$

The remaining error term is

$$E^{0,t}_{i,q} g = \psi_{i,q} L^{r}_{i,q} K^{\boldsymbol{b},t}_{i,q}[\chi_{i,q} g], \tag{10.139}$$

which is a bounded map of $\mathscr{C}^{k,\gamma}_{WF}$ to itself, for any $k \in \mathbb{N}_0$. We need to estimate both $\|E^{0,t}_{i,q} g\|_{L^\infty}$ and $\left[\!\left[E^{0,t}_{i,q} g\right]\!\right]_{WF,0,\gamma}$. The vanishing properties of the coefficients of $L^{r}_{i,q}$, Proposition 9.2.1 and Lemmas 10.1.2 and 10.1.3 imply that the L^∞-term satisfies

$$\|E^{0,t}_{i,q} g\|_{L^\infty} \leq C\epsilon \|\chi_{i,q} g\|_{WF,0,\gamma} \leq C\epsilon^{1-\gamma} \|g\|_{WF,0,\gamma}. \tag{10.140}$$

The second inequality follows from Lemmas 10.1.2 and 10.1.3.

To estimate the Hölder semi-norm we need to consider a variety of terms, much like those in (10.94) and (10.95). For the case at hand we have the terms

$$\left[\!\left[\psi_{i,q} b'_{i,q}(\boldsymbol{x}, \boldsymbol{y})\partial_{x_i} K^{\boldsymbol{b},t}_{i,q}[\chi_{i,q} g]\right]\!\right]_{WF,0,\gamma}, \quad \left[\!\left[\psi_{i,q} c'_{kl,q}(\boldsymbol{x}, \boldsymbol{y})\partial_{y_k}\partial_{y_l} K^{\boldsymbol{b},t}_{i,q}[\chi_{i,q} g]\right]\!\right]_{WF,0,\gamma},$$
$$\left[\!\left[\psi_{i,q}\sqrt{x_i x_j} a'_{ij}(\boldsymbol{x}, \boldsymbol{y})\sqrt{x_i x_j}\partial_{x_i}\partial_{x_j} K^{\boldsymbol{b},t}_{i,q}[\chi_{i,q} g]\right]\!\right]_{WF,0,\gamma},$$
$$\left[\!\left[\psi_{i,q}\sqrt{x_i} b'_{il}(\boldsymbol{x}, \boldsymbol{y})\sqrt{x_i}\partial_{x_i}\partial_{y_l} K^{\boldsymbol{b},t}_{i,q}[\chi_{i,q} g]\right]\!\right]_{WF,0,\gamma}, \tag{10.141}$$

each of which is estimated by using the Leibniz formula in Lemma 10.1.2. Lemma 10.1.3 shows that there is a constant C so that the terms

$$\|\psi_{i,q} b'_{i,q}(\boldsymbol{x}, \boldsymbol{y})\|_{L^\infty}\left[\!\left[\partial_{x_i} K^{\boldsymbol{b},t}_{i,q}[\chi_{i,q} g]\right]\!\right]_{WF,0,\gamma},$$
$$\|\psi_{i,q} c'_{kl,q}(\boldsymbol{x}, \boldsymbol{y})\|_{L^\infty}\left[\!\left[\partial_{y_k}\partial_{y_l} K^{\boldsymbol{b},t}_{i,q}[\chi_{i,q} g]\right]\!\right]_{WF,0,\gamma},$$
$$\|\psi_{i,q}\sqrt{x_i} b'_{il}(\boldsymbol{x}, \boldsymbol{y})\|_{L^\infty}\left[\!\left[\sqrt{x_i}\partial_{x_i}\partial_{y_l} K^{\boldsymbol{b},t}_{i,q}[\chi_{i,q} g]\right]\!\right]_{WF,0,\gamma} \tag{10.142}$$

are all bounded by

$$C\epsilon\left[\chi_{i,q} g\right]_{WF,0,\gamma} \leq C\epsilon^{1-\gamma}\|g\|_{WF,0,\gamma}. \tag{10.143}$$

Similarly, we see that

$$\|\psi_{i,q}\sqrt{x_i x_j} a'_{ij}(\boldsymbol{x}, \boldsymbol{y})\|_{L^\infty}\left[\!\left[\sqrt{x_i x_j}\partial_{x_i}\partial_{x_j} K^{\boldsymbol{b},t}_{i,q}[\chi_{i,q} g]\right]\!\right]_{WF,0,\gamma} \leq C\epsilon^{2-\gamma}\|g\|_{WF,0,\gamma}. \tag{10.144}$$

To complete the estimates for the terms in (10.141) we need to bound

$$\left[\!\left[\psi_{i,q} b'_{i,q}(\boldsymbol{x}, \boldsymbol{y})\right]\!\right]_{WF,0,\gamma}\|\partial_{x_i} K^{\boldsymbol{b},t}_{i,q}[\chi_{i,q} g]\|_{L^\infty},$$
$$\left[\!\left[\psi_{i,q} c'_{kl,q}(\boldsymbol{x}, \boldsymbol{y})\right]\!\right]_{WF,0,\gamma}\|\partial_{y_k}\partial_{y_l} K^{\boldsymbol{b},t}_{i,q}[\chi_{i,q} g]\|_{L^\infty},$$
$$\left[\!\left[\psi_{i,q}\sqrt{x_i x_j} a'_{ij}(\boldsymbol{x}, \boldsymbol{y})\right]\!\right]_{WF,0,\gamma}\|\sqrt{x_i x_j}\partial_{x_i}\partial_{x_j} K^{\boldsymbol{b},t}_{i,q}[\chi_{i,q} g]\|_{L^\infty},$$
$$\left[\psi_{i,q}\sqrt{x_i} b'_{il}(\boldsymbol{x}, \boldsymbol{y})\right]_{WF,0,\gamma}\|\sqrt{x_i}\partial_{x_i}\partial_{y_l} K^{\boldsymbol{b},t}_{i,q}[\chi_{i,q} g]\|_{L^\infty}. \tag{10.145}$$

Proposition 9.2.1 shows that for any $0 < \gamma' \leq \gamma < 1$ the sup-norms appearing in (10.145) are bounded by

$$C_{\gamma'}\|\chi_{i,q} g\|_{WF,0,\gamma'} \leq C_{\gamma'}\epsilon^{-\gamma'}\|g\|_{WF,0,\gamma'}. \tag{10.146}$$

We therefore fix a $0 < \gamma' \leq \gamma$ so that

$$\gamma' + \gamma < 1. \tag{10.147}$$

To complete this estimate we only need to bound the Hölder semi-norms of the coefficients. Lemma 10.1.4 shows that all of these terms are bounded by $C\epsilon^{1-\gamma}$, for a constant independent of $\epsilon > 0$. Together these estimates show that there is a constant C independent of ϵ, i and q so that

$$\|E^{0,t}_{i,q} g\|_{WF,0,\gamma,T} \leq C\epsilon^{1-\gamma-\gamma'}\|g\|_{WF,0,\gamma,T}. \tag{10.148}$$

Once again we use the fact that for any point in P at most a fixed finite number of terms in the sum defining E^0_ϵ is non-zero to conclude that there is a constant S so that

$$\|E^{0,t}_{\epsilon} g\|_{WF,0,\gamma,T} \leq SC\epsilon^{1-\gamma-\gamma'}\|g\|_{WF,0,\gamma,T}. \tag{10.149}$$

We can therefore choose $\epsilon > 0$ so that

$$SC\epsilon^{1-\gamma-\gamma'} \leq \delta. \tag{10.150}$$

With this fixed choice of ϵ we let

$$\varphi = \sum_{i=1}^{p} \sum_{q\in F_{i,\epsilon}} \chi_{i,q}; \tag{10.151}$$

this function equals 1 in a neighborhood of Σ_{M+1}. Using the definition for $\widehat{Q}^t_b$ with this choice of ϵ, we see that with

$$E^t_\epsilon = E^{0,t}_\epsilon + E^{1,t}_\epsilon + E^{\infty,t}_\epsilon \tag{10.152}$$

we have that

$$(\partial_t - L)\widehat{Q}^t_b - \varphi g = E^t_\epsilon g, \tag{10.153}$$

and therefore

$$\|(\partial_t - L)\widehat{Q}^t_b - \varphi g\|_{WF,0,\gamma,T} \leq [\delta + C\epsilon^{-\mu(0,\gamma)}T^{\frac{\gamma}{2}}]\|g\|_{WF,0,\gamma,T}. \tag{10.154}$$

This estimate completes the construction of the boundary parametrix for the case of arbitrary maximal codimension between 1 and $\dim P$.

It only remains to verify the small time localization property for the error term. As before, the operator E^t_ϵ is built from a finite combination of terms of the form $GK^t\theta$, where G is a differential operator, θ is a smooth function, and K^t is the heat kernel of a model operator. Precisely the same argument as given in the maximal codimension case shows that if φ and ψ are smooth functions with disjoint supports, then $\varphi GK^t\theta\psi$ is a family of smoothing operators tending to zero as $T \to 0^+$ as a map from $\mathscr{C}^0(P\times[0,T])$ to $\mathscr{C}^j(P \times [0,T])$, for any $j \in \mathbb{N}$. This in turn completes the proof, in case $k = 0$, of the existence of a solution to the inhomogeneous problem up to a time $T_0 > 0$. In the next section we show how to use this result to demonstrate the existence of solutions to the Cauchy problem, which in turn allows us to prove a global in time existence result for the inhomogeneous problem.

10.5 SOLUTION OF THE HOMOGENEOUS PROBLEM

Assuming the existence of a solution to the inhomogeneous problem for data belonging to $\mathscr{C}^{k,\gamma}_{WF}(P \times [0, T_0])$ for a fixed $T_0 > 0$, a very similar parametrix construction is used to show the existence of v, the solution for all time, to the homogeneous Cauchy problem, with initial data $f \in \mathscr{C}^{k,2+\gamma}_{WF}(P)$. Assume that $\mathfrak{Q}^t$, the solution operator for the inhomogeneous problem, is defined for $t \in [0, T_0]$. As above, we use Proposition 9.1.1 to build a boundary parametrix for the homogeneous Cauchy problem, which we then glue to the exact solution operator for P_U. This gives an operator

$$\begin{gathered}\widehat{Q}^t_0 : \mathscr{C}^{k,2+\gamma}_{WF}(P) \longrightarrow \mathscr{C}^{k,2+\gamma}_{WF}(P \times [0, \infty)) \\ (\partial_t - L)\widehat{Q}^t_0 f = E^t_0 f \text{ and } \widehat{Q}^t_0 f \restriction_{t=0} = f,\end{gathered} \tag{10.155}$$

where

$$E^t_0 : \mathscr{C}^{k,2+\gamma}_{WF}(P) \longrightarrow \mathscr{C}^{k,\gamma}_{WF}(P \times [0, \infty)) \tag{10.156}$$

is a bounded map. A slightly stronger statement is true.

PROPOSITION 10.5.1. *Given $\delta > 0$, we can make*

$$\lim_{T \to 0+} \|E^t_0\|_{\mathscr{C}^{k,2+\gamma}_{WF}(P) \to \mathscr{C}^{k,\gamma}_{WF}(P \times [0,\infty))} \leq \delta. \tag{10.157}$$

The existence of the operator $\widehat{Q}^t_0$ is a simple consequence of the induction hypothesis and the properties of the solution operators for the model homogeneous Cauchy problems established in Proposition 9.1.1. Suppose that the maximal codimension of bP is M. Let $\{W_j : j = 1, \ldots, J\}$ be an NCC cover of Σ_M, and W_0 a relatively compact subset of int P, which covers $P \setminus \cup_{j=1}^J W_j$, and has a smooth boundary. Let $\{\varphi_j\}$ be a partition of unity subordinate to this cover of P, and $\{\psi_j\}$ smooth functions of compact support in W_j, with $\psi_j \equiv 1$ on supp φ_j. For each $j \in \{1, \ldots, J\}$ let $\widehat{Q}^t_{j0}$ be the solution operator for the homogeneous Cauchy problem defined by the model operator in W_j. As above, we let $\widehat{Q}^t_{00}$ be the exact solution operator for the Cauchy problem $(\partial_t - L)u = 0$ on W_0 with Dirichlet data on $bW_0 \times [0, \infty)$. We then define

$$\widehat{Q}^t_0 = \sum_{j=0}^{J} \psi_j \widehat{Q}^t_{j0} \varphi_j. \tag{10.158}$$

From the mapping properties of the component operators it follows that, for any $0 < \gamma < 1$, and $k \in \mathbb{N}_0$, this operator defines bounded maps:

$$\begin{gathered}\widehat{Q}^t_0 : \mathscr{C}^{k,\gamma}_{WF}(P) \longrightarrow \mathscr{C}^{k,\gamma}_{WF}(P \times [0, \infty)) \\ \widehat{Q}^t_0 : \mathscr{C}^{k,2+\gamma}_{WF}(P) \longrightarrow \mathscr{C}^{k,2+\gamma}_{WF}(P \times [0, \infty)).\end{gathered} \tag{10.159}$$

As $t \to 0^+$, the operator $\widehat{Q}^t_0$ tends strongly to the identity, with respect to the topologies $\mathscr{C}^{k,\widetilde{\gamma}}_{WF}(P)$, $\mathscr{C}^{k,2+\widetilde{\gamma}}_{WF}(P)$, respectively, for any $\widetilde{\gamma} < \gamma$.

If we set $\widetilde{Q}_0^t f = \mathfrak{Q}^t E_0^t f$, and $\mathfrak{Q}_0^t f = (\widehat{Q}_0^t - \widetilde{Q}_0^t) f$, then

$$\begin{aligned} \mathfrak{Q}_0^t : \mathscr{C}_{WF}^{k,2+\gamma}(P) &\longrightarrow \mathscr{C}_{WF}^{k,2+\gamma}(P \times [0, T_0]) \text{ is bounded} \\ (\partial_t - L)\mathfrak{Q}_0^t f &= 0. \end{aligned} \tag{10.160}$$

For any $0 < \widetilde{\gamma} < \gamma$, the solution $\mathfrak{Q}_0^t f$ tends to f in $\mathscr{C}_{WF}^{k,2+\widetilde{\gamma}}(P)$. From the induction hypothesis and the properties of the boundary terms this is certainly true of $\widehat{Q}_0^t f$. To treat the correction term we observe that $\mathfrak{Q}^t$ defines a bounded map from the space $\mathscr{C}_{WF}^{k,\gamma}(P \times [0, T])$ to $\mathscr{C}_{WF}^{k,2+\gamma}(P \times [0, T])$. For a fixed $\delta > 0$, by constructing the partition of unity $\{\varphi_j\}$ as in Section 10.4, and choosing $\epsilon > 0$ sufficiently small, we can arrange to have

$$\lim_{T \to 0^+} \| E_0^t f \|_{WF,k,\gamma,T} \leq \delta \| f \|_{WF,k,2+\gamma}.$$

Hence, for any $\delta > 0$,

$$\lim_{t \to 0^+} \| \mathfrak{Q}^t f - f \|_{WF,k,2+\widetilde{\gamma}} \leq C\delta \| f \|_{WF,k,2+\gamma}. \tag{10.161}$$

To show that the solution to the homogeneous problem exists for all $t > 0$, we observe that the time of existence T_0 already obtained is independent of the initial data, and there is a constant C so that, with $v = \mathfrak{Q}_0^t f$,

$$\| v(\cdot, T_0) \|_{WF,k,2+\gamma} \leq C \| f \|_{WF,k,2+\gamma}. \tag{10.162}$$

We can therefore apply this argument again, with data $v(\cdot, T_0)$ specified at $t = T_0$, to obtain a solution on $[0, 2T_0]$. We have the same estimate on $[0, 2T_0]$ with C replaced by C^2. This can be repeated ad libitum to show that there is are constants C_1, M and a solution v to the homogeneous Cauchy problem, belonging to $\mathscr{C}_{WF}^{k,2+\gamma}(P \times [0, T])$, for any $T > 0$, which satisfies the estimate

$$\| v \|_{WF,k,2+\gamma,T} \leq M \exp(C_1 T) \| f \|_{WF,k,2+\gamma}. \tag{10.163}$$

To verify that $\mathfrak{Q}_0^t$ satisfies the small time localization property (condition (3) in the induction hypothesis) we recall that $\mathfrak{Q}_0^t = \widehat{Q}_0^t - \mathfrak{Q}^t E_0^t$. The induction hypothesis and the properties of the model heat kernels show that $\widehat{Q}_0^t$ has this property. We have established this for the operator $\mathfrak{Q}^t$. The error term is again of the form $GK_0^t\theta$, where G is a differential operator and K_0^t is either a model heat kernel, or the heat kernel from the interior. As before, if φ and ψ have disjoint support, then we can choose φ' satisfying (10.110). From this it is immediate that, as maps from $\mathscr{C}_{WF}^{k,2+\gamma}(P)$ to $\mathscr{C}_{WF}^{k,\gamma}(P)$, the operators

$$\varphi G K_0^t \theta \psi = \varphi G \varphi' K_0^t \theta \psi \tag{10.164}$$

have the small time localization property. Using the arguments in the proof of Lemma 10.3.1 it follows easily that, as maps from $\mathscr{C}_{WF}^{k,2+\gamma}(P)$ to itself, the operator $\mathfrak{Q}^t E_0^t$ also has the small time localization property.

This completes the proof of the following theorem, which is part of Theorem 10.0.2, in the $k = 0$ case.

THEOREM 10.5.2. *Let P be a manifold with corners and L a generalized Kimura diffusion operator defined on P. There is an operator*

$$\mathfrak{Q}_0^t : \mathscr{C}_{WF}^{k,2+\gamma}(P) \longrightarrow \mathscr{C}_{WF}^{k,2+\gamma}(P \times [0,\infty)), \tag{10.165}$$

so that

$$(\partial_t - L)\mathfrak{Q}_0^t f = 0 \text{ for all } t > 0; \tag{10.166}$$

moreover, for any $\widetilde{\gamma} < \gamma$, $\mathfrak{Q}_0^t f$ converges to f in $\mathscr{C}_{WF}^{k,2+\widetilde{\gamma}}(P)$. There are constants $C_{k,\gamma}$, $M_{k,\gamma}$ so that

$$\|\mathfrak{Q}_0^t f\|_{WF,k,2+\gamma,T} \leq M_{k,\gamma} \exp(C_{k,\gamma} T)\|f\|_{WF,k,2+\gamma}. \tag{10.167}$$

Contingent upon verification of the convergence of the Neumann series for $k \in \mathbb{N}$, and the proof of Theorem 10.2.1, this completes the proof of Theorem 10.5.2.

The maximum principle has a useful corollary about the point spectrum of L on the spaces $\mathscr{C}_{WF}^{k,2+\gamma}(P)$.

COROLLARY 10.5.3. *If there is a non-trivial solution $f \in \mathscr{C}_{WF}^{0,2+\gamma}(P)$ to the equation $(L - \mu)f = 0$, then* $\operatorname{Re}\mu \leq 0$.

PROOF. Suppose there were a solution $f_\mu \neq 0$, for a complex number $\mu = \mu_1 + i\mu_2$ with $\mu_1 > 0$. The maximum principle, Proposition 3.3.5, shows that $\mu_2 \neq 0$. The unique solution to the initial value problem $(\partial_t - L)v = 0$, with $v(x,0) = f_\mu(x)$, is $v(x,t) = e^{\mu t} f_\mu(x)$. The $\operatorname{Re}(v(x,t))$ and $\operatorname{Im}(v(x,t))$ are also solutions. The sup-norms of these solutions at times of the form $\frac{2\pi j}{\mu_2}$ grow exponentially, which contradicts Proposition 3.3.1. □

We also observe that the solution of the homogeneous problem can be used to extend the time of existence for the inhomogeneous problem. Contingent upon proving the convergence of the Neumann series for $(\mathrm{Id} + E^t)^{-1}$, we have proved the existence of a solution, $u \in \mathscr{C}_{WF}^{k,2+\gamma}(P \times [0,T_0])$ to

$$(\partial_t - L)u = g \in \mathscr{C}_{WF}^{k,\gamma}(P \times [0,T]) \text{ with } u(w,0) = 0, \tag{10.168}$$

where we assume that $T > T_0$. We now let v_1 denote the solution to the Cauchy problem with initial data $u(w,T_0) \in \mathscr{C}_{WF}^{k,2+\gamma}(P)$, which exists on the interval $[0,T_0]$, and let u_1 denote the solution to (10.168), with g replaced by $g(w,t+T_0)$. We see that setting

$$u(w,t) = v_1(w,t-T_0) + u_1(w,t-T_0) \text{ for } t \in [T_0, 2T_0] \tag{10.169}$$

extends u as a solution of (10.168) to the interval $[0,2T_0]$. This process is repeated n times until $nT_0 \geq T$, or infinitely often if $T = \infty$. It is clear that the solution $u \in \mathscr{C}_{WF}^{k,2+\gamma}(P \times [0,T])$, and its norm grows at most exponentially in T.

To complete the proof of Theorem 10.0.2 we need to prove 10.2.1, and the convergence of the Neumann series for $(\mathrm{Id} + E^t)^{-1}$ in the topologies defined by $\mathscr{C}_{WF}^{k,\gamma}(P \times [0,T_0])$, for $k > 0$. Theorem 10.2.1 is proved in the next section, and the higher regularity is established at the end of this chapter.

10.6 PROOF OF THE DOUBLING THEOREM

Let P be a manifold with corners up to codimension M and L a generalized Kimura diffusion operator on P. Let $\Sigma = \Sigma_M$ denote the corner of maximal codimension M. This is a closed manifold without boundary; for simplicity we assume here that it is connected, although this is not important. We first examine the geometry of P near Σ and use this to indicate how to perform the doubling construction for P itself. Once we have accomplished this, we show how to extend L to an operator of the same type on the doubled space.

A key property of manifolds with corners is that Σ possesses a neighborhood $\tilde{U}$ which is diffeomorphic in the category of manifolds with corners to a bundle over Σ, where each fiber is the positive unit ball $B^M_+ = \{x \in \mathbb{R}^M : x_j \geq 0 \; \forall\, j, \; ||x|| < 1\}$ in the positive orthant in $\mathbb{R}^M$. Indeed, the existence of this fibration is just the correct global version of the fact that near any point $q \in \Sigma$ there is an adapted coordinate chart $(x_1, \ldots, x_M, y_1, \ldots, y_\ell)$ for P with each $x_j \in [0, 1)$ and $y_i \in (-1, 1)$. The point we do not belabor is that one can choose a coherent set of coordinate charts of this type so that in the overlaps of these charts, the fibers $\{y = \text{const.}\}$ are the same and the transition maps induce diffeomorphisms of the positive orthant fibers. In fact, we need a slightly more refined version of this. Use polar coordinates $0 \leq r < 1$ and $\omega \in S^M_+ = \{x \in B^M_+ : ||x|| = 1\}$ to identify each fiber with a truncated cone $C_1(S^{M-1}_+)$. Then it is possible to choose the atlas of coordinate charts so that the transition maps preserve the radial coordinate r. In other words, each hypersurface $\{r = \text{const.}\}$ is globally defined, and is itself a manifold with corners up to codimension $M - 1$. In particular, set $\Sigma^o = \{r = 1\}$. Note that Σ^o is the total space of a fibration over Σ with fiber S^{M-1}_+.

We next define the doubled space $\widetilde{P}$. Let P^o denote the open manifold with corners $P \setminus \Sigma$. As a set, define

$$\widetilde{P} = \big((-P^o) \sqcup P^o \sqcup (-1, 1) \times \Sigma^o\big) / \sim,$$

where $-P^o$ denotes P^o with the opposite orientation. The identification is the obvious one between $P^o \cap \mathcal{U} \cong (0, 1) \times \Sigma^o$ and the corresponding portion of the cylinder, with the analogous identification between $-P^o$ and the other side of the cylinder. This space has the structure of a smooth manifold with corners only up to codimension $M - 1$.

For the second step of the proof, we must define an extension of the operator L to $\widetilde{P}$. It is most convenient now to express the restriction of L to the neighborhood $\mathcal{U}$ of Σ in polar coordinate form. For this we recall that in these coordinates,

$$\partial_{x_j} = \omega_j \partial_r + \frac{1}{r} V_j,$$

where V_j is tangent to each hypersurfaces $r = \text{const.}$ and transversal to $\{\omega_i = 0\}$. On the other hand, each ∂_{y_i} lifts to a vector field of precisely the same form. Therefore,

$$L = r\partial_r^2 + \sum_{i=1}^{M}\left(\frac{1}{r}\omega_i V_i^2 - \frac{1}{r}\omega_i^2 V_i + \omega_i V_i(\omega_i)\partial_r\right) + \sum c'_{kl}\partial_{y_k}\partial_{y_l}$$
$$+ \sum_{i\neq j} a'_{ij}\left(\omega_i^2\omega_j^2(r\partial_r)^2 + \omega_i\omega_j V_i V_j + \omega_i V_i(\omega_j^2) r\partial_r + \omega_i V_i(\omega_j) V_j\right)$$
$$+ \sum b'_{il}\left(r\omega_i^2\partial_r\partial_{y_l} + \omega_i V_i \partial_{y_l}\right) + \sum d_l \partial_{y_l}.$$

The coefficients $a'_{ij}, b'_{il}, c'_{kl}, d_l$ are smooth in (y, r, ω). Notice that the first term, $r\partial_r^2$, and the operator in the first parenthetic expression are both homogeneous of degree -1 and odd in r and are independent of y. All of the other operators are homogeneous of degree 0 and even in r provided we neglect the smooth dependence of their coefficients in r. We can obviously regard this as an operator on the cylinder, at least away from $r = 0$, so we must simply define a modification of the coefficients which extends smoothly and in the same class of Kimura-type operators across $r = 0$. Recall that we wish to make this modification in any fixed but arbitrarily small region $|r| < \eta$. To this end, choose a smooth non-negative cutoff function $\chi(r)$ which equals 1 in $r \geq \eta$ and vanishes when $r \leq \eta/2$. Now replace a'_{ij}, for example, by

$$a''_{ij} := \chi(r)a'_{ij}(y, r, \omega) + (1 - \chi(r))a'_{ij}(y, 0, \omega),$$

and similarly for all the other coefficients. These modified terms are now exactly homogeneous of degree 0 in $r \leq \eta/2$ and extend by even reflection across $r = 0$. It remains only to define the extensions of the first two terms. For this, let $\rho(r)$ be a smooth function defined when $|r| < 1$ with the following properties: $\rho(r) = \rho(-r)$, $\rho(r) \geq \eta/4$ for all $|r| < 1$, $\rho(r) = r$ when $|r| \geq \eta$ and $\rho(r) \leq \eta$ for $|r| < \eta$. We then replace these first two terms in L by

$$\rho(r)\partial_r^2 + \sum_{i=1}^{M}\left(\frac{1}{\rho(r)}\omega_i V_i^2 - \frac{1}{\rho(r)}\omega_i^2 V_i + \omega_i V_i(\omega_i)\partial_r\right).$$

We have now defined the full extension of L to an operator $\widetilde{L}$ of Kimura type on the doubled space $\widetilde{P}$. This completes the proof of Theorem 10.2.1.

10.7 THE RESOLVENT OPERATOR AND C_0-SEMI-GROUP

The existence of a solution to the Cauchy problem with initial data in $\mathscr{C}^{0,2+\gamma}_{WF}(P)$ suffices to establish the existence of a contraction semi-group on $\mathscr{C}^0(P)$, generated by the $\mathscr{C}^0$-graph closure of L acting on $\mathscr{C}^{0,2+\gamma}_{WF}(P)$, for any $0 < \gamma < 1$. Though these results suffice to establish the uniqueness of the solution to the stochastic differential equation associated to L and therefore the existence of a strong Markov process with support in P, they are not optimal as regards the smoothing properties of the resolvent $(\mu - L)^{-1}$. We revisit the question of the resolvent operator in the next chapter.

If $f \in \mathscr{C}^{0,2+\gamma}(P)$ then Theorem 10.5.2 shows that there is a unique solution

$$v \in \mathscr{C}^{0,2+\gamma}(P \times [0,\infty))$$

to the initial value problem

$$(\partial_t - L)v = 0 \text{ with } v(\cdot, 0) = f. \tag{10.170}$$

The maximum principle shows that

$$\|v\|_{L^\infty(P\times[0,\infty))} \leq \|f\|_{L^\infty(P)}, \tag{10.171}$$

and Theorem 10.5.2 gives the estimate

$$\|v\|_{WF,0,2+\gamma,T} \leq M\exp(CT)\|f\|_{WF,0,2+\gamma}. \tag{10.172}$$

The estimate in (10.171) easily implies that, so long as $\operatorname{Re}\mu > 0$, the limit

$$\lim_{\epsilon\to 0+} \int_{\epsilon}^{\frac{1}{\epsilon}} v(\cdot, t)e^{-\mu t}dt \tag{10.173}$$

exists as a $\mathscr{C}^0(P)$-valued integral, and, if $\operatorname{Re}\mu > C$, as a $\mathscr{C}^{0,2+\gamma}_{WF}(P)$-valued integral. We denote this limit by $R(\mu)f$. The estimates on v given above imply that

$$\begin{aligned} \|R(\mu)f\|_{L^\infty} &\leq \frac{1}{\operatorname{Re}\mu}\|f\|_{L^\infty} \\ \|R(\mu)f\|_{WF,0,2+\gamma} &\leq \frac{M}{[\operatorname{Re}\mu - C]}\|f\|_{WF,0,2+\gamma}. \end{aligned} \tag{10.174}$$

Using the same integration by parts argument as was used in Section 4.3 we establish that

$$(\mu - L)R(\mu)f = f. \tag{10.175}$$

The maximum principle shows that the operator L with domain $\mathscr{D}^2_{WF}(P)$, considered as an unbounded operator on $\mathscr{C}^0(P)$, is dissipative (see Lemma 3.2.2). As $\mathscr{C}^{0,2+\gamma}_{WF}(P)$ is a dense subset of $\mathscr{C}^0(P)$, we can apply a theorem of Lumer and Phillips [30] to conclude the existence of a $\mathscr{C}_0$-semi-group of operators $e^{tL} : \mathscr{C}^0(P) \to \mathscr{C}^0(P)$, with domain given by the $\mathscr{C}^0$-graph closure of $(L, \mathscr{C}^{0,2+\gamma}_{WF}(P))$. The maximum principle implies that this semi-group is contractive.

This establishes, for example, the uniqueness of the solution to the martingale problem, supported on $\mathscr{C}^0([0,\infty); P)$ and the uniqueness-in-law for the solution to the stochastic differential equation formally defined by this second order operator. The fact that the paths of this process are confined, almost surely, to P follows using an argument like that in [6, 7, 8]. We will return to these questions in a later publication.

10.8 HIGHER ORDER REGULARITY

In the earlier sections of this chapter we constructed a boundary parametrix with an error term E^t_ϵ defined in (10.107) or (10.152). These operators define bounded maps from $\mathscr{C}^{k,\gamma}_{WF}(P\times[0,T])$ to itself for any $k\in\mathbb{N}_0$ and $0<\gamma<1$. To complete the proof of Theorems 10.0.2 and 10.5.2 we need only establish the convergence of the Neumann series for $(\mathrm{Id}+E^t_\epsilon)^{-1}$ in the operator norm topology defined by $\mathscr{C}^{k,\gamma}_{WF}(P\times[0,T_0])$, for some $\epsilon>0$ and $T_0>0$. We accomplish this by using a general result about the convergence of Neumann series in higher norms proved in [14]. We begin by recalling the main result of that paper.

Suppose that we have a ladder of Banach spaces $X_0\supset X_1\supset X_2\supset\cdots$, with norms $\|\cdot\|_k$, satisfying

$$\|x\|_{k-1}\leq\|x\|_k \text{ for all } x\in X_k. \tag{10.176}$$

THEOREM 10.8.1. *Fix any $K\in\mathbb{N}$. Assume that A is a linear map so that $AX_k\subset X_k$ for every $k\in\mathbb{N}_0$, and that there are non-negative constants $\{\alpha_j: j=0,1,\dots,K\}$ and $\{\beta_j: j=1,\dots,K\}$, with*

$$\alpha_j<1\quad \text{for } 0\leq j\leq K, \tag{10.177}$$

for which we have the estimates

$$\begin{aligned}&\|Ax\|_0\leq\alpha_0\|x\|_0 \text{ for } x\in X_0, \text{ and}\\ &\|Ax\|_k\leq\alpha_k\|x\|_k+\beta_k\|x\|_{k-1} \text{ for } x\in X_k.\end{aligned} \tag{10.178}$$

In this case the Neumann series

$$(\mathrm{Id}-A)^{-1}=\sum_{j=0}^{\infty}A^j \tag{10.179}$$

converges in the operator norm topology defined by $(X_k,\|\cdot\|_k)$ for all $k\in\{0,\dots,K\}$.

To apply this theorem we need to show that for any $K\in\mathbb{N}$ and $1<\gamma<0$, we can choose $0<\epsilon$, $0<T_0$ so that there are constants $\{\alpha_0,\dots,\alpha_K\}$ and $\{\beta_0,\dots,\beta_K\}$ with $\beta_0=0$, $\alpha_j<1$, for $0\leq j\leq K$, and we have the estimates

$$\|E^t_\epsilon g\|_{WF,k,\gamma,T_0}\leq\alpha_k\|g\|_{WF,k,\gamma,T_0}+\beta_k\|g\|_{WF,k-1,\gamma,T_0} \text{ for } 0\leq k\leq K. \tag{10.180}$$

Recalling the definitions of the norms on the function spaces $\mathscr{C}^{k,\gamma}_{WF}(P\times[0,T])$ and $\mathscr{C}^{k,2+\gamma}_{WF}(P\times[0,T])$, we see that the proofs of such estimates follow quite easily from what is done in Chapter 10. Equivalent norms can be defined inductively by starting at $k=0$ with the definitions in (5.59) and (5.67) and then setting

$$\begin{aligned}\|g\|_{WF,k,\gamma,T}&=\|g\|_{WF,k-1,\gamma,T}+\sup_{|\boldsymbol{\alpha}|+|\boldsymbol{\beta}|+2l=k}\|\partial_t^l\partial_{\boldsymbol{x}}^{\boldsymbol{\alpha}}\partial_{\boldsymbol{y}}^{\boldsymbol{\beta}}g\|_{WF,0,\gamma,T}\\ \|g\|_{WF,k,2+\gamma,T}&=\|g\|_{WF,k-1,2+\gamma,T}+\sup_{|\boldsymbol{\alpha}|+|\boldsymbol{\beta}|+2l=k}\|\partial_t^l\partial_{\boldsymbol{x}}^{\boldsymbol{\alpha}}\partial_{\boldsymbol{y}}^{\boldsymbol{\beta}}g\|_{WF,0,2+\gamma,T}.\end{aligned} \tag{10.181}$$

The operators appearing in the sum that defines the boundary contributions to E^t_ϵ are of the form

$$GK^{b,t}_{i,q}\chi_\epsilon, \tag{10.182}$$

where G is a differential operator. From the form of this operator it is clear that we can regard it as acting on functions with support in a compact subset of the coordinate chart, independent of ϵ. This allows the application of the higher order estimates proved in Chapters 6–9 with constants that are independent of ϵ. The higher order estimates for the contributions from the interior are covered by the induction hypothesis.

The part of the estimate for $\|E^t_\epsilon g\|_{WF,k,\gamma,T}$ which cannot be subsumed into a large multiple of $\|E^t_\epsilon g\|_{WF,k-1,\gamma,T}$ will be called

$$\|E^t_\epsilon g\|_{WF,k,\gamma,T} \text{ rel } \|E^t_\epsilon g\|_{WF,k-1,\gamma,T}. \tag{10.183}$$

This arises only from terms of the form

$$\|\partial_t^l\partial_{\boldsymbol{x}}^{\boldsymbol{\alpha}}\partial_{\boldsymbol{y}}^{\boldsymbol{\beta}}E^t_\epsilon g\|_{WF,0,\gamma,T}, \text{ where } 2l+|\boldsymbol{\alpha}|+|\boldsymbol{\beta}|=k. \tag{10.184}$$

The structure of the operators that make up E^t_ϵ shows that the parts of these terms that cannot be estimated by a multiple of $\|E^t_\epsilon g\|_{WF,k-1,\gamma,T}$ arise from one of two sources. The simpler terms to estimate are of the form

$$\|\widetilde{E}^t_\epsilon\partial_t^l\partial_{\boldsymbol{x}}^{\boldsymbol{\alpha}}\partial_{\boldsymbol{y}}^{\boldsymbol{\beta}}g\|_{WF,0,\gamma,T}, \tag{10.185}$$

where $\widetilde{E}^t_\epsilon$ is the error term in the parametrix construction for a generalized Kimura diffusion operator $\widetilde{L}_{\boldsymbol{\alpha}}$ derived in a straightforward manner from L. The other "new" terms arise from $\boldsymbol{x}$-derivatives being applied to the coefficients of terms in E^t_ϵ involving $x_j\partial_{x_j}$, $x_i\partial_{x_i}\partial_{y_l}$ and $x_ix_j\partial_{x_i}\partial_{x_j}$. These terms are not of lower order, but applying a derivative to the coefficients of one of these terms leaves one less derivative to apply to g. Terms of the type appearing in (10.185) are controlled by choosing a small $\epsilon>0$, whereas this latter type of term is controlled by taking T_0 sufficiently small.

10.8.1 The 1-Dimensional Case

We explain this first in the 1-dimensional case, where P is the interval $[0,1]$. The operator takes the form $L=x(1-x)\partial_x^2+b(x)\partial_x$, with $b(x)\partial_x$ inward pointing at each boundary component. We can introduce coordinates x_0, x_1, respectively, so that $j\leftrightarrow x_j=0$, $j=0,1$ and, in these coordinates

$$L=x_j\partial_{x_j}^2+(b_j+\widetilde{b}_j(x))\partial_{x_j}, \text{ where } b_j\geq 0 \text{ and } \widetilde{b}(0)=0. \tag{10.186}$$

We let $L^b=x\partial_x^2+b\partial_x$ denote the model operators, and K^b_t the solution operators for $(\partial_t-L^b)u=g$, $u(x,0)=0$. The boundary parametrix has the form

$$\widehat{Q}^t_{b\epsilon}=\sum_{j=0}^1\psi_\epsilon(x_j)K^{b_j}_t\varphi_\epsilon(y_j). \tag{10.187}$$

Here $\varphi(x)$ is a smooth function equal to 1 in $[0, \frac{1}{8}]$, and supported in $[0, \frac{1}{4}]$, and ψ is a smooth function equal to 1 in $[0, \frac{1}{2}]$, and supported in $[0, \frac{3}{4}]$. As usual we let $f_\epsilon(x) = f(x/\epsilon^2)$. We observe that for any smooth function θ

$$[L, \theta]u = 2x(1-x)\partial_x\theta\partial_x u + [b(x)\partial_x\theta + x(1-x)\partial_x^2\theta]u, \tag{10.188}$$

which consists entirely of lower order terms, and

$$(L - \partial_t)\widehat{Q}^t_{b\epsilon} = \sum_{j=0}^{1}\left[\varphi_{j,\epsilon} + ([\psi_{j,\epsilon}, L] + \psi_{j,\epsilon}(\widetilde{b}_j(x_j)\partial_{x_j}))K_t^{b_j}\varphi_{j,\epsilon}\right]. \tag{10.189}$$

The error terms are

$$\begin{aligned} E_\epsilon^{\infty,t} &= [\psi_{0,\epsilon}, L]K_t^{b_0}\varphi_{0,\epsilon} + [\psi_{1,\epsilon}, L]K_t^{b_1}\varphi_{1,\epsilon} \\ E_\epsilon^{0,t} &= [\psi_{0,\epsilon}(\widetilde{b}_0(x_0)\partial_{x_0})]K_t^{b_0}\varphi_{0,\epsilon} + [\psi_{1,\epsilon}(\widetilde{b}_1(x_1)\partial_{x_1})]K_t^{b_1}\varphi_{1,\epsilon}. \end{aligned} \tag{10.190}$$

Together $E^t_{b\epsilon} = E_\epsilon^{0,t} + E_\epsilon^{\infty,t}$.

We want to give an estimate $\|E^t_{b\epsilon}g\|_{WF,k,\gamma,T}$ of the form

$$\|E^t_{b\epsilon}g\|_{WF,k,\gamma,T} \leq \alpha_k\|g\|_{WF,k,\gamma,T} + \beta_k\|g\|_{WF,k-1,\gamma,T}, \tag{10.191}$$

where $\alpha_k < 1$. The new terms in going from $k-1$ to k are of the form

$$\|\partial_t^m\partial_x^l E^t_{b\epsilon}g\|_{WF,0,\gamma,T}, \tag{10.192}$$

where $2m + l = k$. Any derivatives that fall onto the coefficients of $K_t^{b_j}\varphi_{j,\epsilon}g$, other than $\widetilde{b}_j(x_j)$, will lead to terms that can be estimated by multiples (possibly depending on ϵ) of $\|g\|_{WF,k-1,\gamma,T}$, which are of no consequence. From Lemma 4.1.2 it follows that

$$\partial_t^m\partial_x^l K_t^b g \equiv K_t^{b+l}[L^m_{b+l}\partial_y^l g] + \sum_{q=0}^{m-1} L^q_{b+l}\partial_t^{m-q-1}g. \tag{10.193}$$

We write that

$$\partial_t^m\partial_x^l K_t^b g \equiv K_t^{b+l}[L^m_{b+l}\partial_y^l g] + \mathbb{O}(k-2). \tag{10.194}$$

Here $\mathbb{O}(k-2)$ denotes terms for which $(WF, 0, \gamma, T)$-norms are estimated by multiples of $\|g\|_{WF,k-2,\gamma,T}$, which are also of no consequence.

The new contributions to $\|E^t_\epsilon g\|_{WF,k,\gamma,T}$ come from terms like

$$\begin{aligned} &\|[\psi_{j,\epsilon}, L]K_t^{b_j+l}L^m_{b+l}\partial_y^l(\varphi_{j,\epsilon}g)\|_{WF,0,\gamma,T}, \\ &\|\psi_{j,\epsilon}(\widetilde{b}_j(x_0)\partial_{x_j})K_t^{b_j+l}L^m_{b+l}\partial_y^l(\varphi_{j,\epsilon}g)\|_{WF,0,\gamma,T}, \end{aligned} \tag{10.195}$$

and

$$\psi_{j,\epsilon}[\partial_{x_j}\widetilde{b}_j]\partial_{x_j}K_t^{b_j+l-1}L^m_{b+l-1}\partial_y^{l-1}(\varphi_{0,\epsilon}g), \tag{10.196}$$

for $j = 0, 1$. The terms in (10.195) are precisely the sorts of terms estimated earlier in the chapter, with exactly the same coefficients. All that has changed is that we have replaced $K_t^{b_j}$ with $K_t^{b_j+l}$ and $\varphi_{j,\epsilon}g$ with $L_{b+l}^m \partial_y^l(\varphi_{j,\epsilon}g)$. From the Leibniz formula, it is again clear that the only terms that cannot be subsumed into $\|g\|_{WF,k-1,\gamma,T}$ are those of the form

$$\|[\psi_{j,\epsilon}, L]K_t^{b_j+l}\varphi_{j,\epsilon}L_{b+l}^m\partial_y^l(g)\|_{WF,0,\gamma,T},$$
$$\|\psi_{j,\epsilon}(\widetilde{b}_j(x_0)\partial_{x_j})K_t^{b_j+l}\varphi_{j,\epsilon}L_{b+l}^m\partial_y^l(g)\|_{WF,0,\gamma,T}.$$

These terms can all be estimated by

$$C\epsilon^{2-\gamma}\|g\|_{WF,k,\gamma,T}, \tag{10.197}$$

where the constant C is uniformly bounded for $k \leq K$.

To complete this case we need to consider the terms in (10.196); these are not a priori of lower order because $\partial_{x_j}\widetilde{b}_j$ may not vanish at $x_j = 0$. On the other hand, the estimate given in (6.104) shows that there are constants C_k, $\mu(k,\gamma)$ so that

$$\begin{aligned}\|\psi_{j,\epsilon}[\partial_{x_j}\widetilde{b}_j]\partial_{x_j}K_t^{b_j+l-1}L_{b+l-1}^m\partial_y^{l-1}(\varphi_{0,\epsilon}g)\|_{WF,0,\gamma,T} &= \\ \|\psi_{j,\epsilon}[\partial_{x_j}\widetilde{b}_j]K_t^{b_j+l}\partial_y L_{b+l-1}^m\partial_y^{l-1}(\varphi_{0,\epsilon}g)\|_{WF,0,\gamma,T} & \\ \leq C_k\epsilon^{-\mu(k,\gamma)}T^{1-\frac{\gamma}{2}}\|g\|_{WF,k,\gamma,T}. &\end{aligned} \tag{10.198}$$

If we fix any $\delta > 0$, then we can choose an $\epsilon > 0$ so that, $C\epsilon^{2-\gamma} < \delta$ and therefore for some constants $\{\beta_k'\}$, the estimates

$$\|E_{b\epsilon}^t g\|_{WF,k,\gamma,T} \leq (\delta + C_k\epsilon^{-\mu(k,\gamma)}T^{1-\frac{\gamma}{2}})\|g\|_{WF,k,\gamma,T} + \beta_k'\|g\|_{WF,k-1,\gamma,T} \tag{10.199}$$

hold for $k \leq K$. Fix a $\delta < 1/2$, which thereby fixes an $\epsilon > 0$. Let $\varphi_b = \varphi_{0,\epsilon} + \varphi_{1,\epsilon}$, and choose ψ_i with compact support $[a,b] \subset (0,1)$ and equal to 1 on a neighborhood of $\text{supp}(1-\varphi_b)$. Finally we let $\widehat{Q}_i^t$ be the exact solution operator to the Dirichlet problem

$$(\partial_t - L)u = g \text{ on } [a,b] \text{ with } u(x,0) = u(a,t) = u(b,t) = 0. \tag{10.200}$$

With the global parametrix given by

$$\widetilde{Q}^t = \widehat{Q}_{b\epsilon}^t + \psi_i\widehat{Q}_i^t(1-\varphi_b), \tag{10.201}$$

we see that

$$(\partial_t - L)\widetilde{Q}^t g = g + E_{b\epsilon}^t g + [\psi_i, L]\widehat{Q}_i^t(1-\varphi_b)g. \tag{10.202}$$

Since the support of $1-\psi_i$ and $1-\varphi_b$ do not overlap, the induction hypothesis shows that there is a constant $C(T,k,\gamma)$, which tends to 0 as $T \to 0^+$, so that

$$\|[\psi_i, L]\widehat{Q}_i^t(1-\varphi_b)g\|_{WF,k,\gamma,T} \leq C(T,k,\gamma)\|g\|_{WF,k,\gamma,T}. \tag{10.203}$$

Note that $\epsilon > 0$ has already been fixed.

If we let $E^t g = E^t_{b\epsilon} g + [\psi_i, L]\widehat{Q}^t_i(1-\varphi_b)g$, then for some $T_0 > 0$, there are constants $\{\beta_k : k = 0, \ldots, K\}$ so that we have the estimates

$$\|E^t g\|_{WF,k,\gamma,T_0} \leq 2\delta\|g\|_{WF,k,\gamma,T_0} + \beta_k\|g\|_{WF,k-1,\gamma,T_0}, \text{ for } 0 \leq k \leq K. \quad (10.204)$$

Theorem 10.8.1 applies to show that the Neumann series for $(\mathrm{Id}+E^t)^{-1}$ converges in the operator norm topologies defined by $\mathscr{C}^{k,\gamma}_{WF}([0,1]\times[0,T_0])$ for $0 \leq k \leq K$. The argument at the end of Section 10.3 applies to show that this operator has the small time localization property as a map from $\mathscr{C}^{k,\gamma}_{WF}([0,1]\times[0,T_0])$ to $\mathscr{C}^{k,2+\gamma}_{WF}([0,1]\times[0,T_0])$. As K is arbitrary we see that this completes the proof, in dimension 1, of the induction step for the inhomogeneous case and any k.

10.8.2 The Higher Dimensional Case

The argument in the general case is quite similar to the 1-dimensional case, though there are more terms analogous to those appearing in (10.196). We now briefly describe it. As above the key point is to show that estimates like those in (10.191) and (10.199) hold for the error terms coming from the boundary parametrix. This fixes a choice of $\epsilon > 0$, and then we can apply the induction hypothesis to obtain similar estimates for the contribution of the interior parametrix to the error term, which, along with the contributions of terms like those in (10.196), is made as small as we like by taking $0 < T_0$ small enough. The boundary contributions to the error term are enumerated in (10.134).

It is immediate that the only contributions to

$$\|[E^{\infty,t}_\epsilon + E^{1,t}_\epsilon]g\|_{WF,k,\gamma,T} \text{ rel } \|[E^{\infty,t}_\epsilon + E^{1,t}_\epsilon]g\|_{WF,k-1,\gamma,T}$$

are of the terms of the types

$$\|\psi_{i,q} d_l(\boldsymbol{x},\boldsymbol{y})\partial_{y_l} K^{\boldsymbol{b}+\boldsymbol{\alpha},t}_{i,q}[L_{\boldsymbol{b}+\boldsymbol{\alpha},m}\partial^{\boldsymbol{\alpha}}_{\boldsymbol{w}}\partial^{\boldsymbol{\beta}}_{\boldsymbol{z}}]\chi_{i,q} g\|_{WF,0,\gamma,T} \quad (10.205)$$

and

$$\|[\psi_{i,q}, L]K^{\boldsymbol{b}+\boldsymbol{\alpha},t}_{i,q}[L_{\boldsymbol{b}+\boldsymbol{\alpha},m}\partial^{\boldsymbol{\alpha}}_{\boldsymbol{w}}\partial^{\boldsymbol{\beta}}_{\boldsymbol{z}}]\chi_{i,q} g\|_{WF,0,\gamma,T}. \quad (10.206)$$

These are types of terms that we have estimated earlier (see (10.138)); hence for $0 \leq k \leq K$, there are constants $C(T,k,\gamma)$, $\mu(k,\gamma)$ and $\{\beta'_k\}$ so that

$$\|[E^{\infty,t}_\epsilon + E^{1,t}_\epsilon]g\|_{WF,k,\gamma,T} \leq C(T,k,\gamma)\epsilon^{-\mu(k,\gamma)}\|g\|_{WF,k,\gamma,T} + \beta'_k\|g\|_{WF,k-1,\gamma,T}. \quad (10.207)$$

Moreover $C(T,k,\gamma)$ tend to zero as $T \to 0^+$.

This leaves only terms of the form

$$\|\partial^m_t \partial^{\boldsymbol{\beta}}_{\boldsymbol{y}} \partial^{\boldsymbol{\alpha}}_{\boldsymbol{x}} \left(\psi_{i,q} L^r_{i,q} K^{\boldsymbol{b},t}_{i,q}[\chi_{i,q} g]\right)\|_{WF,0,\gamma,T}, \quad (10.208)$$

with $2m + |\boldsymbol{\alpha}| + |\boldsymbol{\beta}| = k$. As in the 1-dimensional case, there are two types of terms that now need to be estimated. The first type arises by passing all derivatives through to $\chi_{i,q} g$, which are of the form

$$\|\psi_{i,q} L^r_{i,q} K^{\boldsymbol{b}+\boldsymbol{\alpha},t}_{i,q} \left(L_{\boldsymbol{b}+\boldsymbol{\alpha},m} \partial^{\boldsymbol{\beta}}_{\boldsymbol{y}} \partial^{\boldsymbol{\alpha}}_{\boldsymbol{x}} [\chi_{i,q} g] \right) \|_{WF,0,\gamma,T}. \tag{10.209}$$

These are precisely the sorts of terms estimated in the $k = 0$ case. As before we choose $0 < \gamma'$ so that $\gamma + \gamma' < 1$. For $0 \leq k \leq K$, there are constants $C(k, \gamma)$ for which

$$\|\psi_{i,q} L^r_{i,q} K^{\boldsymbol{b}+\boldsymbol{\alpha},t}_{i,q} \left(L_{\boldsymbol{b}+\boldsymbol{\alpha},m} \partial^{\boldsymbol{\beta}}_{\boldsymbol{y}} \partial^{\boldsymbol{\alpha}}_{\boldsymbol{x}} [\chi_{i,q} g] \right) \|_{WF,0,\gamma,T} \leq C(k, \gamma) \epsilon^{1-\gamma-\gamma'} \|g\|_{WF,k,\gamma,T}. \tag{10.210}$$

The only terms that remain result from differentiation of the coefficients of terms appearing in $L^r_{i,q}$ of the forms $x_j \partial_{x_j}$, $x_j \partial_{x_j} \partial_{y_l}$, or $x_i x_j \partial_{x_i} \partial_{x_j}$. The parts of terms of these types that cannot be subsumed by a large multiple of $\|g\|_{WF,k-1,\gamma,T}$ are

$$\|\psi_{i,q} \partial_{x_j} K^{\boldsymbol{b}+\boldsymbol{\alpha}',t}_{i,q} \left(L_{\boldsymbol{b}+\boldsymbol{\alpha}',m} \partial^{\boldsymbol{\beta}}_{\boldsymbol{y}} \partial^{\boldsymbol{\alpha}'}_{\boldsymbol{x}} [\chi_{i,q} g] \right) \|_{WF,0,\gamma,T} = \\ \|\psi_{i,q} K^{\boldsymbol{b}+\boldsymbol{\alpha},t}_{i,q} \partial_{x_j} \left(L_{\boldsymbol{b}+\boldsymbol{\alpha}',m} \partial^{\boldsymbol{\beta}}_{\boldsymbol{y}} \partial^{\boldsymbol{\alpha}'}_{\boldsymbol{x}} [\chi_{i,q} g] \right) \|_{WF,0,\gamma,T}, \tag{10.211}$$

$$\|\psi_{i,q} \partial_{x_j} \partial_{y_l} K^{\boldsymbol{b}+\boldsymbol{\alpha}',t}_{i,q} \left(L_{\boldsymbol{b}+\boldsymbol{\alpha}',m} \partial^{\boldsymbol{\beta}}_{\boldsymbol{y}} \partial^{\boldsymbol{\alpha}'}_{\boldsymbol{x}} [\chi_{i,q} g] \right) \|_{WF,0,\gamma,T} = \\ \|\psi_{i,q} K^{\boldsymbol{b}+\boldsymbol{\alpha},t}_{i,q} \partial_{x_j} \partial_{y_l} \left(L_{\boldsymbol{b}+\boldsymbol{\alpha}',m} \partial^{\boldsymbol{\beta}}_{\boldsymbol{y}} \partial^{\boldsymbol{\alpha}'}_{\boldsymbol{x}} [\chi_{i,q} g] \right) \|_{WF,0,\gamma,T}, \tag{10.212}$$

where $\boldsymbol{\alpha}' = \boldsymbol{\alpha} - e_j$; and

$$\|\psi_{i,q} x_i \partial_{x_j} \partial_{x_i} K^{\boldsymbol{b}+\boldsymbol{\alpha}'',t}_{i,q} \left(L\boldsymbol{b} + \boldsymbol{\alpha}'', m \partial^{\boldsymbol{\beta}}_{\boldsymbol{y}} \partial^{\boldsymbol{\alpha}''}_{\boldsymbol{x}} [\chi_{i,q} g] \right) \|_{WF,0,\gamma,T} = \\ \|\psi_{i,q} x_i K^{\boldsymbol{b}+\boldsymbol{\alpha},t}_{i,q} \partial_{x_i} \partial_{x_j} \left(L_{\boldsymbol{b}+\boldsymbol{\alpha}'',m} \partial^{\boldsymbol{\beta}}_{\boldsymbol{y}} \partial^{\boldsymbol{\alpha}''}_{\boldsymbol{x}} [\chi_{i,q} g] \right) \|_{WF,0,\gamma,T}, \tag{10.213}$$

where $\boldsymbol{\alpha}'' = \boldsymbol{\alpha} - e_i - e_j$; and

$$\|\psi_{i,q} \partial_{x_j} \partial_{x_i} K^{\boldsymbol{b}+\boldsymbol{\alpha}'',t}_{i,q} \left(L_{\boldsymbol{b}+\boldsymbol{\alpha}'',m} \partial^{\boldsymbol{\beta}}_{\boldsymbol{y}} \partial^{\boldsymbol{\alpha}''}_{\boldsymbol{x}} [\chi_{i,q} g] \right) \|_{WF,0,\gamma,T} = \\ \|\psi_{i,q} K^{\boldsymbol{b}+\boldsymbol{\alpha},t}_{i,q} \partial_{x_j} \partial_{x_i} \left(L_{\boldsymbol{b}+\boldsymbol{\alpha}'',m} \partial^{\boldsymbol{\beta}}_{\boldsymbol{y}} \partial^{\boldsymbol{\alpha}''}_{\boldsymbol{x}} [\chi_{i,q} g] \right) \|_{WF,0,\gamma,T}. \tag{10.214}$$

It follows from (9.50) and the foregoing argument that for $0 \leq k \leq K$, there are constants $C(k, \gamma)$, $\mu(k, \gamma)$ so that, for $T < 1$, each of the terms in (10.211)–(10.214) is bounded by

$$C(k, \gamma) \epsilon^{-\mu(k,\gamma)} T^{\frac{\gamma}{2}} \|g\|_{WF,0,\gamma,T}. \tag{10.215}$$

Combining (10.207) with (10.210) and (10.215), we see that there are constants $\{\beta_k\}$ so that

$$\|E^t_{b\epsilon} g\|_{WF,k,\gamma,T} \leq \\ [C(T, k, \gamma) \epsilon^{-\mu(k,\gamma)} + C(k, \gamma) \epsilon^{1-\gamma-\gamma'}] \|g\|_{WF,k,\gamma,T} + \beta_k \|g\|_{WF,k-1,\gamma,T},$$

where $C(T, k, \gamma) \to 0$ as $T \to 0^+$. If we fix $0 < \delta < 1/2$, then by first choosing $\epsilon > 0$ and then $T_0 > 0$ we can arrange to have

$$\|E^t_{b\epsilon} g\|_{WF,k,\gamma,T} \leq \delta \|g\|_{WF,k,\gamma,T} + \beta_k \|g\|_{WF,k-1,\gamma,T}, \tag{10.216}$$

for $0 \leq k \leq K$. As in the 1-dimensional case, the argument is finished by augmenting the boundary parametrix with an interior term, obtaining

$$\mathcal{Q}^t = \widehat{Q}^t_{b\epsilon} + \psi_i \widehat{Q}^t_i (1 - \varphi_b). \tag{10.217}$$

Possibly decreasing T_0, we obtain an error term E^t that satisfies

$$\|E^t g\|_{WF,k,\gamma,T_0} \leq 2\delta \|g\|_{WF,k,\gamma,T_0} + \beta_k \|g\|_{WF,k-1,\gamma,T_0}, \tag{10.218}$$

for $0 \leq k \leq K$. This completes the proof of (10.180) for an arbitrary $K \in \mathbb{N}$ and $0 < \gamma < 1$.

Chapter Eleven

The Resolvent Operator

We have shown that e^{tL}, the formal solution operator for the Cauchy problem

$$(\partial_t - L)v = 0,\ v(p,0) = f(p),$$

makes sense for initial data $f \in \mathscr{C}_{WF}^{0,2+\gamma}(P)$, and that the solution belongs to $\mathscr{C}_{WF}^{0,2+\gamma}(P\times[0,\infty))$. Of course, much more is true, but the extension to merely continuous data seems to entail rather different techniques from those employed herein.

The Laplace transform of e^{tL} is formally the resolvent operator:

$$(\mu - L)^{-1} = \int_0^\infty e^{-\mu t} e^{tL} f dt. \tag{11.1}$$

Using the Laplace transform of a parametrix for the heat kernel and a perturbative argument, we construct below an operator $R(\mu)$, which depends analytically on μ lying in the complement of a set $E \subset \mathbb{C}$ which lies in a conic neighborhood of $(-\infty, 0]$. This means that for any $\alpha > 0$ there exists an $0 < R_\alpha$ so that

$$E \subset \{|\arg \mu| > \pi - \alpha \text{ or } |\mu| < R_\alpha\}. \tag{11.2}$$

See Figure 11.1.

If $f \in \mathscr{C}_{WF}^{0,\gamma}(P)$, then $R(\mu)f \in \mathscr{C}_{WF}^{0,2+\gamma}(P)$, satisfies

$$(\mu - L)R(\mu)f = f. \tag{11.3}$$

Hence $R(\mu)$ is a right inverse for $(\mu - L)$. As a map from $\mathscr{C}_{WF}^{0,\gamma}(P)$ to itself the operator $R(\mu)$ is compact. In fact for any $0 < \gamma < 1$ and $k \in \mathbb{N}$, $R(\mu)$ defines a bounded map from $\mathscr{C}_{WF}^{k,\gamma}(P)$ to $\mathscr{C}_{WF}^{k,2+\gamma}(P)$. In Hölder spaces, these are the natural elliptic estimates for generalized Kimura diffusions. Coupling this with Corollary 10.5.3 shows that the operator $\mu - L$, acting on the spaces $\mathscr{C}_{WF}^{k,2+\gamma}(P)$, is injective for μ in the right half plane. Since equation (11.3) shows that for such μ,

$$(\mu - L) : \mathscr{C}_{WF}^{0,2+\gamma}(P) \to \mathscr{C}_{WF}^{0,\gamma}(P)$$

is surjective, the open mapping theorem implies that $R(\mu)$ is also a left inverse, and hence equals the resolvent operator $(\mu - L)^{-1}$.

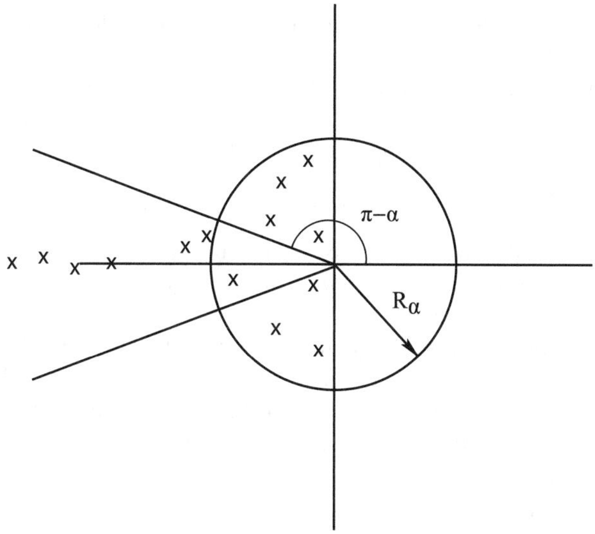

Figure 11.1: A discrete set lying in a conic neighborhood of $(-\infty, 0)$.

Since the domain $\mathscr{C}^{0,2+\gamma}_{WF}(P)$ is *not* dense in $\mathscr{C}^{0,\gamma}_{WF}(P)$ a few more remarks are in order. Suppose that μ is a value for which $(\mu - L)$ is invertible. We can rewrite

$$(\mu + \nu - L) = [\mathrm{Id} + \nu(\mu - L)^{-1}](\mu - L). \tag{11.4}$$

The map $(\mu - L) : \mathscr{C}^{0,2+\gamma}_{WF}(P) \to \mathscr{C}^{0,\gamma}_{WF}(P)$ is an isomorphism. The maps

$$[\mathrm{Id} + \nu(\mu - L)^{-1}] : \mathscr{C}^{0,\gamma}_{WF}(P) \to \mathscr{C}^{0,\gamma}_{WF}(P)$$

depend analytically on ν and are Fredholm of index zero. From this we conclude that the set of ν for which $(\mu + \nu - L)$ fails to be invertible is discrete and coincides with the set

$$\{\nu : \ker(\mu + \nu - L) \neq 0\}. \tag{11.5}$$

Thus $L : \mathscr{C}^{0,2+\gamma}_{WF}(P) \to \mathscr{C}^{0,\gamma}_{WF}(P)$ has a compact resolvent, with discrete spectrum lying in a conic neighborhood of the negative real axis. Moreover, the elliptic estimates show that all eigenfunctions belong to $\mathscr{C}^{\infty}(P)$, so the spectrum of L acting on the spaces $\mathscr{C}^{k,2+\gamma}_{WF}(P)$ does not depend on k or γ.

Using standard functional analytic techniques this allows us to show that the solution to the Cauchy problem

$$(\partial_t - L)v = 0 \text{ with } v(p, 0) = f(p) \tag{11.6}$$

is defined for $f \in \mathscr{C}^{0,\gamma}_{WF}(P)$ and, in fact, extends analytically in t to the right half plane. The solution belongs to $\mathscr{C}^{0,2+\gamma}_{WF}(P)$ for any time with positive real part. Indeed we also show that, for any $k \in \mathbb{N}$, if $f \in \mathscr{C}^{k,\gamma}_{WF}(P)$, then the solution belongs to $\mathscr{C}^{k,2+\gamma}_{WF}(P)$, for t in the right half plane.

The solution operator, $\mathscr{Q}^t_0$, defines a semi-group; thus, for any $N \in \mathbb{N}$

$$\mathscr{Q}^t_0 f = [\mathscr{Q}^{\frac{t}{N}}_0]^N f. \tag{11.7}$$

We have the obvious inclusions $\mathscr{C}^{1,\gamma}_{WF}(P) \subset \mathscr{C}^{0,2+\gamma}_{WF}(P)$, and in fact for any $k \in \mathbb{N}$, we have $\mathscr{C}^{k+1,\gamma}_{WF}(P) \subset \mathscr{C}^{k,2+\gamma}_{WF}(P)$. These inclusions, the semi-group property, and these regularity results show that the solution v to the Cauchy problem, with Hölder initial data, belongs to $\mathscr{C}^\infty(P \times (0,\infty))$.

In the next section we construct the resolvent kernel, using an induction over the maximal codimension of bP, similar to that employed in the previous chapter to construct the heat kernel. We also prove various estimates on it and corresponding estimates for the solution operator for the Cauchy problem.

11.1 CONSTRUCTION OF THE RESOLVENT

To construct the resolvent operator we proceed very much as for the construction of the heat kernel. We use an induction over the maximal codimension of bP, which allows us to construct an approximate solution operator for the Cauchy problem of the form

$$\widehat{Q}^t = \widehat{Q}^t_i + \widehat{Q}^t_b, \tag{11.8}$$

with $\widehat{Q}^t_i$ the "interior" and $\widehat{Q}^t_b$ the "boundary" contributions, respectively. We then analyze the operator:

$$\begin{aligned}\widehat{R}(\mu) &= \int\limits_0^\infty e^{-t\mu}\widehat{Q}^t dt \\ &= \widehat{R}_b(\mu) + \widehat{R}_i(\mu).\end{aligned} \tag{11.9}$$

The operator $\widehat{Q}^t_b$ extends analytically to $\operatorname{Re} t > 0$, and from its form we see that $\widehat{R}_b(\mu)$ extends analytically to the complement of $(-\infty, 0]$. From the induction hypothesis it follows that $\widehat{R}_i(\mu)$ is analytic in the complement of a discrete set lying in a conic neighborhood of $(-\infty, 0]$.

We show that

$$(\mu - L)\widehat{R}(\mu) = (\mathrm{Id} - E_\mu), \tag{11.10}$$

where the operator $E_\mu : \mathscr{C}^{k,\gamma}_{WF}(P) \to \mathscr{C}^{k,\gamma}_{WF}(P)$ is bounded for arbitrary $k \in \mathbb{N}_0$ and $0 < \gamma < 1$. We show that for a given k, $0 < \gamma$, and $0 < \alpha$ there is an R_α so that for μ satisfying

$$|\arg \mu| \leq \pi - \alpha \text{ and } |\mu| > R_\alpha, \tag{11.11}$$

the norm of this operator is less than 1 and therefore, for μ in this domain, we can define the analytic family of operators:

$$R(\mu) = \widehat{R}(\mu)(\mathrm{Id} - E_\mu)^{-1}. \tag{11.12}$$

This operator is a right inverse

$$(\mu - L)R(\mu)f = f \text{ for all } f \in \mathcal{C}^{k,\gamma}_{WF}(P). \tag{11.13}$$

We then verify the estimates in the induction hypothesis.

As noted above this allows us to construct the solution operator for the Cauchy problem for the heat equation via the contour integral:

$$\mathfrak{Q}_0^t = \frac{1}{2\pi i}\int\limits_{\Gamma_\alpha} R(\mu)e^{\mu t}d\mu. \tag{11.14}$$

The contour Γ_α is the boundary of the complement of the region described in (11.11). This defines a semi-group, analytic in $\operatorname{Re} t > 0$, acting on the spaces $\mathcal{C}^{k,\gamma}_{WF}(P)$.

The theorem we prove is the following:

THEOREM 11.1.1. *Let P be a manifold with corners of codimension n and L a generalized Kimura diffusion operator. Fix $k \in \mathbb{N}$ and $0 < \gamma < 1$. There is a discrete subset E, independent of (k, γ), contained in* $\operatorname{Re}\mu \leq 0$ *and lying in a conic neighborhood of $(-\infty, 0]$, such that the spectrum of L acting on $\mathcal{C}^{k,2+\gamma}_{WF}(P)$ is contained in the set E. The resolvent operator $R(\mu)$ is analytic in $\mathbb{C} \setminus E$. For $0 < \alpha$ there is an R_α so that for μ satisfying* (11.11) *there are constants $C_\alpha, C_{k,\alpha}$ so that $R(\mu)$ satisfies the following estimates:*

$$\begin{aligned}
\|R(\mu)f\|_{L^\infty} &\leq \frac{C_\alpha}{\mu}\|f\|_{L^\infty} \text{ for } \mu \in (0,\infty)\\
\|R(\mu)f\|_{WF,k,\gamma} &\leq \frac{C_{k,\alpha}}{|\mu|}\|f\|_{WF,k,\gamma}\\
\|R(\mu)f\|_{WF,k,2+\gamma} &\leq C_{k,\alpha}\|f\|_{WF,k,\gamma}.
\end{aligned} \tag{11.15}$$

Let V be a vector field defined in P so that, in the neighborhood of a boundary point of codimension l, V takes the form

$$V(\boldsymbol{x}, \boldsymbol{y}) = \sum_{j=1}^{l} b_j(\boldsymbol{x}, \boldsymbol{y})x_j\partial_{x_j} + \sum_{l=1}^{n-l} d_l(\boldsymbol{x}, \boldsymbol{y})\partial_{y_l}. \tag{11.16}$$

For μ satisfying (11.11) *there are constants $C_{k,\alpha}$ so that, if $|\mu| > 1$, then*

$$\|VR(\mu)f\|_{WF,k,\gamma} \leq \frac{C_{k,\alpha}}{|\mu|^{\frac{\gamma}{2}}}\|f\|_{WF,k,\gamma}. \tag{11.17}$$

PROOF. The proof is very similar to that of Theorem 10.0.2, and so many details are left to the reader. The construction of the resolvent is done by induction over the maximal codimension of bP. The verification of the induction hypothesis in this proof is actually somewhat simpler, as we do not use the weak localization property. We begin with the case that P is compact manifold without boundary, i.e., the maximum codimension of bP is zero, and L is a non-degenerate elliptic operator without constant term. The Hölder spaces are simply the classical Hölder spaces, and the statement of the theorem is more or less contained in [29], though this text does not address the compact manifold case explicitly. As the detailed estimates for $R(\mu)$ stated in (11.15) and (11.17) also do not seem to be available in the literature, we start by briefly outlining this case.

As before we begin with the $k = 0$ case. The case of $k > 0$ follows by applying 10.8.1, very much like in the proof of Theorem 10.0.2. The details of this argument are also left to the reader. We continue to use the notation and constructions from Sections 10.2–10.4.

11.1.1 The Compact Manifold Case

For $\epsilon > 0$ we cover P by open balls of radius ϵ, $\{B_\epsilon(\boldsymbol{x}_q) : q = 1, \ldots, N_\epsilon\}$, so that any point lies in at most S of the balls $\{B_{3\epsilon}(\boldsymbol{x}_q) : q = 1, \ldots, N_\epsilon\}$. As noted earlier, S can be taken to be independent of $\epsilon > 0$. Let $\{\varphi_q\}$ denote a partition of unity subordinate to this cover and $\{\psi_q\}$ smooth functions, such that

$$\psi_q(x) = 1 \text{ in } B_{2\epsilon}(\boldsymbol{x}_q), \text{ and } \operatorname{supp} \psi_q \subset B_{3\epsilon}(\boldsymbol{x}_q). \tag{11.18}$$

We let L^q denote the constant coefficient operator obtained by freezing the coefficients of the second order part of L at the point $\boldsymbol{x}_q$. If $(y_1, \ldots, y_n)$ are local coordinates near to $\boldsymbol{x}_q$, then

$$L^q = \sum_{l,m=1}^{n} c_{lm}(\boldsymbol{x}_q)\partial_{y_l}\partial_{y_m}. \tag{11.19}$$

We let Q_q^t be the heat kernel defined by L^q. This is obtained from the Euclidean heat kernel by a linear change of variables.

The parametrix for the heat kernel is defined, for t in the right half plane, by

$$\widehat{Q}^t = \sum_{q=1}^{N_\epsilon} \psi_q Q_q^t \varphi_q, \tag{11.20}$$

and the resolvent, for $\mu \in \mathbb{C} \setminus (-\infty, 0]$, by

$$\widehat{R}(\mu) = \int_0^\infty e^{-se^{i\theta}\mu} \widehat{Q}^{se^{i\theta}} e^{i\theta} ds, \tag{11.21}$$

where $\operatorname{Re}[e^{i\theta}\mu] > 0$. In the sequel we let

$$\Gamma_\theta = \{se^{i\theta} : s \in [0, \infty)\}. \tag{11.22}$$

We now compute the "error term," E_μ

$$\begin{aligned} L\widehat{R}(\mu)f &= \int\limits_{\Gamma_\theta} e^{-t\mu} \sum_{q=1}^{N_\epsilon} L\psi_q Q_q^t \varphi_q f dt \\ &= \int\limits_{\Gamma_\theta} e^{-t\mu} \sum_{q=1}^{N_\epsilon} \left\{[L, \psi_q] + \psi_q(L - L^q) + \psi_q L^q r\right\} Q_q^t \varphi_q f dt. \end{aligned} \tag{11.23}$$

Using that $L^q Q_q^t = \partial_t Q_q^t$, we integrate by parts in the last term and obtain

$$\begin{aligned} (\mu - L)\widehat{R}(\mu)f &= f - \int\limits_{\Gamma_\theta} e^{-t\mu} \sum_{q=1}^{N_\epsilon} \left\{[L, \psi_q] + \psi_q(L - L^q)\right\} Q_q^t \varphi_q f dt \\ &= f - E_\mu f. \end{aligned} \tag{11.24}$$

There are two kinds of error terms: those arising from the commutators $[L, \psi_q]$, which are lower order, and those arising from freezing coefficients $\psi_q(L - L^q)$. The differences $L - L^q$ are of the form

$$\begin{aligned} L - L^q &= \sum_{l,m=1}^{n} (c(\boldsymbol{y}) - c_{lm}(\boldsymbol{x}_q))\partial_{y_l}\partial_{y_m} + \sum_{l=1}^{n} d_l(\boldsymbol{y})\partial_{y_l} \\ &= \sum_{l,m=1}^{n} \Delta c^q(\boldsymbol{y})\partial_{y_l}\partial_{y_m} + V^q f. \end{aligned} \tag{11.25}$$

As in the previous case, the second order terms of this type are controlled by taking ϵ sufficiently small. The contribution of each such term is of the form

$$\|\psi_q \Delta c^q \partial_{y_l}\partial_{y_m} R^q(\mu)\varphi_q f\|_{WF,0,\gamma}, \tag{11.26}$$

where

$$R^q(\mu)f = \int\limits_{\Gamma_\theta} e^{-\mu t} Q_q^t f dt. \tag{11.27}$$

Arguing as in the proof of Theorem 10.0.2 and using the estimates for $R^q(\mu)$ given in Proposition 9.4.1, we see that there is a $\widetilde{\gamma} < 1$, so that

$$\|\psi_q \Delta c^q \partial_{y_l}\partial_{y_m} R^q(\mu)\varphi_q f\|_{WF,0,\gamma} \leq C_\alpha \epsilon^{1-\widetilde{\gamma}} \|f\|_{WF,0,\gamma}. \tag{11.28}$$

As before, for each point in P there is at most a fixed finite number of terms, S, independent of ϵ, contributing to this error term. This gives the estimate

$$\|\sum_{q=1}^{N_\epsilon} \psi_q \sum_{l,m=1}^{n} (c(\boldsymbol{y}) - c_{lm}(\boldsymbol{x}_q))\partial_{y_l}\partial_{y_m} R^q(\mu)\varphi_q f\|_{WF,0,\gamma} \leq S C_\alpha \epsilon^{1-\widetilde{\gamma}} \|f\|_{WF,0,\gamma}. \tag{11.29}$$

We can now fix $\epsilon > 0$ so that the coefficient

$$SC_\alpha \epsilon^{1-\widetilde{\gamma}} \leq \frac{1}{4}.$$

The commutators are first order operators:

$$[L, \psi_q]f = 2 \sum_{l,m=1}^{n} c_{lm}(y)\partial_{y_l}\psi_q\partial_{y_m} f + \sum_{l=1}^{n} d_l(y)(\partial_{y_l}\psi_q)f. \tag{11.30}$$

These terms, along with that defined by the vector fields $\{V^q\}$, are controlled using the estimates in (9.168) and (9.169). These estimates show that, for some positive ν, there is a constant C_α so that

$$\|[L, \psi_q]\varphi_q f\|_{WF,0,\gamma} + \|V^q\varphi_q f\|_{WF,0,\gamma} \leq \frac{\epsilon^{-\nu}C_\alpha}{|\mu|^{\frac{\gamma}{2}}}\|f\|_{WF,0,\gamma}. \tag{11.31}$$

Combining these estimates gives

$$\|E_\mu f\|_{WF,0,\gamma} \leq SC_\alpha \left[\frac{\epsilon^{-\nu}}{|\mu|^{\frac{\gamma}{2}}} + \epsilon^{1-\widetilde{\gamma}}\right]\|f\|_{WF,0,\gamma}. \tag{11.32}$$

This shows that there is an R_0 so that if $|\mu| > R_0$, then the norm of E_μ is less than $\frac{1}{2}$, and therefore $(\mathrm{Id} - E_\mu)$ is well defined as an operator from $\mathscr{C}^{0,\gamma}_{WF}(P)$ to itself. The analytic dependence on μ follows from the analyticity of $\mathrm{Id} - E_\mu$ and the uniform norm convergence of the Neumann series. If we define

$$R(\mu)f = \widehat{R}(\mu)(\mathrm{Id} - E_\mu)^{-1}f, \tag{11.33}$$

then we see that, for any $f \in \mathscr{C}^{0,\gamma}_{WF}(P)$, we have that

$$R(\mu)f \in \mathscr{C}^{0,2+\gamma}_{WF}(P) \text{ and } (\mu - L)R(\mu)f = f. \tag{11.34}$$

Finally, it is a classical result that for $\operatorname{Re}\mu > 0$, on a compact manifold, the only solution of the equation

$$(\mu - L)f = 0 \tag{11.35}$$

is $f \equiv 0$, and therefore $(\mu - L) : \mathscr{C}^{0,2+\gamma}_{WF}(P) \to \mathscr{C}^{0,\gamma}_{WF}(P)$ is a one-to-one and onto mapping. The open mapping theorem implies that $R(\mu)$ is also a left inverse. Hence the identity

$$R(\mu)(\mu - L) = \mathrm{Id} = (\mu - L)R(\mu) \tag{11.36}$$

holds in the connected component where $R(\mu)$ is analytic, which contains the right half plane. We have shown that this set contains the complement of a conic neighborhood of $(-\infty, 0]$.

The first estimate in (11.15) follows from the maximum principle. As a map from $\mathscr{C}^{0,\gamma}_{WF}(P)$ to itself $R(\mu)$ is compact, and therefore the spectrum of L acting on

$\mathscr{C}^{0,2+\gamma}_{WF}(P)$ is a discrete set E. We have shown that E is contained in a conic neighborhood of $(-\infty, 0]$.

Arguing as in Section 10.8 we can show that, for any $0 < \phi$, there are constants $\{\alpha_k, \beta_k, R_k\}$, with $\alpha_k < 1$, so that, if $|\arg\mu| < \pi - \phi$, and $|\mu| > R_k$, then

$$\|E_\mu f\|_{WF,k,\gamma} \leq \alpha_k \|f\|_{WF,k,\gamma} + \beta_k \|f\|_{WF,k-1,\gamma}. \tag{11.37}$$

Applying Theorem 10.8.1 we see that the Neumann series for $(\mathrm{Id} - E_\mu)^{-1}$ converges in the operator norm defined by $\mathscr{C}^{k,\gamma}_{WF}(P)$, thus establishing that these results extend to show that, for any $k \in \mathbb{N}$, the maps

$$R(\mu) : \mathscr{C}^{k,\gamma}_{WF}(P) \longrightarrow \mathscr{C}^{k,2+\gamma}_{WF}(P) \tag{11.38}$$

are also bounded. The estimates in the statement of the theorem, (11.15) and (11.17) for $k > 0$ follow easily since it is simply a matter of establishing these estimates for $\widehat{R}(\mu)$. For example, using that the second estimate in (11.15) holds for $\widehat{R}(\mu)$, we see that

$$\begin{aligned} \|R(\mu) f\|_{WF,k,\gamma} &= \|\widehat{R}(\mu)(\mathrm{Id} - E_\mu^{-1}) f\|_{WF,k,\gamma} \\ &\leq \frac{C_\alpha}{|\mu|} \|(\mathrm{Id} - E_\mu^{-1}) f\|_{WF,k,\gamma} \\ &\leq \frac{C'_\alpha}{|\mu|} \|f\|_{WF,k,\gamma}. \end{aligned} \tag{11.39}$$

As the other estimates hold for $\widehat{R}(\mu)$ it follows by the same sort of argument that they also hold for $R(\mu)$.

Suppose that $\mu \in E$, and $f_\mu \in \mathscr{C}^{0,2+\gamma}_{WF}(P)$ is a non-trivial eigenfunction, with

$$(\mu - L) f_\mu = 0. \tag{11.40}$$

If we select ν so that $\mathrm{Re}(\nu + \mu)$ is sufficiently large, then the eigenvalue equation implies that

$$f_\mu = \nu R(\mu + \nu) f_\mu. \tag{11.41}$$

Since $\mathscr{C}^{k,2+\gamma}_{WF}(P) \subset \mathscr{C}^{k+1,\gamma}_{WF}(P)$, we can use (11.38) in a bootstrap argument to conclude that

$$f_\mu \in \mathscr{C}^{k,\gamma}_{WF}(P) \text{ for all } k. \tag{11.42}$$

From this we conclude that $f_\mu \in \mathscr{C}^\infty(P)$, and the spectrum of L acting on $\mathscr{C}^{k,2+\gamma}_{WF}(P)$, does not depend on k.

11.1.2 The Induction Argument

The proof now proceeds by induction on the maximal codimension of the components of bP. Suppose that the theorem has been proved for all pairs $(\widetilde{P}, \widetilde{L})$ where $\widetilde{P}$ is a manifold with corners, with the maximal codimension of $b\widetilde{P}$ at most M, and $\widetilde{L}$ is a generalized Kimura diffusion on $\widetilde{P}$. We let P be a manifold with corners where the maximal codimension of bP is $M+1$ and L be a generalized Kimura diffusion on P. The parametrix $\widehat{R}(\mu)$ for $R(\mu)$ is constructed as in Section 10.4.2, with $\widehat{R}(\mu) = \widehat{R}_b(\mu) + \widehat{R}_i(\mu)$. As

$$\widehat{R}_i(\mu) = \int_0^\infty e^{-t\mu} \psi e^{t\widetilde{L}}(1-\varphi)dt, \tag{11.43}$$

the induction hypothesis implies that $\widehat{R}_i(\mu)$ has an analytic extension to $\mathbb{C} \setminus F$, where F is a discrete set lying in a conic neighborhood of $(-\infty, 0]$.

The only change is that, instead of (10.130), we let

$$\widehat{R}_b(\mu) = \sum_{i=1}^{p} \sum_{q \in F_{i,\epsilon}} \psi_{i,q} R^{\boldsymbol{b}}_{i,q}(\mu) \chi_{i,q}, \tag{11.44}$$

where

$$R^{\boldsymbol{b}}_{i,q}(\mu) = \int_0^\infty e^{-\mu t} k^{\boldsymbol{b},t}_{i,q} dt, \tag{11.45}$$

with $k^{\boldsymbol{b},t}_{i,q}$ the solution operator for the model problem

$$(\partial_t - L_{i,q})v = 0 \text{ with } v(p,0) = f(p). \tag{11.46}$$

The error terms are quite similar to those arising in the previous case. If

$$w_{i,q} = \psi_{i,q} R^{\boldsymbol{b}}_{i,q}(\mu) \chi_{i,q} f, \tag{11.47}$$

then

$$(\mu - L_{i,q}) w_{i,q} = \chi_{i,q} f + \psi_{i,q}(L_{i,q} - L) R^{\boldsymbol{b}}_{i,q}(\mu)[\chi_{i,q} f] + [\psi_{i,q}, L] R^{\boldsymbol{b}}_{i,q}(\mu)[\chi_i^q f]. \tag{11.48}$$

We again use the decomposition of $L_{i,q} - L$ given in (10.118) to write the error terms as

$$\begin{aligned} E^2_{i,q,\mu} f &= \psi_{i,q} L^r_{i,q} R^{\boldsymbol{b}}_{i,q}(\mu)[\chi_{i,q} f], \\ E^1_{i,q,\mu} f &= \left\{ [\psi_{i,q}, L] + \sum_{l=1}^{m} d_l(\boldsymbol{x}, \boldsymbol{y}) \partial_{y_l} \right\} R^{\boldsymbol{b}}_{i,q}(\mu)[\chi_{i,q} f]. \end{aligned} \tag{11.49}$$

Using the estimates in (9.173) and the argument from Section 10.4.2 we conclude that there is a constant C_α so that if $|\arg \mu| < \pi - \alpha$, then

$$\|E^2_{i,q,\mu} f\|_{WF,0,\gamma} \leq C_\alpha \epsilon^{1-\gamma-\gamma'} \|f\|_{WF,0,\gamma}, \tag{11.50}$$

where $\gamma + \gamma' < 1$. Once again there is an S independent of $\epsilon > 0$, so that

$$\|E^2_{\epsilon,\mu} f\|_{WF,0,\gamma} \leq S C_\alpha \epsilon^{1-\gamma-\gamma'} \|f\|_{WF,0,\gamma}, \tag{11.51}$$

with

$$E_{\epsilon,\mu}^2 = \sum_{i=1}^{p} \sum_{q \in F_{i,\epsilon}} E_{i,q,\mu}^2. \tag{11.52}$$

We can now fix $\epsilon > 0$ so that $SC_\alpha \epsilon^{1-\gamma-\gamma'} < \frac{1}{4}$.

The commutators are of the form

$$[\psi_{i,q}, L] = \sum_{i=1}^{M+1} b_i(\boldsymbol{x}, \boldsymbol{y}) x_i \partial_{x_i} + \sum_{l=1}^{n-(M+1)} d_l'(\boldsymbol{x}, \boldsymbol{y}) \partial_{y_l}. \tag{11.53}$$

The estimates in (9.169) and (9.170), along with the argument in Section 10.4.2, show that there are constants C_α and ν so that

$$\|E_{\epsilon,\mu}^1 f\|_{WF,0,\gamma} \leq SC_\alpha \frac{\epsilon^{-\nu}}{|\mu|^{\frac{\gamma}{2}}} \|f\|_{WF,0,\gamma}. \tag{11.54}$$

With the given choice of $\epsilon > 0$, we again define φ as in (10.151). With this choice we have the estimate

$$\begin{aligned} \|(\mu - L)\widehat{R}_b(\mu) f - \varphi f\|_{WF,0,\gamma} = \|E_{b,\mu} f\|_{WF,0\gamma} \leq \\ SC_\alpha \left[\frac{\epsilon^{-\nu}}{|\mu|^{\frac{\gamma}{2}}} + \epsilon^{1-\gamma-\gamma'} \right] \|f\|_{WF,0,\gamma}. \end{aligned} \tag{11.55}$$

We now proceed exactly as in Section 10.3: let U be a neighborhood of Σ_{M+1} with $\overline{U} \subset\subset \operatorname{int} \varphi^{-1}(1)$. As before we apply Theorem 10.2.1 to find $(\widetilde{P}, \widetilde{L})$ so that the maximal codimension of $b\widetilde{P}$ is M and (P_U, L_U) is embedded into $(\widetilde{P}, \widetilde{L})$. We let $\widetilde{R}(\mu)$ be the resolvent operator for L_U, whose existence and properties follow from the induction hypothesis. Finally we choose ψ, a smooth function equal to 1 on the support of $(1 - \varphi)$, compactly supported in U^c, and let

$$\widehat{R}_i(\mu) f = \psi \widetilde{R}(\mu)[(1 - \varphi) f], \tag{11.56}$$

and

$$\widehat{R}(\mu) f = \widehat{R}_i(\mu) f + \widehat{R}_b(\mu) f. \tag{11.57}$$

We see that

$$(\mu - L)\widehat{R}(\mu) f = f - [L, \psi]\widetilde{R}(\mu)[(1 - \varphi) f] + E_{b,\mu} f. \tag{11.58}$$

The commutator $[L, \psi]$ is a vector field of the form (11.16) in each adapted coordinate frame. Hence the induction hypothesis implies that there is a constant C_α so that

$$\|[L, \psi]\widetilde{R}(\mu)[(1 - \varphi) f]\|_{WF,0,\gamma} \leq \frac{C_\alpha \epsilon^{-\nu}}{|\mu|^{\frac{\gamma}{2}}} \|f\|_{WF,0,\gamma}. \tag{11.59}$$

Altogether this shows that with

$$-E_\mu f = (\mu - L)\widehat{R}(\mu) f - f \tag{11.60}$$

we have

$$\|E_\mu f\|_{WF,0,\gamma} \le SC_\alpha \left[\frac{\epsilon^{-\nu}}{|\mu|^{\frac{\gamma}{2}}} + \epsilon^{1-\gamma-\gamma'} \right] \|f\|_{WF,0,\gamma}. \tag{11.61}$$

Thus, we can choose R_α so that if $|\mu| > R_\alpha$ then

$$SC_\alpha \left[\frac{\epsilon^{-\nu}}{|\mu|^{\frac{\gamma}{2}}} + \epsilon^{1-\gamma-\gamma'} \right] < \frac{1}{2}, \tag{11.62}$$

so that the Neumann series for $(\mathrm{Id} - E_\mu)^{-1}$ converges in the operator norm topology defined by $\mathscr{C}^{0,\gamma}_{WF}(P)$ in the set

$$|\arg \mu| < \pi - \alpha \text{ and } |\mu| > R_\alpha. \tag{11.63}$$

It is clear that the family of operators $(\mathrm{Id} - E_\mu)^{-1}$ is analytic in a conic neighborhood of $(-\infty, 0]$.

If we let

$$R(\mu) = \widehat{R}(\mu)(\mathrm{Id} - E_\mu)^{-1}, \tag{11.64}$$

then this is an analytic family of operators, mapping $\mathscr{C}^{0,\gamma}_{WF}(P)$ to $\mathscr{C}^{0,2+\gamma}_{WF}(P)$, which satisfies

$$(\mu - L)R(\mu)f = f \text{ for } f \in \mathscr{C}^{0,\gamma}_{WF}(P). \tag{11.65}$$

As before, Corollary 10.5.3 shows that $(\mu - L)$ is injective for μ in the right half plane. The open mapping theorem then implies that $(\mu - L)$ is actually invertible for $\operatorname{Re} \mu > 0$, and therefore

$$R(\mu)(\mu - L)f = f \text{ for } f \in \mathscr{C}^{0,2+\gamma}_{WF}(P), \tag{11.66}$$

as well. As noted in the compact manifold case, the fact that $R(\mu)$ satisfies all the estimates in (11.15) and (11.17) follows immediately from the boundedness of

$$(\mathrm{Id} - E_\mu)^{-1} : \mathscr{C}^{0,\gamma}_{WF}(P) \longrightarrow \mathscr{C}^{0,\gamma}_{WF}(P), \tag{11.67}$$

and the fact that $\widehat{R}(\mu)$ satisfies these estimates. This latter claim follows from the fact that the model operators satisfy these estimates, and, by the induction hypothesis, so does $\widetilde{R}(\mu)$. This completes the induction step in the $k = 0$ case.

The cases where $k > 0$ are quite similar to that treated in Section 10.8. This case is somewhat simpler, as we do not need to estimate time derivatives. This means that we only need to use the formulæ in (4.27) with $j = 0$. In this case powers of $L_{\boldsymbol{b},m}$ do not appear on the right-hand side, and no hypothesis is required on the support of the data. The only other significant difference concerns the higher order estimates in (11.17). The contributions of the interior terms are estimated, for all k, by using the induction hypothesis and the fact that the commutator $[\psi, L]$ is of the form given in (11.16). The estimates in (9.170) give

$$\|\boldsymbol{x} \cdot \nabla_{\boldsymbol{x}} \widehat{R}_b(\mu) f\|_{WF,k,\gamma} + \|\nabla_{\boldsymbol{y}} \widehat{R}_b(\mu) f\|_{WF,k,\gamma} \le C_\alpha \epsilon^{-\nu} \left[\frac{1}{|\mu|^{\frac{\gamma}{2}}} + \frac{1}{|\mu|} \right] \|f\|_{WF,k,\gamma}. \tag{11.68}$$

For $|\mu| > 1$, this therefore gives the desired estimate, and completes the verification of the induction hypothesis for $M + 1$. This completes the proof of the theorem. □

11.2 HOLOMORPHIC SEMI-GROUPS

Now that we have constructed the resolvent operator for L and demonstrated that it is analytic in the complement of a conic neighborhood of $(-\infty, 0]$, we can use contour integration to construct the solution to the heat equation. Our second pass through this problem represents a distinct improvement over our previous result for several reasons:

1. This time we can work with data belonging to $\mathscr{C}^{0,\gamma}_{WF}(P)$, rather than $\mathscr{C}^{0,2+\gamma}_{WF}(P)$.
2. For such data the solution is shown to belong to $\mathscr{C}^{0,2+\gamma}_{WF}(P)$ for positive times. If the data is in $\mathscr{C}^{k,\gamma}_{WF}(P)$, then the solution belongs to $\mathscr{C}^{k,2+\gamma}_{WF}(P)$.
3. A bootstrapping argument, using the inclusion

$$\mathscr{C}^{k,2+\gamma}_{WF}(P) \subset \mathscr{C}^{k+1,\gamma}_{WF}(P)$$

and the semi-group property, gives that the solution belongs to $\mathscr{C}^{\infty}(P)$, for positive times.

4. The solution extends analytically to t in the right half plane, H_+.

For any $\alpha > 0$, $0 < \gamma < 1$, and $k \in \mathbb{N}$, there is an $0 < R_\alpha$ so that as a map from $\mathscr{C}^{k,\gamma}_{WF}(P)$ to $\mathscr{C}^{k,2+\gamma}_{WF}(P)$ the operator $R(\mu)$, constructed in Theorem 11.1.1, is analytic in a domain containing the set $|\arg\mu| \leq \pi - \alpha$, and $|\mu| \geq R_\alpha$, and satisfies the estimates in the theorem. We let Γ_α denote the boundary of the complement of this region. From these observations, the following theorem follows from standard results in semi-group theory. See, for example, the proof of Theorem 8.2.1 in [29], or that of Theorem 2.34 in [13].

THEOREM 11.2.1. *For $f \in \mathscr{C}^{k,\gamma}_{WF}(P)$ and t with $|\arg t| < \frac{\pi}{2} - \alpha$, define*

$$v(t, p) = T^t f(p) = \frac{1}{2\pi i} \int\limits_{\Gamma_\alpha} e^{t\mu} R(\mu) f(p) d\mu. \tag{11.69}$$

Then:

1. *For any $p \in P$ the function $t \mapsto v(t, p)$ is analytic in the right half plane, and, for s, t in the right half plane:*

$$T^t T^s f = T^{t+s} f. \tag{11.70}$$

2. *For any $t \in H_+$, $v(t, \cdot) \in \mathscr{C}^{k,2+\gamma}_{WF}(P)$.*

3. *For any* $0 < \alpha$, *there is a* C_α *so that for* t *with* $|\arg t| < \frac{\pi}{2} - \alpha$, *we have the estimates*

$$\|T^t f\|_{WF,k,2+\gamma} \leq C_\alpha \left[e^{R_\alpha |t|} + \frac{1}{|t|} \right] \|f\|_{WF,k,\gamma}. \tag{11.71}$$

4. v *satisfies the heat equation in* $\{t : \operatorname{Re} t > 0\} \times P$:

$$\partial_t v = Lv. \tag{11.72}$$

5. *For* t *real, we have*

$$\|T^t f\|_{L^\infty} \leq \|f\|_{L^\infty} \text{ and } \lim_{t \to 0^+} \|T^t f - f\|_{L^\infty} = 0. \tag{11.73}$$

6. *For* $\tilde{\gamma} < \gamma$, *we have*

$$\lim_{t \to 0^+} \|T^t f - f\|_{WF,k,\tilde{\gamma}} = 0. \tag{11.74}$$

Remark. From the higher order regularity results and a simple integration by parts argument, it follows that if $f \in \mathscr{C}^{2(k+1),\gamma}_{WF}(P)$, for a $k \in \mathbb{N}$ and $0 < \gamma < 1$, then $v(p,t) = e^{tL} f(p)$ is given by the Taylor series, with remainder, for the exponential:

$$v(p,t) = \sum_{j=0}^{k} \frac{t^j L^j f(p)}{j!} + \int_0^t \frac{(t-s)^{j+1}}{(j+1)!} \partial_t^{j+1} v(p,s) ds. \tag{11.75}$$

As noted earlier, the regularity statement in this theorem and the fact that $\mathscr{C}^{k+1,\gamma}_{WF}(P) \subset \mathscr{C}^{k,2+\gamma}_{WF}(P)$ have an important corollary:

COROLLARY 11.2.2. *If, for some* $0 < \gamma$, $f \in \mathscr{C}^{0,\gamma}_{WF}(P)$, *then* $v = T^t f$ *belongs to* $\mathscr{C}^\infty(H_+ \times P)$.

11.3 DIFFUSIONS WHERE ALL COEFFICIENTS HAVE THE SAME LEADING HOMOGENEITY

In this brief section we return to an issue which we have alluded to earlier. Namely, from the points of view of both probability and analysis, it is quite natural to study a slightly broader class of Kimura diffusions with generators of the form

$$L = \sum_{i,j=1}^{n} a_{ij}(\boldsymbol{x},\boldsymbol{y}) \sqrt{x_i x_j} \partial^2_{x_i x_j} + \sum_{i=1}^{n} \sum_{k=1}^{m} \sqrt{x_i} b_{ik}(\boldsymbol{x},\boldsymbol{y}) \partial^2_{x_i y_k} + \sum_{k,l=1}^{m} c_{kl}(\boldsymbol{x},\boldsymbol{y}) \partial^2_{y_k y_l} + V, \tag{11.76}$$

where V is an inward pointing vector field, and assuming a degenerate ellipticity condition on the coefficients. The difference between this expression and the one in the original Definition 2.2.1 is that we allow cross-terms which vanish at the rates $\sqrt{x_i x_j}$ instead of $x_i x_j$. This is not an innocuous change, since as we show below, the cross-terms are no longer negligible error terms, which can be easily absorbed.

It is possible to adapt the methods developed in this monograph to study operators of this more general type. The strategy is essentially the same: one first analyzes the model heat problem and establishes appropriate Hölder estimates, and then uses a similar parametrix construction and perturbation argument to obtain the corresponding estimates for solutions of the actual heat operator $\partial_t - L$. The important difference between this and our earlier work is that the model operator is no longer "diagonal," and hence may be substantially more difficult to analyze. We do not claim to have a general method to carry this out at this time, though we hope to return to this elsewhere. If we assume that the model operator is "nearly diagonal," then one can obtain sufficient information using perturbation arguments. This is similar to what was done by Bass and Perkins [4], who treated a more restricted class of operators on the orthant, $\mathbb{R}^n_+$, instead of a manifold with corners. In particular these operators do not involve tangential, or $\boldsymbol{y}$-variables. For simplicity, we assume that the coefficients a_{ij}, b_{ik} and c_{kl} are smooth functions on P. This hypothesis can be relaxed substantially, but, at present, we are concerned with issues connected to the rate of vanishing of the off-diagonal terms.

First let us comment on the ellipticity of L. Under the change of variables $\xi_i = 2\sqrt{x_i}$, we have

$$\sqrt{x_i}\partial_{x_i} = \partial_{\xi_i}, \quad \sqrt{x_i x_j}\,\partial^2_{x_i x_j} = \begin{cases} \partial^2_{\xi_i \xi_j} & i \neq j \\ \partial^2_{\xi_i} + \frac{1}{2\xi_i}\partial_{\xi_i} & \text{otherwise,} \end{cases}$$

hence the second order part of L becomes

$$\sum_{i,j=1}^{n} a_{ij}\partial^2_{\xi_i \xi_j} + \sum_{i=1}^{n}\sum_{k=1}^{m} b_{ik}\partial^2_{\xi_i y_k} + \sum_{k,l=1}^{m} c_{kl}\partial^2_{y_k y_l}.$$

Thus ellipticity corresponds to the positive definiteness of the matrix

$$A = \begin{pmatrix} a_{11} & \dots & a_{1n} & \frac{1}{2}b_{11} & \dots & \frac{1}{2}b_{1m} \\ \vdots & \dots & \vdots & \vdots & \dots & \vdots \\ a_{n1} & \dots & a_{nn} & \frac{1}{2}b_{n1} & \dots & \frac{1}{2}b_{nm} \\ \frac{1}{2}b_{11} & \dots & \frac{1}{2}b_{n1} & c_{11} & \dots & c_{1m} \\ \vdots & \dots & \vdots & \vdots & \dots & \vdots \\ \frac{1}{2}b_{1m} & \dots & \frac{1}{2}b_{nm} & c_{m1} & \dots & c_{mm} \end{pmatrix}.$$

The fact that the $\boldsymbol{x}$ and $\boldsymbol{y}$ directions are no longer orthogonal when $\boldsymbol{x} = 0$ means that any linear change of coordinates which diagonalizes this matrix will not respect the corner product structure.

We must also consider the first order terms. If $V = \sum b_i \partial_{x_i} + \sum d_k \partial_{y_k}$, then in these new coordinates, the first order part of L becomes

$$\sum_{j=1}^{n} (b_j - \frac{1}{2}) \frac{1}{\xi_j} \partial_{\xi_j} + \sum d_k \partial_{y_k}.$$

Thus the singularity of the full equation cannot be transformed away except in the very special case where $b_j \equiv \frac{1}{2}$ for every $j = 1, \ldots, n$.

Let us now revert to the original coordinate system $(\boldsymbol{x}, \boldsymbol{y})$. The model operator for L at the origin in these coordinates is defined to be

$$\begin{aligned} L_A = &\sum_{i,j=1}^{n} a_{ij}(\mathbf{0}, \mathbf{0}) \sqrt{x_i x_j} \partial^2_{x_i x_j} + \sum_{i=1}^{n} \sum_{k=1}^{m} \sqrt{x_i} b_{ik}(\mathbf{0}, \mathbf{0}) \partial^2_{x_i y_k} \\ &+ \sum_{k,l=1}^{m} c_{kl}(\mathbf{0}, \mathbf{0}) \partial^2_{y_k y_l} + \sum_{i=1}^{n} b_i(\mathbf{0}, \mathbf{0}) \partial_{x_i} + \sum_{j=1}^{m} e_k(\mathbf{0}, \mathbf{0}) \partial_{y_k}. \end{aligned} \tag{11.77}$$

Here we no longer think of $\boldsymbol{x}$ and $\boldsymbol{y}$ as local coordinates, but rather as global linear coordinates on $\mathbb{R}^n_+ \times \mathbb{R}^m$. The key fact which should make it possible to obtain detailed information about L_A is that it has a number of symmetries. First, it is translation invariant in $\boldsymbol{y}$. In addition, provided the coefficients $e_k(\mathbf{0}, \mathbf{0})$ all vanish, it is also homogeneous of degree -1 with respect to the dilation $(\boldsymbol{x}, \boldsymbol{y}) \mapsto (\lambda \boldsymbol{x}, \sqrt{\lambda} \boldsymbol{y})$. However, even reducing by these symmetries, we are typically left with a multidimensional problem (at least provided $n > 1$), which may be quite difficult to solve. The one case where the solution is explicit is precisely the case we have already considered, where the b_{ik} all vanish and (a_{ij}) is diagonal. For in this case, the operator is a sum of 1-dimensional operators and hence its heat kernel is a product of the corresponding 1-dimensional heat kernels.

We say that the coefficient matrix A is near-diagonal if there is a linear change of coordinates of $S_{n,m}$ which preserves the orthant structure in $\mathbb{R}^n_+$ such that the transformation A' of A in these new coordinates satisfies

$$||A' - \mathrm{Id}|| \leq \eta \tag{11.78}$$

for some constant $\eta \ll 1$. For simplicity we call this transformed matrix simply A again. We now explain how it is possible, under this hypothesis, to find a solution of the model heat equation

$$(\partial_t - L_A)w = g, \qquad w|_{t=0} = f \tag{11.79}$$

which satisfies the same Hölder estimates as in the diagonal case.

Again to simplify notation, the coefficients in the first order vector field are still denoted b_i after this linear change of coordinates, and we also only consider the case that the coefficients e_k all vanish. This last hypothesis is less important since if some $b_i \neq 0$, then the affine change of coordinates $\tilde{y} = y - (x_i/b_i)e$ removes these tangential coefficients.

PROPOSITION 11.3.1. *Let $k \in \mathbb{N}_0$, $0 < R$, $(b_1, \ldots, b_n) \in \mathbb{R}_+^n$, and $0 < \gamma < 1$. Suppose that* (11.78) *holds. If the initial data f lies in $\mathscr{C}_{WF}^{k,\gamma}(\mathbb{R}_+^n \times \mathbb{R}^m)$, and is supported in $B_R^+(\mathbf{0}) \times B_R(\mathbf{0})$, and similarly the inhomogeneous term g lies in $\mathscr{C}_{WF}^{k,\gamma}(\mathbb{R}_+^n \times \mathbb{R}^m \times \mathbb{R}_+)$ and is also supported in the same product of balls for all $t \leq T$, then the solution w to* (11.79) *can be written as a sum $v + u$ with $v \in \mathscr{C}_{WF}^{k,\gamma}(\mathbb{R}_+^n \times \mathbb{R}^m \times [0, T])$ and $u \in \mathscr{C}_{WF}^{k,2+\gamma}(\mathbb{R}_+^n \times \mathbb{R}^m \times [0, T])$. Moreover, there are constants $C_{\boldsymbol{b},\gamma,R}$ so that*

$$\|v\|_{WF,k,\gamma} + \|u\|_{WF,k,2+\gamma} \leq C_{\boldsymbol{b},\gamma,R}\left[(1+T)\|g\|_{WF,k,2+\gamma} + \|f\|_{WF,k,\gamma}\right]. \quad (11.80)$$

If $f \in \mathscr{C}_{WF}^{k,2+\gamma}(\mathbb{R}_+^n \times \mathbb{R}^m)$, then v belongs to $\mathscr{C}_{WF}^{k,2+\gamma}(\mathbb{R}_+^n \times \mathbb{R}^m \times \mathbb{R}_+)$, and

$$\|v\|_{WF,k,2+\gamma} + \|u\|_{WF,k,2+\gamma} \leq C_{\boldsymbol{b},\gamma,R}\left[(1+T)\|g\|_{WF,k,2+\gamma} + \|f\|_{WF,k,2+\gamma}\right] \quad (11.81)$$

for some constant $C_{\boldsymbol{b},\gamma,R}$. These constants are uniformly bounded for $\boldsymbol{b} \leq B\mathbf{1}$.

PROOF. We consider only the case $f = 0$ and $k = 0$ for simplicity. Write $L_A = L_{\boldsymbol{b},m} + Q_A$, where Q_A is the second order operator with off-diagonal components, and then define

$$u_{j+1} = \int_0^t e^{(t-s)L_{\boldsymbol{b},m}}(g(s,\cdot) + Q_A u_j(s,\cdot))\,ds.$$

We claim that this sequence of functions converges in $\mathscr{C}_{WF}^{0,2+\gamma}$. This is accomplished by a standard contraction mapping argument. If we let

$$\hat{u}_j(t,\cdot) = \int_0^t e^{(t-s)L_{\boldsymbol{b},m}} Q_A u_j(s,\cdot)\,ds, \quad (11.82)$$

then $\hat{u}_j(0) = 0$ and $(\partial_t - L_{\boldsymbol{b},m})\hat{u}_j = Q_A u_j$. We use the estimates in Proposition 9.2.1 and the fact that

$$||\sqrt{x_i x_j}\partial_{x_i x_j}^2 u||_{WF,0,\gamma} \leq C\|u\|_{WF,0,2+\gamma},$$

where C depends on only the constants above, to obtain

$$\|u_{j+1} - u_j\|_{WF,0,2+\gamma} \leq C\eta\|u_j - u_{j-1}\|_{WF,0,2+\gamma},$$

where $\eta = \max\{|a_{ij}| : i \neq j\}$ and the constant C is independent of these coefficients. We may then choose η so that $C\eta < \frac{1}{2}$, which gives convergence of the sequence, and hence a solution to the problem.

The estimates for the problem when $f \not\equiv 0$ and when $k > 0$ are all handled in a similar manner and we leave details to the reader. □

Once we have established these estimates for the model problem, it is then possible to follow the arguments from the previous two chapters to establish the existence of the heat kernel for the generalized Kimura diffusion problem with generator L having sufficiently small off-diagonal terms near each point of bP. The exact smallness condition depends on the geometry of P and the "size" of the diagonal terms in normalized coordinates. The full details needed to verify all these facts are very lengthy. As before, the main point, which we leave the reader to verify, is that one may use perturbation arguments in the WF-Hölder spaces to pass from the solution operator for the model problem to the solution operator for the general problem on manifold with corners. This in turn only requires the estimates for the model problems that we have just generalized to the near-diagonal case, and general facts above these function spaces established in Section 10.1.1. As before, we may also use the parametrix for the heat kernel and the Laplace transform to perturbatively construct the resolvent for L.

The reader may well wonder if there is an essential difference between the theory for the operator L in this more general setting and the simpler version we have been studying in the rest of this monograph. The answer is not obvious, but it is likely that the key new phenomena involve the regularity of solutions of $Lu = 0$ (or $(\partial_t - L)u = 0$) near the corners where some combination of the x_i vanish. We expect that the regularity of solutions near the boundary hypersurfaces where only one of the x_i vanish does not change much.

Chapter Twelve

The Semi-group on $\mathscr{C}^0(P)$

In the previous chapters we have dealt almost exclusively with solutions to (10.1) with inhomogeneous terms f and g in the WF-Hölder spaces. As explained early in this monograph, the reason for working in Hölder spaces in the first place is to handle the perturbation theory in passing from the model operator to the actual one. The original problem, suggested by applications to population genetics, is to study (10.1) with $g = 0$ and $f \in \mathscr{C}^0(P)$. As noted earlier, the existence theory we have developed suffices to prove that the $\mathscr{C}^0(P)$-graph closure of L with domain $\mathscr{C}^{0,2+\gamma}_{WF}(P)$, for any $0 < \gamma$, is the generator of a C_0-semi-group on $\mathscr{C}^0(P)$. We let $\overline{L}$ denote this operator. As noted earlier this suffices to establish the uniqueness of the solution to the martingale problem, and the weak uniqueness of the solution to associated stochastic differential equation, which leads to the existence of a strong Markov process, whose paths are confined to P, almost surely.

Perhaps surprisingly, the refined regularity of solutions with initial data in $\mathscr{C}^0(P)$ does not seem to follow easily from all that we have accomplished thus far. In fact, if the Cauchy data f is continuous but has no better regularity, then it is not clear that the solution u gains any smoothness along bP at times $t > 0$. Of course, we do know that u becomes smooth if $f \in \mathscr{C}^{0,\gamma}_{WF}(P)$; we also know that solutions to the model problem $(\partial_t - L_{b,m})$ on $\mathbb{R}^n_+ \times \mathbb{R}^m$ with continuous initial data become smooth. While it seems quite likely that this also holds for continuous initial data, it does not seem easy to prove this for general Kimura diffusions using the present methods. There are related difficulties concerning the graph closure $\overline{L}$ of L on $\mathscr{C}^0$. For example, it is not clear that the resolvent of $\overline{L}$ is compact. We will return to these questions in a later publication. In this chapter we establish several properties of the elements of $\mathrm{Dom}(\overline{L})$ and features of the adjoint operator that can be deduced from the analysis above. We also consider the existence of irregular solutions to the equations $Lu = f$, for functions that do not belong to the range of $\overline{L}$.

The graph closure $\overline{L}$ on $\mathscr{C}^0(P)$ is the unique linear operator defined on the dense subspace $\mathrm{Dom}(\overline{L}) \subset \mathscr{C}^0(P)$ characterized by the condition that $u \in \mathrm{Dom}(\overline{L})$ and $\overline{L}u = f \in \mathscr{C}^0(P)$ if there exists a sequence $< u_j > \subset \mathscr{C}^{0,2+\gamma}_{WF}(P)$ such that $Lu_j = f_j$ and

$$u_j \to u \text{ and } f_j \to f \text{ in } \mathscr{C}^0(P). \tag{12.1}$$

Since L is a non-degenerate elliptic operator away from bP, it is standard that any $u \in \mathrm{Dom}(\overline{L})$ is "almost" twice differentiable in int P in the sense that

$$\mathrm{Dom}(\overline{L}) \subset \bigcap_{1<p<\infty} W^{2,p}_{\mathrm{loc}}(\mathrm{int}\, P) \tag{12.2}$$

(see [31]). As is well-known, there is no completely explicit way to characterize the regularity of elements of this domain in the interior, but this is not particularly important here. The more interesting difficulties are connected with describing the boundary behavior of elements of $\text{Dom}(\overline{L})$, and we turn to this now.

We recall from Chapter 3 that under the assumption that L meets bP cleanly, L is either tangent to a hypersurface boundary of P or uniformly transverse. We let $bP^T(L)$ denote the components of bP to which L is tangent. We also recall the notion of the minimal and terminal boundary components, $bP_{\text{min}}(L)$ and $bP_{\text{ter}}(L)$: $bP_{\text{min}}(L)$ consists of boundary components that are themselves manifolds without boundary, and $bP^T_{\text{min}}(L)$, elements of $bP_{\text{min}}(L)$ to which L is tangent. The terminal boundary, $bP_{\text{ter}}(L)$, consists of $bP^T_{\text{min}}(L)$, and boundary components, Σ, to which L is tangent, such that L_Σ is transverse to $b\Sigma$.

Even without the cleanness assumption, if Σ is a component of a stratum of bP to which L is tangent, then L_Σ, the restriction of L to Σ, defines a Kimura diffusion operator on $\mathcal{C}^2(\Sigma)$. We can then say something about the behavior of elements of $\text{Dom}(\overline{L})$ near to Σ.

PROPOSITION 12.0.2. *Suppose that L is tangent to Σ, a component of a stratum of bP. If $w \in \text{Dom}(\overline{L})$, then $w\restriction_\Sigma$ lies in* $\text{Dom}(\overline{L}_\Sigma)$, *and*

$$L_\Sigma[w\restriction_\Sigma] = [Lw]\restriction_\Sigma . \tag{12.3}$$

PROOF. This is immediate from the fact that $\text{Dom}(\overline{L})$ is the $\mathcal{C}^0(P)$-graph closure of L acting on $\mathcal{C}^{0,2+\gamma}_{WF}(P)$. If $w \in \text{Dom}(\overline{L})$, then there exists a sequence $< w_n >$ contained in $\mathcal{C}^{0,2+\gamma}_{WF}(P)$ such that

$$\|w - w_n\|_{\mathcal{C}^0(P)} + \|Lw - Lw_n\|_{\mathcal{C}^0(P)} \to 0. \tag{12.4}$$

Clearly, for each n,

$$L_\Sigma[w_n\restriction_\Sigma] = [Lw_n]\restriction_\Sigma . \tag{12.5}$$

By assumption, the sequence on the right converges to $[Lw]\restriction_\Sigma$, and hence the sequence on the left also converges. This shows that $w\restriction_\Sigma \in \text{Dom}(\overline{L}_\Sigma)$, and

$$L_\Sigma[w\restriction_\Sigma] = [Lw]\restriction_\Sigma .$$

□

As already observed by Shimakura, this result implies that

$$[e^{t\overline{L}} f]\restriction_\Sigma = e^{t\overline{L}_\Sigma}[f\restriction_\Sigma], \tag{12.6}$$

so that these boundary components are effectively "decoupled" from the rest of P.

This result gives some information about the behavior of $w \in \text{Dom}(\overline{L})$ in directions transverse to hypersurfaces to which L is tangent. Let Σ be such a hypersurface again, then, in adapted coordinates near an interior point of Σ, L takes the form

$$L = x\partial_x^2 + \tilde{b}(x, y)x\partial_x + \sum_{l=1}^{N-1} xa_{1l}(x, y)\partial_x\partial_{y_l} + (1 + O(x))L_\Sigma. \tag{12.7}$$

Proposition 12.0.2 implies that if $w \in \text{Dom}(\overline{L})$, then

$$\lim_{x\to 0^+}\left[x\partial_x^2+\widetilde{b}(x,\mathbf{y})x\partial_x+\sum_{l=1}^{N-1}xa_{1l}(x,\mathbf{y})\partial_x\partial_{y_l}\right]w(x,\mathbf{y})=0. \tag{12.8}$$

A similar result holds at strata of codimension greater than 1 to which L is tangent.

12.1 THE DOMAIN OF THE ADJOINT

The space $\mathscr{C}^0(P)$ is non-reflexive, which means that the semi-group defined by $\overline{L}^*$ on $\mathcal{M}(P)$ (the Borel measures of finite total variation) may not be strongly continuous at $t = 0$. This is a reflection of the fact that $\text{Dom}(\overline{L}^*)$ may fail to be dense in $\mathcal{M}(P)$. A solution to this problem was introduced by Lumer and Phillips, whereby we consider $e^{t\overline{L}^*}$ acting on a smaller space:

$$\mathcal{M}^{\odot}(P)=\overline{\text{Dom}(\overline{L}^*)}. \tag{12.9}$$

The semi-group $e^{t\overline{L}^*}$, acting on this space, is strongly continuous at $t = 0$. We denote its generator by $\overline{L}^{\odot}$. The domain of $\overline{L}^{\odot}$ is then given by

$$\text{Dom}(\overline{L}^{\odot})=\{\nu\in\text{Dom}(\overline{L}^*):\ \overline{L}^*\nu\in\mathcal{M}^{\odot}(P)\}, \tag{12.10}$$

which Lumer and Phillips show is a dense subset of $\mathcal{M}^{\odot}(P)$. See Chapter 14 of [26].

Let dV be a smooth non-degenerate density on P. It is clear that if $g \in \mathscr{C}_c^\infty(\text{int}\, P)$, then $gdV \in \text{Dom}(\overline{L}^*)$. The closure of this subspace in the strong topology on $\mathcal{M}(P)$ equals $L^1(P)dV$, which is therefore a subspace of $\mathcal{M}^{\odot}(P)$. Because $\overline{L}^{\odot}$ is formally a non-degenerate elliptic operator in the interior of P it is clear that at positive times $e^{t\overline{L}^{\odot}}\nu$ is represented at interior points of P by a measure with a smooth density. From this it is apparent that all elements of $\text{Dom}(\overline{L}^{\odot})$ are represented by absolutely continuous measures in the interior of P. As indicated by Proposition 12.0.2, the behavior of elements of $\text{Dom}(\overline{L}^{\odot})$ along a hypersurface in bP depends crucially upon the first order part of L at that hypersurface.

Let H be a boundary hypersurface. In adapted coordinates $(x, \mathbf{y})$, defined near an interior point of p_0 of H, where $p_0 \leftrightarrow (0, \mathbf{0})$, a generalized Kimura diffusion takes the form:

$$L=x\partial_x^2+b(x,\mathbf{y})\partial_x+\sum_{l=1}^{N-1}xa_{1l}(x,\mathbf{y})\partial_x\partial_{y_l}+(1+O(x))L_H, \tag{12.11}$$

where L_H is a generalized Kimura diffusion tangent to H. Integrating by parts, we see that an element $\nu = gdxd\mathbf{y} \in \text{Dom}(\overline{L}^{\odot})$ that is smooth in int P, and has no support, as a measure, on bP, must satisfy the boundary condition:

$$\lim_{x\to 0^+}[b(x,\mathbf{y})g(x,\mathbf{y})-\partial_x(xg(x,\mathbf{y}))]=0. \tag{12.12}$$

This is called the *no-flux* boundary condition. Generally, this condition forces g to have a complicated singularity along $\{x = 0\}$, i.e., $g(x, \mathbf{y}) \sim x^{b(0,\mathbf{y})-1}$.

Let $P_\epsilon = \{p \in P : \operatorname{dist}(p, bP) \geq \epsilon\}$. Because $\nu \in \mathcal{M}^{\odot}(P)$ is represented by an absolutely continuous measure in the interior of P, we can define the *interior* measure ν_i via the equation

$$\langle f, \nu_i \rangle = \lim_{\epsilon \to 0^+} \langle \chi_{P_\epsilon} f, \nu \rangle. \tag{12.13}$$

The interior measure is represented as gdV, for a function $g \in L^1_{\text{loc}}(P)$. In fact it is easy to see that $\|g\|_{L^1(P)} \leq \|\nu\|$, so that the boundary measure, $\nu_b = \nu - \nu_i$, is also in $\mathcal{M}(P)$. From the construction it is clear that the support of ν_b is contained in bP. If ν is a non-negative measure, then evidently ν_b is as well. We also observe that there must exist a sequence of $\{\epsilon_n\}$ tending to zero for which

$$\lim_{n \to \infty} \int\limits_{bP_{\epsilon_n}} \epsilon_n |g(p)| dS_{\epsilon_n}(p) = 0, \tag{12.14}$$

for otherwise the L^1-norm of g would be infinite.

Suppose that $\nu \in \operatorname{Dom}(\overline{L}^{\odot})$, then $L^*\nu = \mu \in \mathcal{M}^{\odot}(P)$, and therefore

$$L^*\nu = \mu_i + \mu_b. \tag{12.15}$$

It is easy to see, in the sense of distributions in the interior of P, $L^*\nu_i = \mu_i$. We now study the boundary behavior of elements of $\operatorname{Dom}(\overline{L}^{\odot})$ at points in the interiors of hypersurface boundary components. Let $\nu_i = gdV$ and $\mu_i = hdV$. For this calculation we assume that $h \in L^p(P)$, for a $p > 1$. In this case a distributional solution to $L^*(gdV) = hdV$ belongs to $W^{2,p}_{\text{loc}}(\operatorname{int} P)$, which allows us to integrate by parts. For $f \in \mathcal{C}^3(P)$ and ν as above we have that

$$\langle Lf, \nu \rangle = \langle f, \mu_i \rangle + \langle f, \mu_b \rangle. \tag{12.16}$$

With P_ϵ as above, we can also do this calculation by observing that

$$\langle Lf, \nu \rangle = \lim_{\epsilon \to 0^+} \langle Lf, \nu_i \rangle_{P_\epsilon} + \langle Lf, \nu_b \rangle. \tag{12.17}$$

Using the coordinates, $(x, \mathbf{y})$, introduced in (12.11), we let

$$f(x, \mathbf{y}) = \psi(x)\varphi(\mathbf{y}). \tag{12.18}$$

Here ψ is a smooth function equal to 1 at $x = 0$, and zero for $x > \delta > 0$, and φ is a smooth function compactly supported in $\operatorname{int} H$ near to p_0. We can assume that $dV = dxd\mathbf{y}$; using the sequence $\{\epsilon_n\}$ from (12.14), we integrate by parts to see that

$$\lim_{n\to\infty} \langle Lf, \nu_i \rangle_{P_{\epsilon_n}} = \lim_{n\to\infty} \langle f, \mu_i \rangle_{P_{\epsilon_n}} + \lim_{n\to\infty} \int\limits_{bP_{\epsilon_n}} \Big[\partial_x \psi(x)\varphi(\mathbf{y}) x g(x, \mathbf{y}) +$$

$$\varphi(\mathbf{y})\psi(x)\left(b(x, \mathbf{y})g(x, \mathbf{y}) - \partial_x(xg(x, \mathbf{y}))\right) - \sum_l a_l(x, \mathbf{y})\partial_{y_l}\varphi(\mathbf{y}) x \psi(x) g(x, \mathbf{y}) \Big] d\mathbf{y}. \tag{12.19}$$

It follows from (12.19) that the contributions of the first and last terms in the integral over bP_{ϵ_n} tend to zero.

If we suppose that $\partial_x\psi(0) = 0$, then, for test functions of the type under consideration, $Lf \restriction_{\{x=0\}} = L_H\varphi$. If $B_\epsilon(g)dy$ denotes the boundary operator applied to g in (12.19), then we see that

$$\langle\varphi, \mu_b\rangle = \lim_{n\to\infty}\langle\varphi, B_{\epsilon_n}(g)dy\rangle_{bP_{\epsilon_n}} + \langle L_H\varphi, \nu_b\rangle. \tag{12.20}$$

This shows that, in the sense of distributions on H, we have the equation:

$$\lim_{n\to\infty}[B_{\epsilon_n}(g)dy + L_H^*\nu_b] = \mu_b. \tag{12.21}$$

Suppose that $\overline{L}^{\odot}\nu = hdV$, for an $h \in L^p(P)$: that is, in int P, $\overline{L}^{\odot}\nu_i = hdV$, in the classical sense, and $\mu_b = 0$. For $f \in \mathscr{C}^\infty(P)$ with compact support in int P we have that

$$\langle Lf, \nu_i\rangle = \langle f, \overline{L}^{\odot}\nu_i\rangle = \langle f, hdV\rangle. \tag{12.22}$$

Let $f \in \mathscr{C}^\infty(P)$ have support near a point $p_0 \in H$, and suppose that f vanishes on bP. The functions $f(\epsilon, \cdot) \in \mathscr{C}^\infty(H)$ tend to zero in the $\mathscr{C}^\infty$-topology, as $\epsilon \to 0^+$. Therefore, a limiting argument, using (12.21) and (12.14), shows that $\langle Lf, \nu_i\rangle = \langle f, hdV\rangle$. If $f \in \mathscr{C}^\infty(P)$, with $f \restriction_{bP} = 0$, and $\partial_x f = 0$ outside a small neighborhood, $U \subset\subset \text{int}\, H$ of p, then it follows from these observations that

$$\begin{aligned}\langle Lf, \nu\rangle &= \langle Lf \restriction_U, \nu_b\rangle + \langle f, hdV\rangle \\ &= \langle b(0, \cdot)\partial_x f, \nu_b\rangle + \langle f, hdV\rangle.\end{aligned} \tag{12.23}$$

In order for ν to belong to $\text{Dom}(\overline{L}^{\odot})$, there must be a constant C so that

$$|\langle Lf, \nu\rangle| \leq C\|f\|_{L^\infty}. \tag{12.24}$$

If L meets bP cleanly, then this along with (12.23) implies that supp ν_b is disjoint from the interior of any face of bP to which L is transverse. We summarize these calculations in the following proposition:

PROPOSITION 12.1.1. *Suppose that L meets bP cleanly, and* $\nu \in \text{Dom}(\overline{L}^{\odot})$ *solves* $\overline{L}^{\odot}\nu = hdV$, *with* $h \in L^p(P)$, *for a* $p > 1$. *The support of* ν_b *is disjoint from the interior of faces in* $bP^{\pitchfork}(L)$. *If* $\nu_i = gdV$, *then along faces to which L is transverse, g satisfies the boundary condition in* (12.12).

Remark. Thus far we have only shown that the support of ν_b is disjoint from the interiors of faces to which L is transverse. It seems quite likely that supp $\nu_b \subset bP^T$. To prove this statement requires a more refined analysis of the regularity of ν_i and a stronger sense of convergence in (12.21) than that of a sequence of distributions.

Below we show that if L is everywhere transverse to bP, then there is a unique solution $\nu \in \text{Dom}(\overline{L}^{\odot})$ to $\overline{L}^{\odot}\nu = 0$, which is a probability measure. As explained in [27], Section 15.2, there are circumstances where there may be multiple solutions to this equation, which are non-negative and normalizable. Evidently our method picks out the solution that satisfies the boundary condition in (12.12). A more detailed analysis of this and related questions will need to wait for a later publication.

12.2 THE NULL-SPACE OF $\overline{L}^{\odot}$

As noted above, we are, at present, missing the compactness of the resolvent of $\overline{L}$. We can nonetheless give a precise description of the null-space of the adjoint, $\overline{L}^{\odot}$, under the hypothesis that L meets bP cleanly. For the following results it suffices to consider the operator acting on $\mathscr{C}^{0,2+\gamma}_{WF}(P)$, for a $0 < \gamma < 1$.

PROPOSITION 12.2.1. *Suppose that L meets bP cleanly. To each element of* $bP_{\text{ter}}(L)$ *there is an element of the null-space of* $\overline{L}^{\odot}$. *Each is represented by a non-negative measure supported on a component of* $\Sigma \in bP_{\text{ter}}(L)$.

PROOF. For any $0 < \gamma < 1$, denote by L_γ the operator

$$L_\gamma : \mathscr{C}^{0,2+\gamma}_{WF}(P) \longrightarrow \mathscr{C}^{0,\gamma}_{WF}(P). \tag{12.25}$$

We have established that this map is Fredholm; in fact, this map has index zero since it can be deformed amongst Fredholm operators to $L_\gamma - 1$, which is invertible. Thus

$$\dim\ker L_\gamma = \dim\ker L^*_\gamma. \tag{12.26}$$

For the remainder of the argument we fix a $0 < \gamma < 1$.

Consider first the extreme case that L is transverse to every boundary hypersurface. It then follows from Lemma 3.2.7 that $\ker(L_\gamma)$ consists of constant functions. Moreover, using this same lemma, if $f \not\equiv 0$ is continuous and non-negative, then the equation

$$L_\gamma w = f \tag{12.27}$$

is not solvable, since any solution would be a subsolution of L_γ.

The adjoint operator L^*_γ acts canonically as a map from $[\mathscr{C}^{0,\gamma}_{WF}(P)]^*$ to $[\mathscr{C}^{0,2+\gamma}_{WF}(P)]^*$. Since we are still assuming that $bP_{\text{ter}}(L) = P$, $\ker(L_\gamma)$ contains only the constant functions, so there is precisely one non-trivial element $\ell \in [\mathscr{C}^{0,\gamma}_{WF}(P)]^*$, unique up to scaling, which satisfies $L^*_\gamma \ell = 0$. By the Fredholm alternative, the equation $L_\gamma w = f$ is solvable for $f \in \mathscr{C}^{0,\gamma}_{WF}(P)$ if and only if

$$\ell(f) = 0.$$

This means that if $f \in \mathscr{C}^{0,\gamma}_{WF}(P)$ is non-negative (and non-zero), then $L_\gamma w = f$ is not solvable, so that $\ell(f) \neq 0$. We may as well assume that

$$\ell(f) > 0 \tag{12.28}$$

on the set of non-negative functions; we further normalize so that $\ell(1) = 1$.

A priori, we only know that ℓ lies in the dual of a Hölder space, and thus could be a distribution of negative order. If $f \in \mathscr{C}^{0,\gamma}_{WF}(P)$, then

$$f^+(p) = \max\{f(p), 0\} \text{ and } f^-(p) = \min\{f(p), 0\}, \tag{12.29}$$

both lie in this same function space, and therefore (12.28) and our normalization imply that

$$\min f \leq \ell(f) \leq \max f, \tag{12.30}$$

and therefore, for $f \in \mathscr{C}^{0,\gamma}_{WF}(P)$, we have

$$|\ell(f)| \leq \|f\|_{L^\infty}. \tag{12.31}$$

The WF-Hölder spaces are dense in $\mathscr{C}^0(P)$, so ℓ has a unique extension as an element of $[\mathscr{C}^0(P)]'$. By the Riesz-Markov theorem, there is a non-negative Borel measure ν so that

$$\ell(f) = \int_P f(p)d\nu(p). \tag{12.32}$$

As $\mathscr{C}^{0,2+\gamma}_{WF}(P)$ is dense in $\text{Dom}(\overline{L})$ this shows that $\overline{L}^{\odot}\nu = 0$.

The adjoint $\overline{L}^{\odot}$ is elliptic in $\text{int}\, P$, so by standard elliptic regularity,

$$\nu_i = \upsilon_0 dV, \tag{12.33}$$

for some smooth, non-negative function υ_0 on $\text{int}\, P$; here dV is a smooth, nowhere zero density on P. Using (12.28) again, we see that the support of υ_0 is all of P. Note, however, that since $\overline{L}^{\odot}$ can have a zero order part, there is no obvious reason that υ_0 is strictly positive. Proposition 12.1.1 shows that $\text{supp}\, \nu_b$ is disjoint from hypersurfaces in $bP^{\pitchfork}(L)$ and, along such faces, ν_i satisfies the boundary condition in (12.12).

Let us now turn to components $\Sigma \in bP^T_{\text{min}}(L)$. If $\dim \Sigma = 0$, so $\Sigma = p_\Sigma$ is a single point, then the fact that L is tangent to Σ simply means that the restriction $L_\Sigma \equiv 0$. Hence if δ_Σ denotes the functional

$$\langle w, \delta_\Sigma \rangle = w(p_\Sigma), \tag{12.34}$$

then clearly, for $w \in \mathscr{C}^{0,2+\gamma}_{WF}(P)$, we have

$$\langle Lw, \delta_\Sigma \rangle = 0, \tag{12.35}$$

and this equation remains true for $w \in \text{Dom}(\overline{L})$. Hence $\delta_\Sigma \in \text{Dom}(\overline{L}^{\odot})$, and

$$\overline{L}^{\odot}\delta_\Sigma = 0. \tag{12.36}$$

Suppose, on the other hand, that $\dim \Sigma > 0$, i.e.,, Σ is a compact manifold without boundary, and L_Σ is a non-degenerate elliptic operator, without constant term, acting on $\mathscr{C}^2(\Sigma)$. Clearly $L_\Sigma 1 = 0$. On the other hand, the strong maximum principle shows that all solutions to $L_\Sigma w = 0$ are constant. Also from the strong maximum principle, the equation $L_\Sigma w = f$ is not solvable whenever $f \in \mathscr{C}^0(\Sigma)$ is non-negative and not identically zero. Arguing as above for the case that $bP_{\text{ter}}(L) = P$ we conclude that there is a non-negative measure with smooth density, $\nu = \upsilon_0 dV_\Sigma$ that spans the null-space of $\overline{L}^{\odot}_\Sigma$. The functional

$$\ell(w) = \int_\Sigma w d\nu \tag{12.37}$$

defines an element of $\ker \overline{L}^{\odot}$. As before, the support of v_0 is all of Σ.

To complete the construction of $\ker \overline{L}^{\odot}$ we need only consider boundary components $\Sigma \in bP_{\text{ter}}(L) \setminus bP^T_{\min}(L)$. In this case L_Σ is a generalized Kimura diffusion on Σ, and $b\Sigma_{\text{ter}}(L_\Sigma) = \Sigma$. The argument above produces a measure ν with support equal to Σ and such that $\overline{L}^{\odot}_{\Sigma}\nu = 0$. If we define

$$\ell(w) = \int_{\Sigma} w d\nu, \tag{12.38}$$

then $\overline{L}^{\odot}\ell = 0$. This completes the proof of the proposition. □

DEFINITION 12.2.2. *We denote by* $\{\delta_\Sigma : \Sigma \in bP_{\text{ter}}(L)\}$, *the measures, belonging to* $\ker \overline{L}^{\odot}$, *constructed in the proof of Proposition 12.2.1.*

These measures define non-trivial functionals on $\mathscr{C}^0(P) \supset \mathscr{C}^{0,\gamma}_{WF}(P)$, and are certainly linearly independent. This argument shows that $\dim\ker L^*_\gamma \geq |bP_{\text{ter}}(L)|$. On the other hand, by Corollary 3.2.10, $\dim\ker L_\gamma \leq |bP_{\text{ter}}(L)|$.

We summarize all of this in a proposition:

PROPOSITION 12.2.3. *If L meets bP cleanly, then, for any* $0 < \gamma < 1$,

$$\dim\ker L^*_\gamma = \dim\ker L_\gamma = |bP_{\text{ter}}(L)|. \tag{12.39}$$

The $\ker L_\gamma$ *is contained in* $\mathscr{C}^\infty(P)$; *on the other hand,* $\ker L^*_\gamma$ *is spanned by a finite collection of non-negative Borel measures, each of which has a smooth non-negative density supported on one of the terminal boundary components of P. The operator L has no generalized eigenvectors at* 0, *i.e.,, functions* $w \in \mathscr{C}^{0,2+\gamma}_{WF}(P)$ *with* $Lw \in \ker L$.

Remark. If $bP^T_{\min}(L) = \emptyset$, and $bP_{\text{ter}}(L) = P$, then

$$\dim\ker L^*_\gamma = \dim\ker L_\gamma = 1. \tag{12.40}$$

The $\ker L^*_\gamma$ is spanned by a non-negative measure with support all of P. This is the equilibrium measure. If $|bP_{\text{ter}}(L)| > 1$, then instead of a single equilibrium measure, there is a collection of such measures, each supported on one of the terminal components of bP. Zero-dimensional components of $bP_{\text{ter}}(L)$ are classical absorbing states of the underlying Markov process. Higher dimensional components correspond to generalized absorbing states; these are again characterized by an equilibrium measure.

PROOF. Only the last statement still requires proof. If $bP_{\text{ter}}(L) = P$, then $\ker L_\gamma$ consists of constant functions. We observe that $L_\gamma w = 1$ is not solvable, for otherwise w would be a non-trivial subsolution.

Suppose that $bP_{\text{ter}}(L) \neq P$ and that $w \in \mathscr{C}^{0,2+\gamma}_{WF}(P)$ satisfies $Lw = f$, where $Lf = 0$. Then necessarily

$$\langle f, \delta_\Sigma \rangle = 0 \text{ for all } \Sigma \in bP_{\text{ter}}(L). \tag{12.41}$$

However, any $f \in \ker L \cap \mathscr{C}^{0,2+\gamma}_{WF}(P)$ is constant on each component of $bP_{\text{ter}}(L)$. Since each of the measures $\{\delta_\Sigma\}$ is non-negative and non-trivial, Proposition 3.2.8 shows that $f \equiv 0$. □

We have not proved that all elements of $\ker \overline{L}$ belong to $\mathscr{D}^2_{WF}(P)$, nor have we established the Hopf maximum principle for elements of $\text{Dom}(\overline{L})$, hence we cannot presently conclude that $\dim \ker \overline{L} = \dim \ker \overline{L}^{\odot}$. On the other hand, elements of $\ker \overline{L}^{\odot}$ are represented by Borel measures, and furthermore $\ker \overline{L}^{\odot} \subset \ker L^*_\gamma$. Since we have shown that $\ker L^*_\gamma$ is also spanned by Borel measures, we obtain:

PROPOSITION 12.2.4. *If L meets bP cleanly, then*

$$\dim \ker \overline{L}^{\odot} = |bP_{\text{ter}}(L)|. \tag{12.42}$$

The null-space is spanned by Borel measures with support on the components of $bP_{\text{ter}}(L)$.

12.3 LONG TIME ASYMPTOTICS

These observations have several interesting consequences.

PROPOSITION 12.3.1. *If f in $\mathscr{C}^0(P)$, and $e^{t\overline{L}}f$ denotes the action of the semi-group, then the functions*

$$t \mapsto \langle e^{t\overline{L}} f, \delta_\Sigma \rangle \tag{12.43}$$

are constant for every $\Sigma \in bP_{\text{ter}}(L)$.

PROOF. Indeed, this is clear when $f \in \text{Dom}(\overline{L})$ since

$$\begin{aligned} \partial_t \langle e^{t\overline{L}} f, \delta_\Sigma \rangle &= \langle \overline{L} e^{t\overline{L}} f, \delta_\Sigma \rangle \\ &= \langle e^{t\overline{L}} f, \overline{L}^* \delta_\Sigma \rangle \\ &= 0. \end{aligned} \tag{12.44}$$

However, the domain $\text{Dom}(\overline{L})$ is dense in $\mathscr{C}^0(P)$, so for any $f \in \mathscr{C}^0(P)$ we can choose a sequence $< f_n >$ in $\text{Dom}(\overline{L})$ which converges to f in $\mathscr{C}^0(P)$. Then

$$\langle e^{t\overline{L}} f, \delta_\Sigma \rangle = \lim_{n\to\infty} \langle e^{t\overline{L}} f_n, \delta_\Sigma \rangle. \tag{12.45}$$

The right-hand side is independent of t for each n, hence so is the limit. If $bP_{\text{ter}}(L) = P$, then we can also conclude that

$$\langle e^{t\overline{L}} f, \delta_P \rangle \tag{12.46}$$

is constant. □

Remark. A similar observation, for a special case, appears in [9].

We now show that 0 is the only element in the spectrum of L_γ on the imaginary axis, $\text{Re}\,\mu = 0$.

LEMMA 12.3.2. *Let P be a compact manifold with corners, and L a generalized Kimura diffusion on P. If $\varphi \in \mathcal{C}^{0,2+\gamma}_{WF}(P)$ is a non-trivial solution to $L\varphi = i\alpha\varphi$, for $\alpha \in \mathbb{R}$, then $\alpha = 0$.*

PROOF. Let $q_t(x, dy)$ denote the Schwartz kernel for e^{tL}. Then for each $x \in P$,

$$\int_P q_t(x, dy) = 1, \tag{12.47}$$

and $q_t(x, dy)$ is a non-negative measure. By the strong maximum principle, e^{tL} is strictly positivity improving within $\operatorname{int} P$. Hence if $U \subset \operatorname{int} P$ is any open subset, then

$$\int_U q_t(x, dy) > 0 \tag{12.48}$$

for each $x \in \operatorname{int} P$.

Now,

$$e^{it\alpha}\varphi(x) = e^{tL}\varphi = \int_P q_t(x, dy)\varphi(y), \tag{12.49}$$

so by the non-negativity of $q_t(x, dy)$,

$$|\varphi(x)| \leq \int_P q_t(x, dy)|\varphi(y)| = [e^{tL}|\varphi|](x). \tag{12.50}$$

Note also that $|\varphi|$ lies in $\mathcal{C}^{0,\gamma}_{WF}(P)$, so $e^{tL}|\varphi| \in \mathcal{C}^{\infty}(P)$ for $t > 0$. The estimate in (12.50) implies that for any $s > 0$,

$$0 \leq \frac{[e^{(s+t)L}|\varphi|](x) - [e^{sL}|\varphi|](x)}{t}. \tag{12.51}$$

Since $e^{sL}|\varphi| \in \operatorname{Dom}(L_\gamma)$, we can let $t \to 0^+$ to conclude that

$$Le^{sL}|\varphi|(x) \geq 0 \text{ for all } x \in P. \tag{12.52}$$

Choose any non-negative $\psi \in \mathcal{C}^{\infty}_c(\operatorname{int} P)$. Integrating by parts with respect to some smooth non-degenerate density on P gives

$$0 \leq \langle e^{sL}|\varphi|, L^*\psi\rangle, \tag{12.53}$$

so letting $s \to 0^+$, we obtain that

$$0 \leq \langle |\varphi|, L^*\psi\rangle. \tag{12.54}$$

If the support of ψ is further constrained to lie in a set where $|\varphi|$ is smooth, then we can integrate by parts again to conclude that

$$0 \leq \langle L|\varphi|, \psi\rangle. \tag{12.55}$$

In particular, $L|\varphi| \geq 0$ in the open subset of int P where $|\varphi| > 0$.

There are now several cases to consider. If $bP^T(L) = \emptyset$, then Lemma 3.2.6 shows that $|\varphi(x)| \equiv 1$. We have

$$1 = |\varphi(x)| = \left| \int q_t(x, dy)\varphi(y) \right| \leq \int q_t(x, dy)|\varphi(y)| = \int q_t(x, dy) = 1, \quad (12.56)$$

so the inequality in the middle is an equality, and since $\int q_t(x, dy) = 1$, this can only happen if φ has constant phase. This shows that $\alpha = 0$ in this case.

If $bP^T(L) \neq \emptyset$, then we use an induction on the dimension of P. The result has been proved when $\dim P = 1$ in [20] and [15], so we now assume that it is true whenever $\dim P \leq N - 1$. Let P have dimension N and assume that $bP^T(L) \neq \emptyset$. Suppose that φ is a non-trivial solution, as above, and that $\|\varphi\|_{\mathscr{C}^0(P)} = 1$. For each $\Sigma \in bP^T(L)$ we know that

$$L_\Sigma \, \varphi\restriction_\Sigma = i\alpha\varphi\restriction_\Sigma . \quad (12.57)$$

By induction, either $\alpha = 0$ or $\varphi \restriction_\Sigma = 0$. In the former case we are done, so we can reduce to the case that $\varphi\restriction_\Sigma = 0$ for every $\Sigma \in bP^T(L)$.

If L is tangent to every face of bP, then $|\varphi|$ attains its maximal value 1 at some point $x_0 \in \operatorname{int} P$. It follows directly from (12.47) and (12.48) that $|\varphi(x)| \equiv 1$ in int P. For if this were false, then the fact that e^{tL} is strictly positivity improving in int P would show that

$$\int_P q_t(x_0, dy)|\varphi(y)| < 1, \quad (12.58)$$

which contradicts (12.50). By induction φ vanishes on bP, which is clearly impossible, as $\varphi \in \mathscr{C}^0(P)$.

We are left to consider the case where $bP^T(L) \neq bP$. If $|\varphi(x)|$ assumes its maximum in the interior of P then we conclude as above that $|\varphi(x)| \equiv 1$ in P, which leads to the same contradiction as before. Thus $|\varphi(x)|$ must assume the value 1 at $x_0 \in bP \setminus bP^T(L)$. Indeed $x_0 \in bP^{\pitchfork}(L)$, for otherwise x_0 would belong to the closure of $bP^T(L)$ and hence would vanish. Applying Lemma 3.2.6 gives that $|\varphi|$ is identically equal to 1 in a neighborhood U of x_0. Using the previous argument at $x_1 \in U \cap \operatorname{int} P$ gives $|\varphi(x)| \equiv 1$ in P, which contradicts that $\varphi \restriction_{bP^T(L)} = 0$.

Thus the only tenable case is that $\alpha = 0$, as claimed. □

Combining this lemma with Theorem 11.1.1 and Lemma 12.3.2 gives the following corollary:

COROLLARY 12.3.3. *For any* $0 < \gamma < 1$, *the* $\operatorname{spec}(L_\gamma) \setminus \{0\}$ *lies in a half plane* $\operatorname{Re}\mu < \eta < 0$.

Write $N_0 = |bP_{\text{ter}}(L)|$ and let $\{\Sigma_j : j = 1, \ldots, N_0\}$ enumerate the components of $bP_{\text{ter}}(L)$. Also, denote by $\ell_j = \delta_{\Sigma_j}$ the probability measures which span $\ker \overline{L}^{\odot}$, constructed above. From the support properties of these measures, we can choose smooth functions $\{f_j : j = 1, \ldots, N_0\}$ so that

$$\ell_k(f_j) = \delta_{jk}. \quad (12.59)$$

Choose a smooth basis $\{w_j : j = 1, \ldots, N_0\}$ for $\ker L_\gamma$, for any $0 < \gamma < 1$. Corollary 12.3.3 and the fact that L_γ has no generalized eigenvectors at 0, shows that $\operatorname{spec} L_\gamma \setminus \{0\}$ lies in $\operatorname{Re} \mu < \eta$, for an $\eta < 0$. Thus, as the spectrum lies in a conic neighborhood of the negative real axis, we can deform the contour in (11.69) to show there exist continuous linear functionals $\{a_j\}$ such that, for any $f \in \mathscr{C}^{0,\gamma}_{WF}(P)$, we have

$$e^{tL} f = \sum_{j=1}^{N_0} a_j(f) w_j + O(e^{\eta t}). \tag{12.60}$$

Proposition 12.3.1 shows that the quantities $\{\ell_k(e^{tL} f)\}$ are independent of t, so letting $t \to \infty$, we conclude that

$$\ell_k(f) = \sum_{j=1}^{N_0} a_j(f) \ell_k(w_j). \tag{12.61}$$

In light of (12.59), $\ell_k(w_j)$ is an invertible matrix, so we can find a new basis $\{\tilde{w}_j\}$ for $\ker L_\gamma$ so that $\ell_k(\tilde{w}_j) = \delta_{jk}$ and therefore

$$e^{tL} f = \sum_{j=1}^{N_0} \ell_j(f) \tilde{w}_j + O(e^{\eta t}). \tag{12.62}$$

By duality we can conclude that if ν is a Borel measure belonging to $\mathcal{M}^{\odot}(P)$, then

$$e^{tL^*} \nu = \sum_{j=1}^{N_0} \langle \tilde{w}_j, \nu \rangle \ell_j + O(e^{\eta t}). \tag{12.63}$$

The $\{\ell_j\}$ are non-negative measures with disjoint supports. Since the forward Kolmogorov equation maps non-negative measures to non-negative measures, it follows that the eigenfunctions $\{\tilde{w}_j\}$ must be non-negative. As a special case of (12.62) note that

$$1 = \sum_{j=1}^{N_0} \ell_j(1) \tilde{w}_j. \tag{12.64}$$

We summarize these results in a proposition.

PROPOSITION 12.3.4. *For $0 < \gamma < 1$, there is a basis for* $\ker L_\gamma$ *consisting of non-negative smooth functions $\{\tilde{w}_j\}$. There is an $\eta < 0$, so that for initial data $f \in \mathscr{C}^{0,\gamma}(P)$, the asymptotic formula* (12.62) *holds. For initial data $\nu \in \mathcal{M}^{\odot}(P)$, the asymptotic formula* (12.63) *holds.*

Remark. In the classical case, with $L = L_{\text{Kim}}$ and $P = \mathcal{S}_N$, a basis for $\ker L_{\text{Kim}}$ is given by the functions $\{1, x_1, \ldots, x_N\}$. It is very likely that (12.62) also holds for data in $\mathscr{C}^0(P)$.

12.4 IRREGULAR SOLUTIONS OF THE INHOMOGENEOUS EQUATION

In many applications to probability one would like to solve equations of the form

$$Lw = f, \tag{12.65}$$

where f is a non-negative function. From the analysis in Section 12.2 it is clear that there is often no regular solution to this equation, for $\langle f, \delta_\Sigma \rangle$ will, in general, not vanish for all $\Sigma \in bP_{\text{ter}}(L)$. Nonetheless, probabilistic considerations show that this equation should sometimes have a solution. For example, if $f = 1$, then $-w(p)$ is the expected time for a path of the process defined by $\overline{L}$, starting at p, to arrive at bP.

In this section we show that, under certain hypotheses on L, there is an *irregular* solution to this equation (12.12), by which we mean a solution that does not belong to $\text{Dom}(\overline{L})$. We begin by noting that, since $\overline{L}$ defines a contraction semi-group on $\mathscr{C}^0(P)$, it automatically satisfies the maximum principle:

LEMMA 12.4.1. *If $w \in \text{Dom}(\overline{L})$ assumes its maximum at $p \in P$, then $Lw(p) \leq 0$.*

See Theorem 7.22 in [13]. By choosing a smooth non-degenerate density dV on P we can represent the formal adjoint operator L^* to L as a sum

$$L^* = \widetilde{L} + c, \tag{12.66}$$

where $c = L^*1$ is a smooth function, and $L - \widetilde{L} = X$ is a vector field. Denote by $\{\rho_j : j = 1, \ldots, K\}$ non-negative defining functions for the hypersurface components of bP. If

$$L^*\rho_j \restriction_{\rho_j=0} \geq 0 \text{ for } j = 1, \ldots, K, \tag{12.67}$$

then $\widetilde{L}$ is also a generalized Kimura diffusion operator. We call a generalized Kimura diffusion operator with this property *adversible*. We use this slightly odd neologism as the more natural term "reversible" already has a well established meaning in the context of Markov processes.

For a adversible Kimura diffusion we can apply the analysis above to study the operator $\widetilde{L}$ on the spaces $\mathscr{C}^{k,2+\gamma}_{WF}$, and then consider its $\mathscr{C}^0(P)$-graph closure. We denote this closed operator by $\widetilde{L}_r$; it generates a contraction semi-group of operators on $\mathscr{C}^0(P)$. If the scalar function c is non-positive, then the closed operator $L^*_r = \widetilde{L}_r + c$ also generates a contraction semi-group on $\mathscr{C}^0(P)$, with the same domain as $\widetilde{L}_r$. Using Lemma 12.4.1 we can easily establish the following form of the maximum principle:

PROPOSITION 12.4.2. *If L is an adversible generalized Kimura diffusion, with c non-positive in P and strictly negative on bP, then, for $w \in \text{Dom}(L^*_r)$, the equation*

$$L^*_r w = 0 \tag{12.68}$$

has only the trivial solution.

PROOF. Suppose that there is a non-trivial solution. Evidently w is non-constant, and we can therefore assume that w has a positive maximum. The strong maximum principle implies that the maximum must occur at a point $p_0 \in bP$. Lemma 12.4.1 shows that $\widetilde{L}_r w(p_0) \leq 0$, and therefore $L^*_r w(p_0) < 0$, which is a contradiction. □

If we assume that $c(p) < -\lambda < 0$ throughout P, then

$$L_r^* = \widetilde{L}_r + (c + \lambda) - \lambda. \tag{12.69}$$

The operator $\widetilde{L}_r + (c + \lambda)$ generates a contraction semi-group, and therefore L_r^* is invertible. The formal symbol of the operator $(L_r^*)^*$ agrees with L. Because $\mathscr{C}^0(P)$ is not reflexive, we need to use the operator $(L_r^*)^{\odot}$ on the space $\mathcal{M}^{\odot}(P)$ introduced above. With these preliminaries we can show the existence of irregular solutions.

PROPOSITION 12.4.3. *Suppose that P is a manifold with corners and L is an adversible generalized Kimura diffusion on P such that $L^*1 < 0$ throughout P. If $g \in L^1(P)$ then there is a unique solution $w \in \operatorname{Dom}((L_r^*)^{\odot})$ to the equation*

$$(L_r^*)^{\odot} w = g. \tag{12.70}$$

Remark. In the interior equation (12.70) is equivalent to the classical equation $Lw = g$, which explains why we say that this equation has an irregular solution for any $g \in L^1(P)$.

PROOF. Under the hypotheses of the proposition the operator L_r^* is invertible. Theorem 14.3.4 of [26] shows that $(L_r^*)^{\odot}$ is also invertible. As noted above, $L^1(P)$ is a subspace of $\mathcal{M}^{\odot}(P)$, from which it follows that (12.70) has a unique solution, for any $g \in L^1(P)$. □

Remark. As the operator L is non-degenerate elliptic in the interior of P, the solution w has roughly two more derivatives than g in this subset. On the other hand, the best one can hope for is that w is an absolutely continuous measure, satisfying the no-flux boundary conditions (12.12), along bP. Let $\{\rho_i\}$ be defining functions for the faces of bP. These conditions force w to have leading singularities of the form $\rho_i^{\beta(y)}$ along the sets $\{\rho_i = 0\}$.

In the classical population genetic case, P is a simplex $\Delta_N \subset \mathbb{R}^N$, and

$$L = L_{\text{Kim}} + \sum_{j=1}^{N} b_j(x)\partial_{x_j}. \tag{12.71}$$

A simple calculation shows that, with respect to Lebesgue measure,

$$L^* = L_{\text{Kim}} + \sum_{j=1}^{N} [2 - b_i(x) - 2(N+1)x_j]\partial_{x_j} - [N(N+1) + \sum_{j=1}^{N} \partial_{x_j} b_j(x)]. \tag{12.72}$$

The conditions required for L to be adversible are: the functions $b_j(x) \leq 2$, where $x_j = 0$, for $j = 1, \dots, N$ and

$$\left[\sum_{j=1}^{N} b_j(x)\right]_{\{x_1+\cdots+x_N=1\}} \geq -2. \tag{12.73}$$

That is we are in a "weak mutation" regime. The hypothesis that $L^*1 < 0$ becomes the requirement:

$$\sum_{j=1}^{N} \partial_{x_j} b_j(x) > -N(N+1). \tag{12.74}$$

This inequality involves the strength of both mutation and selection, but holds when both are sufficiently weak. Along $\{x_j = 0\}$ the no-flux condition would force a solution to behave like $x_j^{1-b_j(x)}$. Along faces where $0 < b_j(x) < 1$, the solution will tend to zero. Where $1 < b_j(x) < 2$, the solution will tend to infinity. Where $b_j = 0$, we expect the solution to behave like $x_j \log x_j$, as it is precluded from being regular.

Appendix A

Proofs of Estimates for the Degenerate 1-d Model

This Appendix contains proofs of estimates, used throughout the paper, of the 1-dimensional solution operator $k_t^b(x, y)$, which we recall is given by

$$k_t^b(x, y) = \frac{y^{b-1}}{t} e^{-\frac{(x+y)}{t}} \psi_b\left(\frac{xy}{t^2}\right), \tag{A.1}$$

where

$$\psi_b(z) = \sum_{j=0}^{\infty} \frac{z^j}{j!\Gamma(j+b)}. \tag{A.2}$$

This function has the following asymptotic development, as $z \to \infty$:

$$\psi_b(z) \sim \frac{z^{\frac{1}{4}-\frac{b}{2}} e^{2\sqrt{z}}}{\sqrt{4\pi}} \left[1 + \sum_{j=1}^{\infty} \frac{c_{b,j}}{z^{\frac{j}{2}}}\right]. \tag{A.3}$$

In several of the arguments below we need a suitable replacement for the mean value theorem that is valid for complex valued functions.

LEMMA A.0.4. *Let f be a continuously differentiable, complex valued function defined on the interval $[a, b]$. There is a point $c \in (a, b)$ such that*

$$|f(b) - f(a)| \leq (b - a)|f'(c)|. \tag{A.4}$$

PROOF. As an immediate consequence of the fundamental theorem of calculus and the triangle inequality we see that

$$|f(b) - f(a)| \leq \int_a^b |f'(y)|dy. \tag{A.5}$$

The estimate in the lemma now follows from the standard mean value theorem applied to the differentiable function

$$F(x) = \int_a^x |f'(y)|dy. \tag{A.6}$$

□

The kernel functions $k_t^b(x, y)$ extend to be analytic for $\text{Re}\, t > 0$, and we prove estimates for the spatial derivatives of this analytic continuation. These are needed to study the resolvent kernel, which for the 1-dimensional model problem is defined in the right half plane by

$$R(\mu) = \int_0^\infty e^{-\mu t} e^{tL_b} dt. \tag{A.7}$$

The contour of integration can be deformed to lie along any ray $\arg t = \theta$ with $|\theta| < \frac{\pi}{2}$. This provides an analytic continuation of $R(\mu)$ to $\mathbb{C} \setminus (-\infty, 0]$. In these arguments we let $t = \tau e^{i\theta}$, where $\tau = |t|$.

Notational Convention. To simplify the notation in the ensuing arguments we let

$$e_\phi \overset{d}{=} e^{-i\phi}. \tag{A.8}$$

A.1 BASIC KERNEL ESTIMATES

LEMMA A.1.1. *[LEMMA 6.1.4] For $b > 0$, $0 < \gamma < 1$, and $0 < \phi < \frac{\pi}{2}$, there are constants $C_{b,\phi}$ uniformly bounded with b, so that for $t \in S_\phi$*

$$\int_0^\infty |k_t^b(x, y) - k_t^b(0, y)| y^{\frac{\gamma}{2}} dy \leq C_{b,\phi} x^{\frac{\gamma}{2}}. \tag{A.9}$$

PROOF. We let $t = \tau e^{i\theta}$, where $|\theta| < \frac{\pi}{2} - \phi$. Using the formula for k_t^b we see that

$$\begin{aligned} \int_0^\infty |k_t^b(x, y) - k_t^b(0, y)| y^{\frac{\gamma}{2}} dy &= \int_0^\infty \left(\frac{y}{\tau}\right)^b e^{-\cos\theta \frac{y}{\tau}} \left| e^{-e_\theta \frac{x}{\tau}} \psi_b\left(\frac{xye_{2\theta}}{\tau^2}\right) - \psi_b(0) \right| y^{\frac{\gamma}{2}} \frac{dy}{y} \\ &\leq \int_0^\infty w^b e^{-\cos\theta w} \left| e^{-\lambda e_\theta} \psi_b(\lambda w e_{2\theta}) - \psi_b(0) \right| (w\tau)^{\frac{\gamma}{2}} \frac{dw}{w}. \end{aligned} \tag{A.10}$$

On the second line we let $w = y/\tau$ and $\lambda = x/\tau$. We split the integral into a part, $I_1(t, \lambda)$, from 0 to $1/\lambda$, and the rest, $I_2(t, \lambda)$. We estimate the compact part first; using the FTC we see that

$$\begin{aligned} \psi_b(\lambda w e_{2\theta}) - \psi_b(0) &= \lambda w e_{2\theta} \int_0^1 \psi_b'(s\lambda w e_{2\theta}) ds \\ &= M\lambda w e_{2\theta}, \end{aligned} \tag{A.11}$$

where M is a complex number satisfying:

$$|M| \leq \sup_{z:\,|z|\leq 1} |\psi_b'(z)|. \tag{A.12}$$

This gives

$$I_1(t,\lambda) \leq C_b \tau^{\frac{\gamma}{2}} \int_0^{\frac{1}{\lambda}} w^{b+\frac{\gamma}{2}-1} e^{-\cos\theta w}\left[e^{-\cos\theta\lambda}\lambda w + |1-e^{-e_\theta\lambda}|\right] dw. \tag{A.13}$$

The constant C_b is uniformly bounded for $0 < b < B$. For λ bounded, and $|\theta| < \frac{\pi}{2}$ we can estimate $|1-e^{-e_\theta\lambda}|$ by λ and therefore the integral can also be estimated by a constant times λ. Altogether we get

$$I_1(t,\lambda) \leq C_{b,\theta}\tau^{\frac{\gamma}{2}}\lambda = C_{b,\theta}x^{\frac{\gamma}{2}}\lambda^{1-\frac{\gamma}{2}}. \tag{A.14}$$

As λ is bounded, this is the desired estimate. Now we turn to $\lambda \to \infty$. In this case it is easy to see that the integral tends to zero. As $\lambda > 1$, this implies that

$$I_1(t,\lambda) \leq C_{b,\theta}x^{\frac{\gamma}{2}}. \tag{A.15}$$

We are left to estimate I_2. In this case there is no cancellation between the terms on the right-hand side of (A.10). As $\psi_b(0) = 1/\Gamma(b)$, it is elementary to see that, in all cases,

$$\tau^{\frac{\gamma}{2}} \int_{\frac{1}{\lambda}}^{\infty} w^b e^{-\cos\theta w} |\psi_b(0)| w^{\frac{\gamma}{2}} \frac{dw}{w} \leq C_{b,\theta}x^{\frac{\gamma}{2}}. \tag{A.16}$$

To complete the proof, for this case, we need to estimate the other term, which we denote $I_2'(t,\lambda)$. To that end we use the asymptotic expansion to estimate $\psi_b(w\lambda)$:

$$|\psi_b(w\lambda e_{2\theta})| \leq C_b (w\lambda)^{\frac{1}{4}-\frac{b}{2}} e^{2\cos\theta\sqrt{w\lambda}}. \tag{A.17}$$

Inserting this estimate gives:

$$I_2'(t,\lambda) \leq C_b \tau^{\frac{\gamma}{2}} \int_{\frac{1}{\lambda}}^{\infty} \left(\frac{w}{\lambda}\right)^{\frac{b}{2}-\frac{1}{4}} e^{-\cos\theta(\sqrt{w}-\sqrt{\lambda})^2} w^{\frac{\gamma}{2}} \frac{dw}{\sqrt{w}}. \tag{A.18}$$

Applying Lemma 6.1.23 it is a simple matter to see that this is uniformly bounded by $C_{b,\theta}x^{\frac{\gamma}{2}}\|f\|_{WF,0,\gamma}$, for a constant bounded when b is bounded, and therefore

$$I_2(t,\lambda) \leq C_{b,\theta}x^{\frac{\gamma}{2}}. \tag{A.19}$$

Combining (A.14), (A.15), (A.16) and (A.19) completes the proof of the lemma. □

LEMMA A.1.2. *[LEMMA 6.1.5] For $b > 0$, there is a constant $C_{b,\phi}$ so that, for $t \in S_\phi$,*

$$\int_0^\infty |k_t^b(x,z) - k_t^b(0,z)|dz \leq C_{b,\phi}\frac{x/|t|}{1+x/|t|}. \tag{A.20}$$

For $0 < c < 1$, there is a constant $C_{b,c,\phi}$ so that if $cx_2 < x_1 < x_2$ and $t \in S_\phi$, then

$$\int_0^\infty |k_t^b(x_2,z) - k_t^b(x_1,z)|dz \leq C_{b,c,\phi}\left(\frac{\frac{|\sqrt{x_2}-\sqrt{x_1}|}{\sqrt{|t|}}}{1+\frac{|\sqrt{x_2}-\sqrt{x_1}|}{\sqrt{|t|}}}\right). \tag{A.21}$$

PROOF. First observe that Lemma 6.1.3 implies that, for $t \in S_\phi$, the integrals in (A.20) and (A.21) are always bounded by a constant C_ϕ. We start with the proof of (A.20). We let $t = \tau e^{i\theta}$, with $|\theta| < \frac{\pi}{2}$, and set $w = z/\tau$, and $\lambda = x/\tau$, then we see that the expression on the right-hand side of (A.20) equals

$$\int_0^\infty w^{b-1}e^{-\cos\theta w}|e^{-e_\theta\lambda}\psi_b(\lambda w e_{2\theta}) - \psi_b(0)|dw. \tag{A.22}$$

We split this into an integral over $[0, \frac{1}{\lambda}]$ and the rest. As $\lambda \to \infty$ it is clear that the compact part remains bounded, and as $\lambda \to 0$, it is $O(\lambda)$. In the non-compact part we use the trivial bound when $\lambda \to \infty$, and the asymptotic expansion when $\lambda \to 0$. This latter term is easily seen to be bounded by $O(e^{-\frac{\cos\theta}{2\lambda}})$, completing the proof in this case.

To prove (A.21), we assume that $cx_2 < x_1 < x_2$, and use the formula for k_t^b; setting $w = z/\tau$, $\lambda = x_1/\tau$, and $\mu = x_2/x_1$, we obtain:

$$\int_0^\infty |k_t^b(x_2,z) - k_t^b(x_1,z)|dz =$$

$$\int_0^\infty w^{b-1}e^{-\cos\theta w}|e^{-e_\theta\lambda}\psi_b(\lambda w e_{2\theta}) - e^{-e_\theta\mu\lambda}\psi_b(\mu\lambda w e_{2\theta})|dw. \tag{A.23}$$

We let $F(\mu) = e^{-e_\theta\mu\lambda}\psi_b(\mu\lambda w e_{2\theta})$; from Lemma A.0.4 it follows that

$$|F(\mu) - F(1)| \leq (\mu-1)\lambda e^{-\cos\theta\xi\lambda}|\psi_b(\xi\lambda w e_{2\theta}) - e_\theta w\psi_b'(\xi\lambda w e_{2\theta})|, \tag{A.24}$$

for a $\xi \in (1,\mu) \subset (1, \frac{1}{c})$. We split the integral into the part from 0 to $1/\lambda$, I_-, and the rest, I_+. Using the Taylor expansion we see that

$$I_- \leq C(\mu-1)\lambda\int_0^{\frac{1}{\lambda}} w^{b-1}e^{-\cos\theta(w+\lambda)}\left[\frac{1}{\Gamma(b)} + w(1+\lambda)\right]dw. \tag{A.25}$$

As $\lambda \to \infty$ this is bounded by a constant times $\lambda^{1-b}(\mu - 1)e^{-\lambda}$. This in turn satisfies

$$I_{-} \leq e^{-\frac{\lambda}{2}} \left(\frac{\sqrt{x_2} - \sqrt{x_1}}{\sqrt{|t|}} \right). \tag{A.26}$$

As $\lambda \to 0$ the integral in (A.25) remains bounded and therefore

$$I_{-} \leq C_{b,\theta} \left(\frac{x_2 - x_1}{|t|} \right) = C_{b,\theta} \left(\frac{\sqrt{x_2} - \sqrt{x_1}}{\sqrt{|t|}} \right) \left(\frac{\sqrt{x_2} + \sqrt{x_1}}{\sqrt{|t|}} \right), \tag{A.27}$$

which shows that

$$I_{-} \leq C_{b,\theta} \sqrt{\lambda} (\sqrt{\mu} + 1) \left(\frac{\sqrt{x_2} - \sqrt{x_1}}{\sqrt{|t|}} \right), \tag{A.28}$$

thus completing this case.

Using the asymptotic expansions for ψ_b and ψ_b' we see that

$$I_{+} \leq C_b (\mu - 1) \lambda \int\limits_{\frac{1}{\lambda}}^{\infty} \left(\frac{w}{\xi \lambda} \right)^{\frac{b}{2} - \frac{1}{4}} e^{-\lambda \cos\theta (\sqrt{\frac{w}{\lambda}} - \sqrt{\xi})^2} \left| 1 - \sqrt{\frac{w}{\xi \lambda}} + O(\frac{1}{\sqrt{w \lambda}} + \frac{1}{\lambda}) \right| \frac{dw}{\sqrt{w}}. \tag{A.29}$$

As $\lambda \to 0$ this satisfies an estimate of the form

$$I_{+} \leq C_{b,\theta} \left(\frac{x_2 - x_1}{|t|} \right) e^{-\frac{1}{2\lambda}}. \tag{A.30}$$

To analyze the non-compact part as $\lambda \to \infty$, we first note that we are only interested in the case that

$$\frac{\sqrt{x_2} - \sqrt{x_1}}{\sqrt{|t|}} \leq 1, \tag{A.31}$$

for otherwise we use the trivial estimate. Dividing by $\sqrt{\lambda}$ we see that this constraint is equivalent to

$$\sqrt{\mu} - 1 \leq \frac{1}{\sqrt{\lambda}}, \tag{A.32}$$

which clearly implies that $\mu \to 1$ as $\lambda \to \infty$.

We change variables in (A.29) letting $z = \sqrt{w/\lambda}$, to obtain:

$$I_{+} \leq C_b (\mu - 1) \lambda \int\limits_{\frac{1}{\lambda}}^{\infty} z^{b - \frac{1}{2}} e^{-\lambda \cos\theta (z - \sqrt{\xi})^2} \left| 1 - \frac{z}{\sqrt{\xi}} + O(\frac{1}{z\lambda} + \frac{1}{\lambda}) \right| \sqrt{\lambda} dz. \tag{A.33}$$

To estimate this integral, we split the domain into three pieces: $[\frac{1}{\lambda}, 1]$, $[1, \sqrt{\mu}]$, and $[\sqrt{\mu}, \infty]$. Recall that $\xi \in [1, \sqrt{\mu}]$, and therefore, in the first segment we see that

$$|z - \sqrt{\xi}| > |z - 1|, \tag{A.34}$$

and in the third segment,

$$|z - \sqrt{\xi}| > |z - \sqrt{\mu}|. \tag{A.35}$$

With these observations we see that

$$I_+ \leq C_b(\mu - 1)\lambda\Bigg[\int_{\frac{1}{\lambda}}^{1} z^{b-\frac{1}{2}} e^{-\lambda\cos\theta(z-1)^2}\left[|1-z| + |\sqrt{\mu} - 1| + O(\frac{1}{z\lambda} + \frac{1}{\lambda})\right]\sqrt{\lambda}dz +$$
$$\int_{1}^{\sqrt{\mu}} z^{b-\frac{1}{2}} e^{-\lambda\cos\theta(z-\sqrt{\xi})^2}\left|1 - \frac{z}{\sqrt{\xi}} + O(\frac{1}{z\lambda} + \frac{1}{\lambda})\right|\sqrt{\lambda}dz +$$
$$\int_{\sqrt{\mu}}^{\infty} z^{b-\frac{1}{2}} e^{-\lambda\cos\theta(z-\sqrt{\mu})^2}\left[|\sqrt{\mu} - z| + |\sqrt{\mu} - 1| + O(\frac{1}{z\lambda} + \frac{1}{\lambda})\right]\sqrt{\lambda}dz\Bigg]. \tag{A.36}$$

Using Laplace's method to estimate the first and third terms, as well as (A.32), we easily show that the sum of the three integrals is bounded by $C_{b,c,\theta}/\sqrt{\lambda}$, which implies that

$$I_+ \leq C_{b,c,\theta}\left(\frac{\sqrt{x_2} - \sqrt{x_1}}{\sqrt{|t|}}\right). \tag{A.37}$$

This completes the proof of the lemma. □

LEMMA A.1.3. *[LEMMA 6.1.6] For $b > 0$, $0 < \gamma < 1$ and $t \in S_\phi$, $0 < \phi < \frac{\pi}{2}$, there is a $C_{b,\phi}$ so that*

$$\int_{0}^{\infty} |k_t^b(x, y)||\sqrt{x} - \sqrt{y}|^\gamma \, dy \leq C_{b,\phi}|t|^{\frac{\gamma}{2}}. \tag{A.38}$$

For fixed $0 < \phi$, and B, these constants are uniformly bounded for $0 < b < B$.

PROOF. We let $t = \tau e^{i\theta}$, with $|\theta| < \frac{\pi}{2} - \phi$, and set $w = y/\tau$, $\lambda = x/\tau$, obtaining:

$$\int_{0}^{\infty} |k_t^b(x, y)||\sqrt{x} - \sqrt{y}|^\gamma \, dy =$$
$$|t|^{\frac{\gamma}{2}} \int_{0}^{\infty} w^b e^{-\cos\theta(w+\lambda)}|\psi_b(w\lambda e_{2\theta})||\sqrt{w} - \sqrt{\lambda}|^\gamma \frac{dw}{w}. \tag{A.39}$$

We split this integral into the part from 0 to $1/\lambda$, I_1, and the rest, I_2. Using the estimate

$$|\psi_b(z)| \leq C_b\left(\frac{1}{\Gamma(b)} + z\right), \tag{A.40}$$

we easily show that the compact part is uniformly bounded by $C_{b,\theta}|t|^{\frac{\gamma}{2}}$, for a constant $C_{b,\theta}$ uniformly bounded for $b < B$ and $|\theta| \leq \frac{\pi}{2} - \phi$.

In the non-compact part use the asymptotic expansion to obtain that

$$I_2 \leq C_b|t|^{\frac{\gamma}{2}} \lambda^{\frac{1}{4}-\frac{b}{2}} \int\limits_{\frac{1}{\lambda}}^{\infty} w^{\frac{1}{2}\left(b-\frac{1}{2}\right)} e^{-\cos\theta(\sqrt{w}-\sqrt{\lambda})^2} |\sqrt{w} - \sqrt{\lambda}|^{\gamma} \frac{dw}{\sqrt{w}}. \tag{A.41}$$

Lemma 6.1.23 shows that as $\lambda \to 0$ this is bounded by $C_{b,\theta}|t|^{\frac{\gamma}{2}} \lambda^{-(b+\gamma-\frac{1}{2})} e^{-\frac{1}{\lambda}}$, showing that in this regime (6.34) holds.

As $\lambda \to \infty$, Lemma 6.1.23 shows that this is bounded by $C_{b,\theta}|t|^{\frac{\gamma}{2}}$, thus completing the argument to show that (6.34) holds for all x, and $t \in S_{\phi}$. □

LEMMA A.1.4. *[LEMMA 6.1.7] We assume that $x_1/x_2 > 1/9$ and $J = [\alpha, \beta]$, as defined in* (6.35). *For $b > 0$, $0 < \gamma < 1$ and $0 < \phi < \frac{\pi}{2}$, there is a $C_{b,\phi}$ so that, for $t \in S_{\phi}$,*

$$\int\limits_{J^c} |k_t^b(x_2, y) - k_t^b(x_1, y)||\sqrt{y} - \sqrt{x_1}|^{\gamma}\, dy \leq C_{b,\phi}|\sqrt{x_2} - \sqrt{x_1}|^{\gamma}. \tag{A.42}$$

PROOF. We let $t = \tau e^{i\theta}$, with $|\theta| < \frac{\pi}{2} - \phi$, and set

$$\begin{aligned} I_- &= \int\limits_0^{\alpha} |k_t^b(x_2, y) - k_t^b(x_1, y)||\sqrt{y} - \sqrt{x_1}|^{\gamma}\, dy \\ I_+ &= \int\limits_{\beta}^{\infty} |k_t^b(x_2, y) - k_t^b(x_1, y)||\sqrt{y} - \sqrt{x_1}|^{\gamma}\, dy. \end{aligned} \tag{A.43}$$

Since $x_1/x_2 > 1/9$, we know that $\alpha > 0$.

$$I_- \leq \int\limits_0^{\alpha} \left(\frac{y}{\tau}\right)^b e^{-\cos\theta\frac{y}{\tau}} \left| e^{-\frac{x_2}{t}} \psi_b\left(\frac{x_2 y}{t^2}\right) - e^{-\frac{x_1}{t}} \psi_b\left(\frac{x_1 y}{t^2}\right)\right| |\sqrt{x_1} - \sqrt{y}|^{\gamma} \frac{dy}{y}. \tag{A.44}$$

As usual we let $y/\tau = w$ and $x_1/\tau = \lambda$, obtaining

$$I_- \leq |t|^{\frac{\gamma}{2}} \int\limits_0^{\frac{\alpha}{|t|}} w^b e^{-\cos\theta w} \left| e^{-\frac{x_2}{x_1} e_{\theta}\lambda} \psi_b\left(\frac{x_2}{x_1} w\lambda e_{2\theta}\right) - e^{-e_{\theta}\lambda} \psi_b\left(w\lambda e_{2\theta}\right)\right| |\sqrt{\lambda} - \sqrt{w}|^{\gamma} \frac{dw}{w}. \tag{A.45}$$

The upper limit of integration can be re-expressed as

$$\frac{\alpha}{|t|} = \lambda \frac{\left(3 - \sqrt{\frac{x_2}{x_1}}\right)^2}{4}. \tag{A.46}$$

As before we use Lemma A.0.4 to obtain:

$$\left|e^{-\frac{x_2}{x_1}e_\theta\lambda}\psi_b\left(\frac{x_2}{x_1}w\lambda e_{2\theta}\right)-e^{-e_\theta\lambda}\psi_b\left(w\lambda e_{2\theta}\right)\right|\leq \lambda e^{-\lambda\xi\cos\theta}\left|we_\theta\psi_b'(\xi\lambda we_{2\theta})-\psi_b(\xi\lambda we_{2\theta})\right|\left(\frac{x_2}{x_1}-1\right), \quad (A.47)$$

where $\xi\in(1,\frac{x_2}{x_1})\subset(1,9)$.

As in an earlier estimate we need to split this integral into the part from 0 to $1/\lambda$ and the rest. In the first part, I_{-1}, we estimate

$$|\psi_b(\xi\lambda we_{2\theta})|\leq\frac{1}{\Gamma(b)}+C_b(\xi\lambda w) \quad (A.48)$$

and $|\psi_b'(\xi\lambda we_{2\theta})|$ by a constant; in the second part, I_{-2}, we will use the asymptotic expansions. The term in (A.45) coming from $1/\Gamma(b)$ is estimated by

$$I_{-1}'\leq C_\theta|t|^{\frac{\gamma}{2}}\lambda^{1+b+\gamma/2}\frac{e^{-\cos\theta\lambda}}{b\Gamma(b)}\left(\frac{x_2}{x_1}-1\right). \quad (A.49)$$

We observe that this is bounded by $C_\theta\|f\|_{WF,0,\gamma}|\sqrt{x_2}-\sqrt{x_1}|^\gamma$ provided that

$$\lambda^{b+\gamma/2}e^{-\cos\theta\lambda}\left(\sqrt{\frac{x_2}{|t|}}-\sqrt{\frac{x_1}{|t|}}\right)^{1-\gamma}\left(\sqrt{\frac{x_2}{|t|}}+\sqrt{\frac{x_1}{|t|}}\right)\leq C. \quad (A.50)$$

As $c<x_1/x_2$, we see that the quantity on the left is bounded by a multiple of $\lambda^{b+1}e^{-\cos\theta\lambda}$, which remains bounded as $\lambda\to\infty$. Thus, there is a constant C_θ, independent of b, so that

$$I_{-1}'\leq C_\theta|\sqrt{x_2}-\sqrt{x_1}|^\gamma. \quad (A.51)$$

The other part of I_{-1} (coming from the $C_b[\xi\lambda w+w]$-terms) is easily seen to satisfy an estimate of the form

$$I_{-1}''\leq C_\theta|t|^{\frac{\gamma}{2}}\lambda^{2+b+\gamma/2}e^{-\cos\theta\lambda}\left(\frac{x_2}{x_1}-1\right), \quad (A.52)$$

for a constant independent of b. Arguing as before shows that this also satisfies (A.51), so that I_{-1} satisfies the desired estimate.

Using the asymptotic expansions we see that the other part, I_{-2}, satisfies

$$I_{-2}\leq C|t|^{\frac{\gamma}{2}}\lambda^{\frac{3}{4}-\frac{b}{2}}\left(\frac{x_2}{x_1}-1\right)\int_{\frac{1}{\lambda}}^{\frac{\alpha}{|t|}}e^{-\cos\theta(\sqrt{w}-\sqrt{\lambda\xi})^2}w^{\frac{b}{2}-\frac{1}{4}}|\sqrt{w}-\sqrt{\xi\lambda}+O((w\lambda)^{-\frac{1}{2}})||\sqrt{w}-\sqrt{\lambda}|^\gamma\frac{dw}{\sqrt{w}}. \quad (A.53)$$

As $\xi > 1$, for $w \in [0, \frac{\alpha}{|t|}]$ the exponential is only increased if we replace $\lambda\xi$ with λ. For a large enough C it is also the case that for w in the domain of integration:

$$\sqrt{\xi\lambda} - \sqrt{w} \leq C(\sqrt{\lambda} - \sqrt{w}). \tag{A.54}$$

Letting $z = \sqrt{\frac{w}{\lambda}} - 1$, we obtain

$$I_{-2} \leq C_b|t|^{\frac{\gamma}{2}}\lambda^{\frac{3}{2}+\frac{\gamma}{2}}\left(\frac{x_2}{x_1} - 1\right) \times \int_{\frac{1}{\lambda}-1}^{\frac{1}{2}\left(1-\sqrt{\frac{x_2}{x_1}}\right)} e^{-\cos\theta\lambda z^2}(1+z)^{b-\frac{1}{2}}|z|^{1+\gamma}\,dz. \tag{A.55}$$

We are interested in the case x_2/x_1 approaches 1, and $\lambda \to \infty$. Even if $b < \frac{1}{2}$, then we see that the part of the integral near to $z = -1$ contributes a term much like I_{-1}. Lemma 6.1.24 shows that

$$I_{-2} \leq C_{b,\theta}I_{-1} + C_{b,\theta}|t|^{\frac{\gamma}{2}}\lambda^{\frac{3}{2}+\frac{\gamma}{2}}\left(\frac{x_2}{x_1} - 1\right)\frac{e^{-\frac{\cos\theta(\sqrt{x_1}-\sqrt{x_2})^2}{4|t|}}\left(\frac{\sqrt{x_2}-\sqrt{x_1}}{2|t|}\right)^{\gamma}}{\lambda^{1+\frac{\gamma}{2}}}. \tag{A.56}$$

The complicated expression on the right-hand side can be rewritten as

$$C_{b,\theta}(\sqrt{x_2} - \sqrt{x_1})^{\gamma}\left(\frac{\sqrt{x_2} - \sqrt{x_1}}{\sqrt{|t|}}\right)\left(\frac{\sqrt{x_2} + \sqrt{x_1}}{\sqrt{x_1}}\right)e^{-\frac{\cos\theta(\sqrt{x_2}-\sqrt{x_1})^2}{4|t|}}, \tag{A.57}$$

showing that

$$I_{-2} \leq C_{b,\theta}I_{-1} + C_{b,\theta}|\sqrt{x_2} - \sqrt{x_1}|^{\gamma}, \tag{A.58}$$

which is precisely the bound that we need. The error term contributes a term of this size times λ^{-1}, completing the analysis of this term.

We now turn to I_+; in this part the lower limit of integration is

$$w = \frac{\beta}{|t|} = \frac{\lambda}{4}\left(3\sqrt{\frac{x_2}{x_1}} - 1\right)^2 \geq \lambda.$$

If $\lambda < 1$, we need to split the integral into the part from β/t to $1/\lambda$, and use the Taylor expansion at zero to estimate the ψ_b- and ψ_b'-terms. If $\lambda > 1$, then we only need to use the asymptotic expansions of ψ_b and ψ_b'. For $\lambda < 1$, we have to estimate

$$\int_{\frac{\beta}{t}}^{\frac{1}{\lambda}} w^b e^{-\cos\theta w}\left|e^{-\frac{x_2}{x_1}\lambda e_\theta}\psi_b\left(\frac{x_2}{x_1}w\lambda e_{2\theta}\right) - e^{-\lambda e_\theta}\psi_b(w\lambda e_{2\theta})\right||\sqrt{\lambda} - \sqrt{w}|^{\gamma}\frac{dw}{w}. \tag{A.59}$$

Using (A.47) and arguing as before we can show that the contribution of this term is estimated by $C_{b,\theta}|t|^{\frac{\gamma}{2}}(x_2 - x_1)/|t|$. This, in turn, is estimated by

$$C_{b,\theta}|\sqrt{x_2} - \sqrt{x_1}|^{\gamma}.$$

If $\lambda < 1$, then the contribution of the integral from $1/\lambda$ to infinity is of the form $e^{-\frac{\cos\theta}{2\lambda}} C_{b,\theta}|\sqrt{x_2} - \sqrt{x_1}|^{\gamma}$.

Assuming now that $\lambda > 1$, we change variables as before, to see that

$$I_+ \leq C_b|t|^{\frac{\gamma}{2}} \lambda^{\frac{3}{4}-\frac{b}{2}} \left(\frac{x_2}{x_1} - 1\right) \times \int\limits_{\frac{\lambda}{4}\left(3\sqrt{\frac{x_2}{x_1}}-1\right)^2}^{\infty} w^{\frac{b}{2}-\frac{1}{4}} e^{-\cos\theta(\sqrt{w}-\sqrt{\lambda\xi})^2} |\sqrt{\xi\lambda} - \sqrt{w} + O((w\lambda)^{-\frac{1}{2}})||\sqrt{\lambda} - \sqrt{w}|^{\gamma} \frac{dw}{\sqrt{w}}. \tag{A.60}$$

In this case $\xi \leq \sqrt{\frac{x_2}{x_1}}$, and so changing variables again, as above, we see that the leading term is bounded by

$$I_+ \leq C_b|t|^{\frac{\gamma}{2}} \lambda^{\frac{3}{2}+\frac{\gamma}{2}} \left(\frac{x_2}{x_1} - 1\right) \times \int\limits_{\frac{3}{2}\sqrt{\frac{x_2}{x_1}}-\frac{1}{2}}^{\infty} e^{-\lambda\cos\theta\left(z-\sqrt{\frac{x_2}{x_1}}\right)^2} z^{b-\frac{1}{2}}|z-1|^{\gamma+1} dz. \tag{A.61}$$

This term, as well as the error term, satisfy the same estimates as those satisfied by I_{2-}, thereby completing the proof of Lemma 6.1.7. □

LEMMA A.1.5. *[LEMMA 6.1.8] For $b > 0$, $0 < \gamma < 1$ and $c < 1$ there is a C_b such that if $c < s/t < 1$, then*

$$\int\limits_0^{\infty} \left|k_t^b(x,y) - k_s^b(x,y)\right| |\sqrt{x} - \sqrt{y}|^{\gamma} dy \leq C_b|t-s|^{\frac{\gamma}{2}}. \tag{A.62}$$

PROOF. If we let $w = y/t$, $\lambda = x/t$ and $\mu = t/s$, then this becomes:

$$\int\limits_0^{\infty} \left|k_t^b(x,y) - k_s^b(x,y)\right| |\sqrt{x} - \sqrt{y}|^{\gamma} dy \leq$$

$$t^{\frac{\gamma}{2}} \int\limits_0^{\infty} \left|w^b e^{-(w+\lambda)} \psi_b(w\lambda) - (\mu w)^b e^{-\mu(w+\lambda)} \psi_b(\mu^2 w\lambda)\right| |\sqrt{w} - \sqrt{\lambda}|^{\gamma} \frac{dw}{w}. \tag{A.63}$$

We denote the quantity on the left by $K(x, s, t)$. If $F(\mu) = (\mu w)^b e^{-\mu(w+\lambda)} \psi_b(\mu^2 w\lambda)$, then the difference in the integral can be written:

$$F(\mu) - F(1) = F'(\xi)(\mu - 1) \text{ for a } \xi \in (1, \mu). \tag{A.64}$$

Computing the derivative, we see that

$$K(x, s, t) \le t^{\frac{\gamma}{2}}(\mu - 1) \int_0^\infty (\xi w)^b e^{-\xi(w+\lambda)} \times \\ \left| \left(\frac{b}{\xi} - (w + \lambda)\right) \psi_b(\xi^2 w\lambda) + 2\xi w\lambda \psi_b'(\xi^2 w\lambda) \right| |\sqrt{w} - \sqrt{\lambda}|^\gamma \frac{dw}{w}. \tag{A.65}$$

As usual, we split this into a part, I_1 from 0 to $1/\lambda$, and the rest, which we denote by I_2.

To bound I_1 we estimate ψ_b' by a constant and use the estimate of ψ_b in (A.48). Arguing exactly as before we see that

$$I_1 \le C_b t^{\frac{\gamma}{2}} \left(\frac{t - s}{s}\right). \tag{A.66}$$

The fact that $t/s < 1/c$, implies that there is a constant C so that

$$t^{\frac{\gamma}{2}} \left(\frac{t - s}{s}\right) \le C(t - s)^{\frac{\gamma}{2}}, \tag{A.67}$$

showing that

$$I_1 \le C_b(t - s)^{\frac{\gamma}{2}}, \tag{A.68}$$

for constants C_b that are uniformly bounded for $0 < \lambda < B$.

To estimate I_2 we use the asymptotic formulæ for ψ_b and ψ_b'. It is straightforward to estimate the $\frac{b}{\xi}$-term. To estimate the other two terms we need to take advantage of cancellations that occur, to leading order, and then use the error terms in the asymptotic expansions to estimate the remainder. The $\frac{b}{\xi}$-term is estimated by

$$Cbt^{\frac{\gamma}{2}}(\mu - 1) \int_{\frac{1}{\lambda}}^\infty w^b e^{-\xi(\sqrt{w}-\sqrt{\lambda})^2} (w\lambda)^{\frac{1}{4}-\frac{b}{2}} |\sqrt{w} - \sqrt{\lambda}|^\gamma \frac{dw}{w}. \tag{A.69}$$

As before we apply Lemma 6.1.23 to show that this integral is uniformly bounded for $\lambda \in (0, \infty)$. This term is again bounded by

$$C_b t^{\frac{\gamma}{2}} \left(\frac{t - s}{s}\right), \tag{A.70}$$

which is handled exactly like I_1.

This leaves only

$$I_2' = t^{\frac{\gamma}{2}}(\mu - 1)\times$$
$$\int_{\frac{1}{\lambda}}^{\infty} w^b e^{-(w+\lambda)} \left|2\xi w\lambda \psi_b'(\xi^2 w\lambda) - (w+\lambda)\psi_b(\xi^2 w\lambda)\right| |\sqrt{w} - \sqrt{\lambda}|^\gamma \frac{dw}{w}. \tag{A.71}$$

Using the asymptotic expansions for ψ_b and ψ_b' this is bounded by

$$I_2' \leq C_b t^{\frac{\gamma}{2}}(\mu - 1)\int_{\frac{1}{\lambda}}^{\infty} w^b e^{-\xi(\sqrt{w}-\sqrt{\lambda})^2}(\xi^2 w\lambda)^{\frac{1}{4}-\frac{b}{2}}\times$$
$$\left|(\sqrt{w} - \sqrt{\lambda})^2 + O((w\lambda)^{-\frac{1}{2}})\right| |\sqrt{w} - \sqrt{\lambda}|^\gamma \frac{dw}{w}. \tag{A.72}$$

This is negligible as $\lambda \to 0$; Lemma 6.1.23 implies that the leading term is bounded by $C_b t^{\frac{\gamma}{2}}\left(\frac{t-s}{s}\right)$, as before, and that the error term is bounded by

$$\lambda^{-\frac{1}{2}} C_b t^{\frac{\gamma}{2}}\left(\frac{t-s}{s}\right).$$

This completes the proof of the lemma. □

The proof of Lemma 6.1.8 also establishes the following simpler result:

LEMMA A.1.6. *[*LEMMA *6.1.9] For $b > 0$, there is a C_b such that if $s < t$, then*

$$\int_0^\infty \left|k_t^b(x,y) - k_s^b(x,y)\right| dy \leq C_b\left(\frac{t/s - 1}{1 + [t/s - 1]}\right). \tag{A.73}$$

We now consider the effects of scaling these kernels by powers of x/y.

LEMMA A.1.7. *[*LEMMA *7.1.2] If $0 \leq \gamma \leq 1$, and $b > \nu - \frac{\gamma}{2} > 0$, then there is a constant $C_{b,\phi}$, bounded for $b \leq B$, and $B^{-1} < b + \frac{\gamma}{2} - \nu$, so that, for $t \in S_\phi$, where $0 < \phi < \frac{\pi}{2}$, we have the estimate*

$$\int_0^\infty \left(\frac{x}{y}\right)^\nu |k_t^b(x,y)| y^{\frac{\gamma}{2}} dy \leq C_{b,\phi} x^{\frac{\gamma}{2}}. \tag{A.74}$$

PROOF. We let $t = \tau e^{i\theta}$, with $|\theta| < \frac{\pi}{2} - \phi$. Using $w = y/|t|$ and $\lambda = x/|t|$, it shows that we need to bound

$$|t|^{\frac{\gamma}{2}} \int_0^\infty \left(\frac{\lambda}{w}\right)^\nu w^b e^{-\cos\theta(w+\lambda)} |\psi_b(w\lambda e_{2\theta})| w^{\frac{\gamma}{2}} \frac{dw}{w}. \tag{A.75}$$

As usual we split this into an integral from 0 to $1/\lambda$ and the rest. The compact part we can estimate by

$$C_b|t|^{\frac{\gamma}{2}}\lambda^{\nu}e^{-\cos\theta\lambda}\int_0^{\frac{1}{\lambda}} w^{b+\frac{\gamma}{2}-\nu-1}e^{-\cos\theta w}dw. \tag{A.76}$$

As $b+\nu-\frac{\gamma}{2}>0$, the integral is clearly bounded uniformly in λ. Because $\nu-\frac{\gamma}{2}\geq 0$, the contribution of this term is bounded by

$$C_{b,\theta}x^{\frac{\gamma}{2}}\lambda^{\nu-\frac{\gamma}{2}}e^{-\cos\theta\lambda}\leq C'_{b,\theta}x^{\frac{\gamma}{2}}. \tag{A.77}$$

We use the asymptotic expansion of ψ_b to see that the non-compact part is bounded by

$$C_b|t|^{\frac{\gamma}{2}}\int_{\frac{1}{\lambda}}^{\infty}\left(\frac{w}{\lambda}\right)^{\frac{b}{2}-\frac{1}{4}-\nu}e^{-\cos\theta(\sqrt{w}-\sqrt{\lambda})^2}w^{\frac{\gamma}{2}}\frac{dw}{\sqrt{w}}. \tag{A.78}$$

The integral tends to zero like $e^{-\frac{\cos\theta}{2\lambda}}$ as $\lambda\to 0$, showing that again this term is bounded by $Cx^{\frac{\gamma}{2}}$. We let $\sqrt{w}-\sqrt{\lambda}=z$, to obtain that, as $\lambda\to\infty$, this is bounded by

$$C_b|t|^{\frac{\gamma}{2}}\lambda^{\nu+\frac{1}{4}-\frac{b}{2}}\int_{\frac{1}{\sqrt{\lambda}}-\sqrt{\lambda}}^{\infty}(z+\sqrt{\lambda})^{b-\frac{1}{2}-2\nu+\gamma}e^{-\cos\theta z^2}dz. \tag{A.79}$$

It is again not difficult to see that, as $\lambda\to\infty$, this is bounded by $C_{b,\theta}x^{\frac{\gamma}{2}}$. □

LEMMA A.1.8. *[LEMMA 7.1.3] Let $J=[\alpha,\beta]$, where α and β are given in* (7.61). *Assuming that x'_1 and x_1 satisfy* (7.60), *for $b>0$, $0<\gamma\leq 1$, and $0<\phi<\frac{\pi}{2}$, there is a $C_{b,\phi}$ so that if $t\in S_\phi$, then,*

$$\int_{J^c}\left|k_t^{b+1}(x_1,z_1)\sqrt{\frac{x_1}{z_1}}-k_t^{b+1}(x'_1,z_1)\sqrt{\frac{x'_1}{z_1}}\right||\sqrt{z_1}-\sqrt{x'_1}|^{\gamma}dz_1\leq C_{b,\phi}|\sqrt{x_1}-\sqrt{x'_1}|^{\gamma}. \tag{A.80}$$

PROOF. We let $t=\tau e^{i\theta}$, with $|\theta|<\frac{\pi}{2}-\phi$. Observe that it suffices to show that

$$I=\int_{J^c}|k_t^{b+1}(x_1,z_1)|\left|\frac{\sqrt{x_1}-\sqrt{x'_1}}{\sqrt{z_1}}\right||\sqrt{z_1}-\sqrt{x'_1}|^{\gamma}dz_1\leq C_{b,\phi}|\sqrt{x_1}-\sqrt{x'_1}|^{\gamma}, \tag{A.81}$$

and

$$II=\int_{J^c}\sqrt{\frac{x'_1}{z_1}}\left|k_t^{b+1}(x_1,z_1)-k_t^{b+1}(x'_1,z_1)\right||\sqrt{z_1}-\sqrt{x'_1}|^{\gamma}dz_1\leq C_{b,\phi}|\sqrt{x_1}-\sqrt{x'_1}|^{\gamma}. \tag{A.82}$$

The integral in I is relatively simple to bound, and we can extend the integral over $[0, \infty)$, rather than just over J^c. Before switching the domain of integration we observe that there is a constant C so that if $z_1 \in J^c$, then

$$C^{-1}|\sqrt{z_1} - \sqrt{x_1}| \leq |\sqrt{z_1} - \sqrt{x_1'}| \leq C|\sqrt{z_1} - \sqrt{x_1}|. \tag{A.83}$$

It therefore suffices to show that

$$I' = \int_0^\infty |k_t^{b+1}(x_1, z_1)| \left| \frac{\sqrt{x_1} - \sqrt{x_1'}}{\sqrt{z_1}} \right| |\sqrt{z_1} - \sqrt{x_1}|^\gamma dz_1 \leq C_{b,\phi}|\sqrt{x_1} - \sqrt{x_1'}|^\gamma. \tag{A.84}$$

To estimate this integral we let $w = z_1/|t|$ and $\lambda = x_1/|t|$, to obtain that

$$I' = |t|^{\frac{\gamma-1}{2}}|\sqrt{x_1} - \sqrt{x_1'}| \int_0^\infty w^b e^{-\cos\theta(w+\lambda)}|\psi_{b+1}(w\lambda e_{2\theta})||\sqrt{w} - \sqrt{\lambda}|^\gamma \frac{dw}{\sqrt{w}}. \tag{A.85}$$

We observe that

$$|t|^{\frac{\gamma-1}{2}}|\sqrt{x_1} - \sqrt{x_1'}| = |\sqrt{x_1} - \sqrt{x_1'}|^\gamma \left[\left| 1 - \sqrt{\frac{x_1'}{x_1}} \right| \sqrt{\lambda} \right]^{\frac{1-\gamma}{2}}. \tag{A.86}$$

For bounded λ it is easy to see that the integral in (A.85) is bounded, so we need to consider what happens as $\lambda \to \infty$. As usual we split the integral into the part over $[0, 1/\lambda]$ and the rest. The compact part is estimated by

$$I_-' \leq C_b|\sqrt{x_1} - \sqrt{x_1'}|^\gamma \lambda^{\frac{1-\gamma}{2}} e^{-\cos\theta\lambda} \int_0^{\frac{1}{\lambda}} w^b e^{-\cos\theta w}|\sqrt{w} - \sqrt{\lambda}|^\gamma \frac{dw}{\sqrt{w}}. \tag{A.87}$$

Whether λ is going to zero or infinity, we see that the contribution of this term is bounded by $C_{b,\theta}|\sqrt{x_1} - \sqrt{x_1'}|^\gamma$.

The non-compact is estimated using the asymptotic expansion for ψ_{b+1} as

$$I_+' \leq C_b|\sqrt{x_1} - \sqrt{x_1'}|^\gamma \lambda^{\frac{1-\gamma}{2}} \int_{\frac{1}{\lambda}}^\infty \left(\frac{w}{\lambda}\right)^{\frac{b}{2}-\frac{1}{4}} e^{-\cos\theta(\sqrt{w}-\sqrt{\lambda})^2}|\sqrt{w} - \sqrt{\lambda}|^\gamma \frac{dw}{\sqrt{w\lambda}}. \tag{A.88}$$

As $\lambda \to 0$, the integral is $O(e^{-\frac{1}{2\lambda}})$. We let $z = \sqrt{w} - \sqrt{\lambda}$, to obtain that

$$I_+' \leq C|\sqrt{x_1} - \sqrt{x_1'}|^\gamma \lambda^{-\frac{\gamma}{2}} \int_{\frac{1}{\sqrt{\lambda}}-\sqrt{\lambda}}^\infty \left(1 + \frac{z}{\sqrt{\lambda}}\right)^{b-\frac{1}{2}} e^{-\cos\theta z^2}|z|^\gamma dz, \tag{A.89}$$

from which it follows easily that

$$I \leq I' \leq C_{b,\theta}|\sqrt{x_1} - \sqrt{x_1'}|^{\gamma}. \tag{A.90}$$

We now turn to II. With $y/|t| = w$, $\lambda = x_1'/|t|$, and $\mu = x_1/x_1'$, we see that

$$II \leq \sqrt{x_1'}|t|^{\frac{\gamma-1}{2}} \left[\int_0^{\frac{\alpha}{|t|}} + \int_{\frac{\beta}{|t|}}^{\infty}\right] w^b e^{-\cos\theta w}|e^{-\mu\lambda e_\theta}\psi_{b+1}(\mu\lambda w e_{2\theta}) - e^{-\lambda e_\theta}\psi_{b+1}(\lambda w e_{2\theta})| \times$$
$$|\sqrt{w} - \sqrt{\lambda}|^{\gamma}\frac{dw}{\sqrt{w}}. \tag{A.91}$$

Note that $1 \leq \mu \leq 4$. To estimate this term we use the Lemma A.0.4 to obtain:

$$|e^{-\mu\lambda e_\theta}\psi_{b+1}(\mu\lambda w e_{2\theta}) - e^{-\lambda e_\theta}\psi_{b+1}(\lambda w e_{2\theta})| \leq (\mu - 1)\lambda e^{-\cos\theta\xi\lambda} \times$$
$$|w\psi_{b+2}(\xi\lambda w e_{2\theta}) - \psi_{b+1}(\xi\lambda w e_{2\theta})|, \text{ where } \xi \in (1, \mu). \tag{A.92}$$

The limits of integration in (A.91) can be re-expressed as

$$\frac{\alpha}{|t|} = \lambda\left(\frac{3 - \sqrt{\mu}}{2}\right)^2 \quad \frac{\beta}{|t|} = \lambda\left(\frac{3\sqrt{\mu} - 1}{2}\right)^2. \tag{A.93}$$

When we use the expression in (A.92) in (A.91), we see that the integral is multiplied by

$$\sqrt{x_1'}|t|^{\frac{\gamma-1}{2}}(\mu - 1) = |\sqrt{x_1} - \sqrt{x_1'}|^{\gamma}[\sqrt{\lambda}(\sqrt{\mu} - 1)]^{1-\gamma}(\sqrt{\mu} + 1). \tag{A.94}$$

We therefore need to show that

$$\lambda[\sqrt{\lambda}(\sqrt{\mu} - 1)]^{1-\gamma} \left[\int_0^{\frac{\alpha}{|t|}} + \int_{\frac{\beta}{|t|}}^{\infty}\right] w^b e^{-\cos\theta(w+\xi\lambda)} \times$$
$$|we_\theta\psi_{b+2}(\xi\lambda w e_{2\theta}) - \psi_{b+1}(\xi\lambda w e_{2\theta})||\sqrt{w} - \sqrt{\lambda}|^{\gamma}\frac{dw}{\sqrt{w}} \tag{A.95}$$

is uniformly bounded.

It is clear that the contribution of the integral from 0 to $\frac{\alpha}{|t|}$ is bounded for λ bounded, so we only need to evaluate the behavior of this term as $\lambda \to \infty$. For this purpose we need to split the integral into a part from 0 to $1/\lambda$ and the rest. The part from 0 to $1/\lambda$ is bounded by a constant times $\lambda^{2+b}e^{-\cos\theta\lambda}$, and is therefore controlled. The remaining contribution is bounded by

$$\lambda[\sqrt{\lambda}(\sqrt{\mu}-1)]^{1-\gamma}\int_{\frac{1}{\lambda}}^{\frac{\alpha}{|t|}} w^b e^{-\cos\theta(w+\xi\lambda)}|we_\theta\psi_{b+2}(\xi\lambda we_{2\theta})-\psi_{b+1}(\xi\lambda we_{2\theta})|\times$$
$$|\sqrt{w}-\sqrt{\lambda}|^\gamma\frac{dw}{\sqrt{w}}. \quad \text{(A.96)}$$

Using the asymptotic expansions for ψ_{b+1} and ψ_{b+2} we see that

$$|we_\theta\psi_{b+2}(\xi\lambda we_{2\theta})-\psi_{b+1}(\xi\lambda we_{2\theta})|\leq$$
$$C\sqrt{w}e^{2\cos\theta\sqrt{\xi\lambda w}}(\xi\lambda w)^{-\frac{b}{2}-\frac{3}{4}}\left[|\sqrt{w}-\sqrt{\lambda\xi}|+O\left(\frac{1}{\sqrt{w}}+\frac{1}{\sqrt{\lambda}}\right)\right]. \quad \text{(A.97)}$$

The integral in (A.96) is bounded by

$$\frac{C}{\lambda}\int_{\frac{1}{\lambda}}^{\frac{\alpha}{|t|}}\left(\frac{w}{\xi\lambda}\right)^{\frac{b}{2}-\frac{1}{4}} e^{-\cos\theta(\sqrt{w}-\sqrt{\xi\lambda})^2}|\sqrt{w}-\sqrt{\lambda\xi}|^\gamma\times$$
$$\left[|\sqrt{w}-\sqrt{\lambda}|+O\left(\frac{1}{\sqrt{w}}+\frac{1}{\sqrt{\lambda}}\right)\right]\frac{dw}{\sqrt{w}}. \quad \text{(A.98)}$$

In the interval of integration $\sqrt{\lambda\xi}-\sqrt{w}>\sqrt{\lambda}-\sqrt{w}$, and

$$\sqrt{\lambda\xi}-\sqrt{w}\leq C(\sqrt{\lambda}-\sqrt{w}), \quad \text{(A.99)}$$

provided that $C>3$. We can therefore estimate the leading term in (A.98) by

$$\frac{C}{\lambda}\int_{\frac{1}{\lambda}}^{\frac{\alpha}{|t|}}\left(\frac{w}{\lambda}\right)^{\frac{b}{2}-\frac{1}{4}} e^{-\cos\theta(\sqrt{w}-\sqrt{\lambda})^2}|\sqrt{w}-\sqrt{\lambda}|^{1+\gamma}\frac{dw}{\sqrt{w}}. \quad \text{(A.100)}$$

We let $z=\sqrt{w}-\sqrt{\lambda}$, to obtain that this is bounded by

$$\frac{C}{\lambda}\int_{\frac{1}{\sqrt{\lambda}}-\sqrt{\lambda}}^{\sqrt{\frac{\alpha}{|t|}}-\sqrt{\lambda}}\left(1+\frac{z}{\sqrt{\lambda}}\right)^{b-\frac{1}{2}} e^{-\cos\theta z^2}|z|^{1+\gamma}dz, \quad \text{(A.101)}$$

with the upper limit of integration given by

$$\sqrt{\frac{\alpha}{|t|}}-\sqrt{\lambda}=-\sqrt{\lambda}\frac{\sqrt{\mu}-1}{2}. \quad \text{(A.102)}$$

When the upper limit of integration is bounded, then the integral is bounded, and the contribution of this term is again bounded by $C|\sqrt{x_1}-\sqrt{x_1'}|^{\gamma}$. If the upper limit tends to $-\infty$, then we easily show that this term is bounded by

$$\frac{C_{b,\theta}}{\lambda}[\sqrt{\lambda}(\sqrt{\mu}-1)]^{\gamma}e^{-\cos\theta\frac{\lambda(\sqrt{\mu}-1)^2}{4}}, \tag{A.103}$$

and therefore the contribution of this term is again bounded by $C|\sqrt{x_1}-\sqrt{x_1'}|^{\gamma}$.

To complete the analysis of (A.98) we need to estimate the contribution of the error terms. Using the same change of variables we see that these terms are bounded by

$$\frac{C_b}{\lambda}\int\limits_{\frac{1}{\sqrt{\lambda}}-\sqrt{\lambda}}^{\sqrt{\frac{\alpha}{|t|}}-\sqrt{\lambda}}\left(1+\frac{z}{\sqrt{\lambda}}\right)^{b-\frac{1}{2}}e^{-\cos\theta z^2}|z|^{\gamma}\,O\left(\frac{1}{z+\sqrt{\lambda}}+\frac{1}{\sqrt{\lambda}}\right)dz. \tag{A.104}$$

These contributions are bounded as before if $\sqrt{\lambda}(\sqrt{\mu}-1)$ remains bounded. If the upper limit tends to $-\infty$, then this expression is bounded by

$$\frac{C_{b,\theta}}{\lambda^{\frac{3}{2}}}[\sqrt{\lambda}(\sqrt{\mu}-1)]^{\gamma-1}e^{-\cos\theta\frac{\lambda(\sqrt{\mu}-1)^2}{4}}, \tag{A.105}$$

completing the proof that the contribution from 0 to $\alpha/|t|$ is altogether bounded by $C_{b,\theta}|\sqrt{x_1}-\sqrt{x_1'}|^{\gamma}$.

We turn now to the part of (A.95) from $\beta/|t|$ to ∞. We need to split this integral into two parts only for $\lambda<1$. In this case we get a term bounded by

$$\lambda[\sqrt{\lambda}(\sqrt{\mu}-1)]^{1-\gamma}\int\limits_{\frac{\beta}{|t|}}^{\frac{1}{\lambda}}w^{b}e^{-\cos\theta w}|\sqrt{w}-\sqrt{\lambda}|^{\gamma}\frac{dw}{\sqrt{w}}, \tag{A.106}$$

which is clearly negligible as $\lambda\to 0$. The other term takes the form

$$\lambda[\sqrt{\lambda}(\sqrt{\mu}-1)]^{1-\gamma}\int\limits_{\max\left\{\frac{1}{\lambda},\frac{\beta}{|t|}\right\}}^{\infty}w^{b}e^{-\cos\theta(w+\xi\lambda)}\times$$
$$|we_{\theta}\psi_{b+2}(\xi\lambda we_{2\theta})-\psi_{b+1}(\xi\lambda we_{2\theta})||\sqrt{w}-\sqrt{\lambda}|^{\gamma}\frac{dw}{\sqrt{w}}. \tag{A.107}$$

The integral is bounded by

$$\frac{C_b}{\lambda}\int_{\max\left\{\frac{1}{\lambda},\frac{\beta}{|t|}\right\}}^{\infty}\left(\frac{w}{\xi\lambda}\right)^{\frac{b}{2}-\frac{1}{4}}e^{-\cos\theta(\sqrt{w}-\sqrt{\xi\lambda})^2}|\sqrt{w}-\sqrt{\lambda\xi}|^{\gamma}\times$$
$$\left[|\sqrt{w}-\sqrt{\lambda}|+O\left(\frac{1}{\sqrt{w}}+\frac{1}{\sqrt{\lambda}}\right)\right]\frac{dw}{\sqrt{w}}. \quad \text{(A.108)}$$

For $w\in[\frac{\beta}{|t|},\infty)$ we have the inequalities

$$\frac{1}{3}(\sqrt{w}-\sqrt{\lambda})\leq(\sqrt{w}-\sqrt{\lambda\xi})\leq(\sqrt{w}-\sqrt{\lambda}), \quad \text{(A.109)}$$

and therefore the expression in (A.108) is bounded by

$$\frac{C_b}{\lambda}\int_{\max\left\{\frac{1}{\lambda},\frac{\beta}{|t|}\right\}}^{\infty}\left(\frac{w}{\lambda}\right)^{\frac{b}{2}-\frac{1}{4}}e^{-\cos\theta\frac{(\sqrt{w}-\sqrt{\lambda})^2}{9}}|\sqrt{w}-\sqrt{\lambda}|^{\gamma}\times$$
$$\left[|\sqrt{w}-\sqrt{\lambda}|+O\left(\frac{1}{\sqrt{w}}+\frac{1}{\sqrt{\lambda}}\right)\right]\frac{dw}{\sqrt{w}}. \quad \text{(A.110)}$$

As before we let $z=\sqrt{w}-\sqrt{\lambda}$. If $1/\lambda>\beta/|t|$, then we obtain:

$$\frac{C_b}{\lambda}\int_{\frac{1}{\sqrt{\lambda}}-\sqrt{\lambda}}^{\infty}\left(1+\frac{z}{\sqrt{\lambda}}\right)^{b-\frac{1}{2}}e^{-\cos\theta\frac{z^2}{9}}|z|^{\gamma}\left[|z|+O\left(\frac{1}{z+\sqrt{\lambda}}+\frac{1}{\sqrt{\lambda}}\right)\right]dz. \quad \text{(A.111)}$$

This is bounded by $C_{b,\theta}e^{-\frac{\cos\theta}{20\lambda}}/\lambda$ as $\lambda\to 0$, so in this case the contribution of the integral from $\beta/|t|$ to infinity is bounded by $C_{b,\theta}|\sqrt{x_1}-\sqrt{x_1'}|^{\gamma}$.

The final case to consider is when $1/\lambda<\beta/|t|$, so that the lower limit of integration in (A.111) would be:

$$\sqrt{\frac{\beta}{|t|}}-\sqrt{\lambda}=\sqrt{\lambda}\frac{3(\sqrt{\mu}-1)}{2}. \quad \text{(A.112)}$$

An analysis, essentially identical to that above, shows that this term is bounded by

$$\frac{C_b}{\lambda}[\sqrt{\lambda}(\sqrt{\mu}-1)]^{\gamma}e^{-\cos\theta\frac{\lambda(\sqrt{\mu}-1)^2}{4}}. \quad \text{(A.113)}$$

As before we conclude that the contribution of this term is bounded by $C|\sqrt{x_1}-\sqrt{x_1'}|^{\gamma}$, which completes the proof of the lemma. □

LEMMA A.1.9. *[LEMMA 7.1.4] For $b > 0$, $\gamma \geq 0$, there is a constant $C_{b,\phi}$, bounded for $b \leq B$, so that for $t \in S_\phi$,*

$$\int_0^\infty |k_t^{b+1}(x,y)||\sqrt{y}-\sqrt{x}|y^{\frac{\gamma-1}{2}}dy \leq C_{b,\phi}|t|^{\frac{\gamma}{2}}. \tag{A.114}$$

PROOF. We let $t = \tau e^{i\theta}$, with $|\theta| < \frac{\pi}{2} - \phi$, and change variables with $w = y/|t|$, and $\lambda = x/|t|$ to obtain:

$$|t|^{\frac{\gamma}{2}} \int_0^\infty w^{b+\frac{\gamma}{2}} e^{-\cos\theta(w+\lambda)} |\psi_{b+1}(w\lambda e_{2\theta})||\sqrt{w}-\sqrt{\lambda}|\frac{dw}{\sqrt{w}}. \tag{A.115}$$

In the part of the integral from 0 to $1/\lambda$, we estimate ψ_{b+1} by a constant, obtaining

$$C_b|t|^{\frac{\gamma}{2}} \int_0^{\frac{1}{\lambda}} w^{b+\frac{\gamma}{2}} e^{-\cos\theta(w+\lambda)}|\sqrt{w}-\sqrt{\lambda}|\frac{dw}{\sqrt{w}}, \tag{A.116}$$

which is easily seen to be bounded by $C_{b,\theta}|t|^{\frac{\gamma}{2}}$.

In the non-compact part we use the asymptotic expansion of ψ_{b+1} to see that this contribution is bounded by

$$C_b|t|^{\frac{\gamma}{2}} \int_{\frac{1}{\lambda}}^\infty \left(\frac{w}{\lambda}\right)^{\frac{b}{2}-\frac{1}{4}} e^{-\cos\theta(\sqrt{w}-\sqrt{\lambda})^2} w^{\frac{\gamma}{2}} \frac{|\sqrt{w}-\sqrt{\lambda}|}{\sqrt{\lambda}}\frac{dw}{\sqrt{w}}. \tag{A.117}$$

As $\lambda \to 0$ this is bounded by $C_{b,\theta}|t|^{\frac{\gamma}{2}}e^{-\frac{\cos\theta}{2\lambda}}$. To estimate this term as $\lambda \to \infty$, we let $z = \sqrt{w} - \sqrt{\lambda}$ to obtain:

$$C_b|t|^{\frac{\gamma}{2}} \int_{\frac{1}{\sqrt{\lambda}}-\sqrt{\lambda}}^\infty \left(1+\frac{z}{\sqrt{\lambda}}\right)^{b-\frac{1}{2}} e^{-\cos\theta z^2}(\sqrt{\lambda}+z)^\gamma \frac{|z|}{\sqrt{\lambda}}dz. \tag{A.118}$$

This term is bounded by $C_{b,\theta}|t|^{\frac{\gamma}{2}}\lambda^{\frac{\gamma-1}{2}}$, thereby completing the proof of the lemma. □

LEMMA A.1.10. *If $b > \nu > 0$, and $0 < \phi < \frac{\pi}{2}$, then there is a constant $C_{b,\phi}$, bounded for $b \leq B$, so that if $t \in S_\phi$, we have*

$$\int_0^\infty \left(\frac{x}{y}\right)^\nu |k_t^b(x,y)||\sqrt{y}-\sqrt{x}|^\gamma dy \leq C_{b,\phi}|t|^{\frac{\gamma}{2}}. \tag{A.119}$$

PROOF. We let $t = \tau e^{i\theta}$, with $|\theta| < \frac{\pi}{2} - \phi$. Using $w = y/|t|$, and $\lambda = x/|t|$, we see that the integral in the lemma equals

$$|t|^{\frac{\gamma}{2}} \int_0^\infty \left(\frac{\lambda}{w}\right)^\nu w^b e^{-\cos\theta(w+\lambda)} \psi_b(w\lambda e_{2\theta}) |\sqrt{w} - \sqrt{\lambda}|^\gamma \frac{dw}{w}. \tag{A.120}$$

It therefore suffices to show that this integral is uniformly bounded for $\lambda \in [0, \infty)$.

The contribution from $[0, 1/\lambda]$ is bounded by

$$C_b \lambda^\nu e^{-\cos\theta\lambda} \int_0^{\frac{1}{\lambda}} w^{b-\nu-1} e^{-\cos\theta w} |\sqrt{w} - \sqrt{\lambda}|^\gamma \, dw. \tag{A.121}$$

Since $b - \nu > 0$ this is uniformly bounded for all λ. The remaining contribution comes from

$$\int_{\frac{1}{\lambda}}^\infty \left(\frac{\lambda}{w}\right)^\nu w^b e^{-\cos\theta(w+\lambda)} \psi_b(w\lambda e_{2\theta}) |\sqrt{w} - \sqrt{\lambda}|^\gamma \frac{dw}{w} \leq$$

$$C_b \int_{\frac{1}{\lambda}}^\infty \left(\frac{w}{\lambda}\right)^{\frac{b}{2}-\nu-\frac{1}{4}} e^{-\cos\theta(\sqrt{w}-\sqrt{\lambda})^2} |\sqrt{w} - \sqrt{\lambda}|^\gamma \frac{dw}{\sqrt{w}}. \tag{A.122}$$

As $\lambda \to 0$, this is bounded by $C_b e^{-\frac{\cos\theta}{2\lambda}}$. Letting $z = \sqrt{w} - \sqrt{\lambda}$, we obtain

$$C_b \int_{\frac{1}{\sqrt{\lambda}}-\sqrt{\lambda}}^\infty \left(1 + \frac{z}{\sqrt{\lambda}}\right)^{b-2\nu-\frac{1}{2}} e^{-\cos\theta z^2} |z|^\gamma \, dz. \tag{A.123}$$

It is again straightforward to see that this remains bounded as $\lambda \to \infty$, thus completing the proof of the lemma. □

LEMMA A.1.11. *[LEMMA 7.1.5] For $0 \leq \gamma < 1$, $1 \leq b$, and $0 < c < 1$, there is a constant C, so that if $ct < s < t$, then*

$$\int_0^\infty \left| k_t^b(x, z) - k_s^b(x, z) \right| |\sqrt{x} - \sqrt{z}| z^{\frac{\gamma-1}{2}} dz \leq C|t - s|^{\frac{\gamma}{2}}. \tag{A.124}$$

PROOF. The proof of this lemma is very similar to that of Lemma 6.1.8. If we let $w = z/t$, $\lambda = x/t$, and $\mu = t/s$, then the integral we need to estimate becomes

$$t^{\frac{\gamma}{2}} \int_0^\infty w^{b+\frac{\gamma}{2}-1} |e^{-(w+\lambda)} \psi_b(w\lambda) - \mu^b e^{-\mu(w+\lambda)} \psi_b(\mu^2 w\lambda)| |\sqrt{w} - \sqrt{\lambda}| \frac{dw}{\sqrt{w}}. \tag{A.125}$$

Proceeding as in the proof of Lemma 6.1.8 we see that

$$|e^{-(w+\lambda)}\psi_b(w\lambda) - \mu^b e^{-\mu(w+\lambda)}\psi_b(\mu^2 w\lambda)| =$$
$$(\mu-1)\xi^b e^{-\xi(w+\lambda)}\left|\left(\frac{b}{\xi}-(w+\lambda)\right)\psi_b(\xi^2 w\lambda) + 2\xi w\lambda\psi_{b+1}(\xi^2 w\lambda)\right|, \quad \text{(A.126)}$$

where $\xi \in (1, \mu) \subset (1, \frac{1}{c})$. Since

$$t^{\frac{\gamma}{2}}\left(\frac{t-s}{s}\right) \le \frac{|t-s|^{\frac{\gamma}{2}}}{c}, \quad \text{(A.127)}$$

it again suffices to show that the integral

$$\int_0^\infty w^{b+\frac{\gamma}{2}-1}e^{-\xi(w+\lambda)}\left|\left(\frac{b}{\xi}-(w+\lambda)\right)\psi_b(\xi^2 w\lambda) + \right.$$
$$\left. 2\xi w\lambda\psi_{b+1}(\xi^2 w\lambda)\right| |\sqrt{w}-\sqrt{\lambda}|\frac{dw}{\sqrt{w}} \quad \text{(A.128)}$$

is uniformly bounded for $\lambda \in (0, \infty)$.

We split the integral into a part from 0 to $1/\lambda$ and the rest. The compact part is bounded by

$$C\int_0^{\frac{1}{\lambda}} w^{b+\frac{\gamma}{2}-1}e^{-(w+\lambda)}|b+w+\lambda+2w\lambda||\sqrt{w}-\sqrt{\lambda}|\frac{dw}{\sqrt{w}}. \quad \text{(A.129)}$$

As $b \ge 1$, it is not difficult to see that this integral is uniformly bounded for $\lambda \in (0, \infty)$. For the non-compact part, we first estimate the contribution from the b/ξ-term. Using the asymptotic expansion we see that this part is bounded by

$$C\int_{\frac{1}{\lambda}}^\infty \left(\frac{w}{\lambda}\right)^{\frac{b}{2}-\frac{1}{4}} w^{\frac{\gamma-1}{2}}e^{-\xi(\sqrt{w}-\sqrt{\lambda})^2}|\sqrt{w}-\sqrt{\lambda}|\frac{dw}{\sqrt{w}}. \quad \text{(A.130)}$$

As $\lambda \to 0$ this is $O(e^{-\frac{1}{2\lambda}})$. To estimate this expression as $\lambda \to \infty$, we let $\sqrt{w}-\sqrt{\lambda} = z$; this integral is then bounded by

$$\int_{\frac{1}{\sqrt{\lambda}}-\sqrt{\lambda}}^\infty \left(1+\frac{z}{\sqrt{\lambda}}\right)^{b-\frac{1}{2}}(\sqrt{\lambda}+z)^{\gamma-1}e^{-z^2}|z|dz, \quad \text{(A.131)}$$

which, as $\lambda \to \infty$, is bounded by $C\lambda^{\frac{\gamma-1}{2}}$.

The other term is estimated by

$$\left|(w+\lambda)\psi_b(\xi^2 w\lambda) - 2\xi w\lambda\psi_{b+1}(\xi^2 w\lambda)\right| =$$
$$(\xi^2 w\lambda)^{\frac{1}{4}-\frac{b}{2}} \frac{e^{2\sqrt{\xi^2 w\lambda}}}{\sqrt{4\pi}} \left[(\sqrt{w}-\sqrt{\lambda})^2 + O\left(1+\sqrt{\frac{w}{\lambda}}+\sqrt{\frac{\lambda}{w}}\right)\right]. \quad \text{(A.132)}$$

It is again easy to see that, as $\lambda \to 0$, the contribution of this term is $O(e^{-\frac{1}{2\lambda}})$. To bound this term as $\lambda \to \infty$, we use the estimate from (A.132) in (A.128), and let $z = \sqrt{w} - \sqrt{\lambda}$ to obtain

$$\int_{\frac{1}{\sqrt{\lambda}}-\sqrt{\lambda}}^{\infty} \left(1+\frac{z}{\sqrt{\lambda}}\right)^{b-\frac{1}{2}} (\sqrt{\lambda}+z)^{\gamma-1} e^{-z^2} |z| \times$$
$$\left[|z|^2 + O\left(1+\frac{z}{\sqrt{\lambda}}+\frac{\sqrt{\lambda}}{z+\sqrt{\lambda}}\right)\right] dz. \quad \text{(A.133)}$$

It is again straightforward to see that all contributions in this integral are $O(\lambda^{\frac{\gamma-1}{2}})$, which completes the proof of the lemma. □

A.2 FIRST DERIVATIVE ESTIMATES

LEMMA A.2.1. *[LEMMA 6.1.10] For* $b > 0$, $0 \leq \gamma < 1$, *and* $0 < \phi < \frac{\pi}{2}$, *there is a* $C_{b,\phi}$ *so that for* $t \in S_\phi$ *we have*

$$\int_0^\infty |\partial_x k_t^b(x,y)||\sqrt{y}-\sqrt{x}|^\gamma \, dy \leq C_{b,\phi} \frac{|t|^{\frac{\gamma}{2}-1}}{1+\lambda^{\frac{1}{2}}}, \quad \text{(A.134)}$$

where $\lambda = x/|t|$.

PROOF. We let $t = \tau e^{i\theta}$, where $|\theta| < \frac{\pi}{2} - \phi$. Arguing as in the proof of Lemma 8.1 in [15] we see that

$$\int_0^\infty |\partial_x k_t^b(x,y)||\sqrt{y}-\sqrt{x}|^\gamma \, dy \leq$$
$$|t|^{\frac{\gamma}{2}-1} \int_0^\infty w^{b-1} e^{-\cos\theta(w+\lambda)} \left|we_\theta \psi_b'(w\lambda e_{2\theta}) - \psi_b(w\lambda e_{2\theta})\right| |\sqrt{w}-\sqrt{\lambda}|^\gamma \, dw.$$
$$\text{(A.135)}$$

Here $\lambda = x/|t|$ and $w = y/|t|$. We now estimate the quantity:

$$I(\lambda) = \int_0^\infty w^{b-1} e^{-\cos\theta(w+\lambda)} \left| w e_\theta \psi_b'(w\lambda e_{2\theta}) - \psi_b(w\lambda e_{2\theta}) \right| |\sqrt{w} - \sqrt{\lambda}|^\gamma \, dw. \tag{A.136}$$

We divide this integral into a part from 0 to $1/\lambda$ and the rest. We write $I(\lambda) = I_0(\lambda) + I_1(\lambda)$. In the first part we can estimate the integral using the Taylor expansions for ψ_b and ψ_b' as

$$I_0(\lambda) \leq C_b \int_0^{1/\lambda} w^{b-1} e^{-\cos\theta(w+\lambda)} \left(w + \frac{1}{\Gamma(b)} \right) |\sqrt{w} - \sqrt{\lambda}|^\gamma \, dw. \tag{A.137}$$

When λ remains bounded, the only difficulty that arises is that as $b \to 0$, the w^{b-1} term introduces a $1/b$, but this is compensated for by the $1/\Gamma(b)$, showing that this expression remains bounded as $b \to 0$. This term is bounded by a constant times

$$e^{-\cos\theta\lambda} \left(\frac{1 + \lambda^{\frac{\gamma}{2}}}{1 + \lambda^b} \right). \tag{A.138}$$

To estimate the other part of the integral, we use the asymptotic expansions for ψ_b and ψ_b', giving

$$I_1(\lambda) \leq C \int_{\frac{1}{\lambda}}^\infty \left(\frac{w}{\lambda} \right)^{\frac{b}{2} - \frac{1}{4}} \left| \sqrt{\lambda} - \sqrt{w} \right|^{1+\gamma} e^{-\cos\theta(\sqrt{\lambda} - \sqrt{w})^2} \frac{dw}{\sqrt{w\lambda}}. \tag{A.139}$$

Applying Lemma 6.1.23, this is $O(e^{-\frac{1}{\lambda}})$, as $\lambda \to 0$, and, as $\lambda \to \infty$,

$$I_1(\lambda) \leq C_{b,\phi} \lambda^{-\frac{1}{2}}. \tag{A.140}$$

Combining this with the estimates above, we can show that

$$I(\lambda) \leq \frac{C}{1 + \sqrt{\lambda}}. \tag{A.141}$$

This proves Lemma 6.1.10. □

LEMMA A.2.2. *[*LEMMA *6.1.11] For* $b > 0$, $0 < \gamma < 1$, $0 < \phi < \frac{\pi}{2}$, *and* $0 < c < 1$, *there is a constant* $C_{b,c,\phi}$ *so that for* $cx_2 < x_1 < x_2$, $t \in S_\phi$,

$$\int_0^\infty |\sqrt{x_1}\partial_x k_t^b(x_1, y) - \sqrt{x_2}\partial_x k_t^b(x_2, y)||\sqrt{x_1} - \sqrt{y}|^\gamma \, dy \leq$$

$$C_{b,c,\phi}|t|^{\frac{\gamma-1}{2}} \frac{\left(\frac{|\sqrt{x_2} - \sqrt{x_1}|}{\sqrt{|t|}} \right)}{1 + \left(\frac{|\sqrt{x_2} - \sqrt{x_1}|}{\sqrt{|t|}} \right)}. \tag{A.142}$$

PROOF. We let $t = \tau e^{i\theta}$, where $|\theta| < \frac{\pi}{2} - \phi$, and note that Lemma 6.1.10 provides a "trivial" estimate,

$$\int_0^\infty |\sqrt{x_1}\partial_x k_t^b(x_1, y) - \sqrt{x_2}\partial_x k_t^b(x_2, y)||\sqrt{x_1} - \sqrt{y}|^\gamma\, dy \leq$$

$$C_{b,\phi}\frac{\sqrt{x_1}|t|^{\frac{\gamma-1}{2}}}{\sqrt{x_1} + \sqrt{|t|}} \leq C_{b,\phi}|t|^{\frac{\gamma-1}{2}}, \quad \text{(A.143)}$$

which is the desired estimate when $|\sqrt{x_2} - \sqrt{x_1}|/\sqrt{|t|} >> 1$. Recalling that

$$\partial_x k_t^b(x, y) = \frac{1}{yt}\left(\frac{y}{t}\right)^b e^{-\frac{x+y}{t}}\left[\left(\frac{y}{t}\right)\psi_b'\left(\frac{xy}{t^2}\right) - \psi_b\left(\frac{xy}{t^2}\right)\right] \quad \text{(A.144)}$$

and setting $w = y/|t|$, $\lambda = x_1/|t|$, and $\mu = x_2/x_1$, we see that

$$\int_0^\infty |\sqrt{x_1}\partial_x k_t^b(x_1, y) - \sqrt{x_2}\partial_x k_t^b(x_2, y)||\sqrt{x_1} - \sqrt{y}|^\gamma\, dy =$$

$$\sqrt{\lambda}|t|^{\frac{\gamma-1}{2}}\int_0^\infty w^{b-1}e^{-\cos\theta w}|F(\mu, \lambda, w) - F(1, \lambda, w)||\sqrt{\lambda} - \sqrt{w}|^\gamma\, dw, \quad \text{(A.145)}$$

where we let

$$F(\mu, \lambda, w) = \sqrt{\mu}e^{-\mu\lambda e_\theta}\left[we_\theta\psi_b'(\mu\lambda we_{2\theta}) - \psi_b(\mu\lambda we_{2\theta})\right]. \quad \text{(A.146)}$$

Using Lemma A.0.4 we see that

$$|F(\mu, \lambda, w) - F(1, \lambda, w)| \leq (\mu - 1)|\partial_\mu F(\xi, \lambda, w)| \quad \text{for a} \quad \xi \in (1, \mu), \quad \text{(A.147)}$$

where $\xi \in (1, \frac{1}{c})$. We therefore need to estimate

$$\sqrt{\lambda}|t|^{\frac{\gamma-1}{2}}(\mu - 1)\int_0^\infty w^{b-1}e^{-\cos\theta w}|\partial_\mu F(\xi, \lambda, w)||\sqrt{\lambda} - \sqrt{w}|^\gamma\, dw. \quad \text{(A.148)}$$

Using the equation satisfied by ψ_b, we see that

$$\partial_\mu F(\xi, \lambda, w) =$$

$$e_\theta\frac{e^{-\xi\lambda e_\theta}}{\sqrt{\xi}}\left[\left(\xi\lambda + w - \frac{e_{-\theta}}{2}\right)\psi_b(\xi\lambda we_{2\theta}) - \left(2\xi\lambda we_\theta + (b - \frac{1}{2})w\right)\psi_b'(\xi\lambda we_{2\theta})\right]. \quad \text{(A.149)}$$

As usual we split the integral in (A.148) into the part from 0 to $1/\lambda$ and the rest. In the compact part we use the estimate

$$|\partial_\mu F(\xi,\lambda,w)| \leq C_b e^{-\cos\theta\xi\lambda}\left[\frac{1+\lambda+w}{\Gamma(b)}+\frac{w}{\Gamma(b+1)}+\lambda w O(1+\lambda+w)\right]. \tag{A.150}$$

Using this estimate we see that

$$\sqrt{\lambda}|t|^{\frac{\gamma-1}{2}}(\mu-1)\int_0^{\frac{1}{\lambda}} w^{b-1}e^{-\cos\theta w}|\partial_\mu F(\xi,\lambda,w)||\sqrt{\lambda}-\sqrt{w}|^\gamma\,dw \leq$$
$$C_{b,\theta}\sqrt{\lambda}|t|^{\frac{\gamma-1}{2}}e^{-\frac{\cos\theta\lambda}{2}}|\mu-1| \tag{A.151}$$

where the constant is uniformly bounded for $b\in(0,B]$, and $|\theta|\leq\frac{\pi}{2}-\phi$.

Now we turn to the non-compact part where we use the asymptotic expansions of ψ_b and ψ_b' to obtain:

$$|\partial_\mu F(\xi,\lambda,w)| =$$
$$(\xi\lambda w)^{\frac{1}{4}-\frac{b}{2}}e^{2\cos\theta\sqrt{\xi\lambda w}}e^{-\cos\theta\xi\lambda}\left[\frac{\lambda}{\xi}\left(1-\sqrt{\frac{w}{\xi\lambda}}\right)^2+O\left(1+\sqrt{\frac{w}{\lambda}}+\sqrt{\frac{\lambda}{w}}\right)\right]. \tag{A.152}$$

Using this expansion in the integral and setting $z=\sqrt{w/\lambda}$, we see that this term is bounded by

$$C_b\lambda^{\frac{\gamma}{2}+1}|t|^{\frac{\gamma-1}{2}}(\mu-1)\int_{\frac{1}{\lambda}}^{\infty} z^{b-\frac{1}{2}}e^{-\cos\theta\lambda(z-\sqrt{\xi})^2}|z-1|^\gamma\times$$
$$\left(\lambda(z-\sqrt{\xi})^2+O(1+z+1/z)\right)dz. \tag{A.153}$$

As $\lambda\to 0$ this term is easily seen to be bounded by $C_{b,\theta}|t|^{\frac{\gamma-1}{2}}(\mu-1)e^{-\frac{\cos\theta}{2\lambda}}$.

In this case, when $|\sqrt{x_2}-\sqrt{x_1}|/\sqrt{|t|}>1/4$, we use the "trivial" estimate in (A.143). We henceforth assume that $|\sqrt{x_2}-\sqrt{x_1}|/\sqrt{|t|}\leq 1/4$, which implies that

$$\sqrt{\mu}-1\leq\frac{1}{4\sqrt{\lambda}}. \tag{A.154}$$

In order to estimate the integral, we need to split it into three parts, with z lying in $[1/\lambda,1]$, $[1,\sqrt{\mu}]$, and $[\sqrt{\mu},\infty)$, respectively. Using the assumption in (A.154) we easily show that the integral over $[1,\sqrt{\mu}]$ is bounded by $C_{b,\theta}\sqrt{\lambda}|t|^{\frac{\gamma-1}{2}}(\mu-1)$, as desired. To treat the other two terms we use Laplace's method. The integral over

$[1/\lambda, 1]$ is bounded by

$$C_b\lambda^{\frac{\gamma}{2}+1}|t|^{\frac{\gamma-1}{2}}(\mu-1)\int\limits_{\frac{1}{\lambda}}^{1} z^{b-\frac{1}{2}}e^{-\cos\theta\lambda(z-1)^2}|z-1|^{\gamma}\times \left[\lambda(z-\sqrt{\mu})^2+O(1+z+1/z)\right]dz. \tag{A.155}$$

Laplace's method, using (A.154), shows that this term is also bounded by

$$C_{b,\theta}\sqrt{\lambda}|t|^{\frac{\gamma-1}{2}}(\mu-1).$$

Finally the integral over $[\sqrt{\mu}, \infty)$ is bounded by

$$C_b\lambda^{\frac{\gamma}{2}+1}|t|^{\frac{\gamma-1}{2}}(\mu-1)\int\limits_{\sqrt{\mu}}^{\infty} z^{b-\frac{1}{2}}e^{-\cos\theta\lambda(z-\sqrt{\mu})^2}|z-1|^{\gamma}\times \left[\lambda(z-1)^2+O(1+z+1/z)\right]dz. \tag{A.156}$$

Applying Laplace's method to this integral shows that it is bounded by

$$C_{b,\theta}\sqrt{\lambda}|t|^{\frac{\gamma-1}{2}}(\mu-1),$$

thereby completing the estimate of the non-compact term. The proof of the lemma is completed by noting that

$$\sqrt{\lambda}(\mu-1)=\frac{\sqrt{x_2}-\sqrt{x_1}}{\sqrt{|t|}}\frac{\sqrt{x_2}+\sqrt{x_1}}{\sqrt{x_1}}, \tag{A.157}$$

and therefore the lemma follows from the assumption that $0 < c < x_1/x_2 < 1$. □

LEMMA A.2.3. *[LEMMA 6.1.12] For $b > 0$, $0 < \gamma < 1$, there is a constant C_b so that for $t_1 < t_2 < 2t_1$, we have:*

$$\int\limits_{t_2-t_1}^{t_1}\int\limits_{0}^{\infty}|\partial_x k^b_{t_2-t_1+s}(x,y)-\partial_x k^b_s(x,y)||\sqrt{x}-\sqrt{y}|^{\gamma}\,dyds < C_b|t_2-t_1|^{\frac{\gamma}{2}}. \tag{A.158}$$

This result follows from the more basic:

LEMMA A.2.4. *[LEMMA 6.1.13] For $b > 0$, $0 \geq \gamma < 1$, and $0 < t_1 < t_2 < 2t_1$, we have for $s \in [t_2-t_1, t_1]$ that there is a constant C so that*

$$\int\limits_{0}^{\infty}|\partial_x k^b_{t_2-t_1+s}(x,y)-\partial_x k^b_s(x,y)||\sqrt{x}-\sqrt{y}|^{\gamma}\,dy < C\frac{(t_2-t_1)s^{\frac{\gamma}{2}-1}}{(t_2-t_1+s)(1+\sqrt{x/s})}. \tag{A.159}$$

PROOF OF LEMMA 6.1.12. The estimate in (A.158) follows by integrating the estimate in (6.42):

$$\int_{t_2-t_1}^{t_1}\int_0^{\infty}|\partial_x k^b_{t_2-t_1+s}(x,y)-\partial_x k^b_s(x,y)||\sqrt{x}-\sqrt{y}|^{\gamma}\,dyds \leq C\int_{t_2-t_1}^{t_1}\frac{(t_2-t_1)s^{\frac{\gamma}{2}-1}ds}{(t_2-t_1)+s}$$
$$\leq C\int_{t_2-t_1}^{\infty}(t_2-t_1)s^{\frac{\gamma}{2}-2}ds=\frac{2C}{2-\gamma}|t_2-t_1|^{\frac{\gamma}{2}}. \quad (A.160)$$

□

We now give the proof of Lemma 6.1.13.

PROOF. This argument is very similar to the proof of Lemma 6.1.11. Set $\tau = t_2 - t_1$, and define

$$G(\mu,\lambda,w)=\mu^{b+1}e^{-\mu(w+\lambda)}\left[\mu w\psi_b'(\mu^2 w\lambda)-\psi_b(\mu^2 w\lambda)\right]. \quad (A.161)$$

Setting $w = y/s$, $\lambda = x/s$, and $\mu = s/(\tau + s)$, we see that

$$\int_0^{\infty}|\partial_x k^b_{t_2-t_1+s}(x,y)-\partial_x k^b_s(x,y)||\sqrt{x}-\sqrt{y}|^{\gamma}\,dy=$$
$$s^{\frac{\gamma}{2}-1}\int_0^{\infty}w^{b-1}|G(\mu,\lambda,w)-G(1,\lambda,w)||\sqrt{\lambda}-\sqrt{w}|^{\gamma}\,dw=$$
$$s^{\frac{\gamma}{2}-1}(\mu-1)\int_0^{\infty}w^{b-1}|\partial_\mu G(\xi,\lambda,w)||\sqrt{\lambda}-\sqrt{w}|^{\gamma}\,dw; \quad (A.162)$$

here $\xi \in [\mu, 1]$. In the last line we use the mean value theorem. The assumption $t_1 < t_2 < 2t_1$ shows that $\mu \in [\frac{1}{2}, 1)$.

A calculation, using the equation satisfied by ψ_b shows that

$$\partial_\mu G(\xi,\lambda,w)=\xi^{b+1}e^{-\xi(w+\lambda)}\times$$
$$\left[\psi_b(\xi^2 w\lambda)(3w+\lambda-\frac{1+b}{\xi})-w\psi_b'(\xi^2 w\lambda)(b-2+\xi(w+3\lambda))\right]. \quad (A.163)$$

As usual, we split the integral into a part from 0 to $1/\lambda$ and the rest. In the compact part we observe that

$$|\partial_\mu G(\xi,\lambda,w)|\leq Ce^{-\frac{w+\lambda}{2}}\left|\frac{w+\lambda+1}{\Gamma(b)}+O\left(w(1+w+\lambda)^2\right)\right|. \quad (A.164)$$

The compact part is therefore bounded by

$$s^{\frac{\gamma}{2}-1}(\mu-1)\int_0^{\frac{1}{\lambda}} w^{b-1}e^{-\frac{w+\lambda}{2}}\left|\frac{w+\lambda+1}{\Gamma(b)}+O\left(w(1+w+\lambda)^2\right)\right||\sqrt{\lambda}-\sqrt{w}|^{\gamma}\,dw. \tag{A.165}$$

As $\lambda\to\infty$ this is bounded by $Cs^{\frac{\gamma}{2}-1}(\mu-1)e^{-\frac{\lambda}{4}}$, and as $\lambda\to 0$, by $Cs^{\frac{\gamma}{2}-1}(\mu-1)$.

For the non-compact part we use the asymptotic expansions of ψ_b and ψ_b' of order 2, given in (A.221), to obtain:

$$\begin{aligned} G(\xi,\lambda,w) \le C\xi^{b+1}e^{-\xi(\sqrt{w}-\sqrt{\lambda})^2}(\xi^2 w\lambda)^{\frac{1}{4}-\frac{b}{2}}\times \\ \left|\lambda\left(1-\sqrt{\frac{w}{\lambda}}\right)^3 - a_1(b)\sqrt{\frac{w}{\lambda}}\left(1-\sqrt{\frac{w}{\lambda}}\right)+\right. \\ \left. a_2(b)\left(1-\sqrt{\frac{w}{\lambda}}\right)+a_3(b)\left(1-\sqrt{\frac{\lambda}{w}}\right)+O\left(1+\frac{1}{\lambda}+\frac{1}{w}+\frac{1}{\sqrt{w\lambda}}\right)\right|; \end{aligned} \tag{A.166}$$

here $a_j(b)$, $j=1,2,3$ are polynomials in b. Using this expression in the integral and letting $z=\sqrt{\lambda}(\sqrt{w/\lambda}-1)$, we see this is bounded by

$$Cs^{\frac{\gamma}{2}-1}(\mu-1)\int_{\frac{1}{\sqrt{\lambda}}-\sqrt{\lambda}}^{\infty}\left(\frac{z}{\sqrt{\lambda}}+1\right)^{b-\frac{1}{2}}e^{-\frac{z^2}{2}}z^{\gamma}\left[\frac{|z|^3+|z|}{\sqrt{\lambda}}+O\left(\frac{1}{\lambda}\right)\right]dz. \tag{A.167}$$

As $\lambda\to 0$ this is bounded by $Cs^{\frac{\gamma}{2}-1}(\mu-1)e^{-\frac{1}{4\lambda}}$. When $\lambda\to\infty$, Laplace's method applies to show that it is bounded by $Cs^{\frac{\gamma}{2}-1}(\mu-1)/\sqrt{\lambda}$. This completes the proof of the lemma. □

A.3 SECOND DERIVATIVE ESTIMATES

LEMMA A.3.1. *[*LEMMA *6.1.14] For $b>0$, $0<\gamma<1$, and $0<\phi<\frac{\pi}{2}$, there is a $C_{b,\phi}$ so that for $t=|t|e^{i\theta}$ with $|\theta|<\frac{\pi}{2}-\phi$,*

$$\begin{aligned} &\int_0^{|t|}\int_0^{\infty}|x\partial_x^2 k^b_{se^{i\theta}}(x,y)||\sqrt{y}-\sqrt{x}|^{\gamma}\,dyds \le C_{b,\phi}x^{\frac{\gamma}{2}} \text{ and} \\ &\int_0^{t}\int_0^{\infty}|x\partial_x^2 k^b_{se^{i\theta}}(x,y)||\sqrt{y}-\sqrt{x}|^{\gamma}\,dyds \le C_{b,\phi}|t|^{\frac{\gamma}{2}}. \end{aligned} \tag{A.168}$$

We deduce this lemma from the following result, of interest in its own right:

LEMMA A.3.2. *[LEMMA 6.1.15] For $b > 0$, $0 \le \gamma < 1$, and $0 < \phi < \frac{\pi}{2}$, there is a $C_{b,\phi}$ so that if $t \in S_\phi$, then*

$$\int_0^\infty |x\partial_x^2 k_t^b(x,y)||\sqrt{x} - \sqrt{y}|^\gamma \, dy \le C_{b,\phi} \frac{\lambda |t|^{\frac{\gamma}{2}-1}}{1+\lambda}, \tag{A.169}$$

where $\lambda = x/|t|$.

We let $t = \tau e^{i\theta}$, where $|\theta| < \frac{\pi}{2} - \phi$, and first show how to deduce 6.1.14 from (A.169).

PROOF OF LEMMA 6.1.14. To prove the first estimate in (A.168), using (A.169), we see that

$$\int_0^{|t|}\int_0^\infty |x\partial_x^2 k^b_{se^{i\theta}}(x,y)||\sqrt{y} - \sqrt{x}|^\gamma \, dyds \le C_{b,\phi} \int_0^{|t|} s^{\frac{\gamma}{2}-1} \frac{x/s}{1+x/s} ds. \tag{A.170}$$

Splitting this into an integral from 0 to x and the rest (if needed), we easily see that the first estimate in (A.168) holds. The second estimate follows from (A.170) and the observation that $x/(x+s) \le 1$. □

Now we prove Lemma 6.1.15.

PROOF OF LEMMA 6.1.15. We denote the left-hand side of (A.169) by I. The formula (4.5) for k_t^b, and the second order equation satisfied by $\psi_b(z)$:

$$z\psi_b'' + b\psi_b' - \psi_b = 0, \tag{A.171}$$

imply that

$$x\partial_x^2 k_t^b(x,y) =$$
$$\frac{1}{yt}\left(\frac{y}{t}\right)^b e^{-\frac{(x+y)}{t}} \left[\left(\frac{x+y}{t}\right)\psi_b\left(\frac{xy}{t^2}\right) - \left(\frac{2xy}{t^2}\right)\psi_b'\left(\frac{xy}{t^2}\right) - \left(\frac{by}{t}\right)\psi_b'\left(\frac{xy}{t^2}\right)\right]. \tag{A.172}$$

We let $w = y/|t|$ $\lambda = x/|t|$ to obtain

$$I = \frac{1}{|t|^{1-\frac{\gamma}{2}}} \int_0^\infty w^b e^{-\cos\theta(w+\lambda)} \times$$
$$\left|(w+\lambda)\psi_b(w\lambda e_{2\theta}) - (2w\lambda e_\theta + bw)\psi_b'(w\lambda e_{2\theta})\right| |\sqrt{w} - \sqrt{\lambda}|^\gamma \frac{dw}{w}, \tag{A.173}$$

which we split into a part from $[0, \frac{1}{\lambda}]$ and the rest. In the compact part, we use the estimate

$$\left|(w+\lambda)\psi_b(w\lambda e_{2\theta}) - (2w\lambda e_\theta + bw)\psi_b'(w\lambda e_{2\theta})\right| \leq C_b\lambda\left[\frac{1}{\Gamma(b)} + w(1+w+\lambda)\right]. \tag{A.174}$$

Applying Lemma 6.1.22 shows that these parts of the w-integral are bounded by

$$C_{b,\theta}e^{-\cos\theta\frac{\lambda}{2}} \text{ as } \lambda \to \infty \text{ and } C_{b,\theta}\lambda \text{ as } \lambda \to 0. \tag{A.175}$$

In the non-compact part of the w-integral, we use the asymptotic expansion to obtain

$$\left|(w+\lambda)\psi_b(w\lambda e_{2\theta}) - (2w\lambda e_\theta + bw)\psi_b'(w\lambda e_{2\theta})\right| =$$
$$(w\lambda)^{\frac{1}{4}-\frac{b}{2}}e^{2\cos\theta\sqrt{w\lambda}}\left[(\sqrt{w}-\sqrt{\lambda})^2 + O\left(\frac{w+\lambda}{\sqrt{w\lambda}}+1\right)\right]. \tag{A.176}$$

Applying Lemma 6.1.23 shows that the principal terms of the non-compact part of the w-integral are bounded by

$$C_{b,\theta}e^{-\frac{1}{2\lambda}}. \tag{A.177}$$

This leaves only the error term in (A.176). Again applying Lemma 6.1.23 shows that these terms are also bounded by the expression in (A.177). □

LEMMA A.3.3. *[LEMMA 6.1.16] For $b > 0$, $0 < \gamma < 1$, $0 < \phi < \frac{\pi}{2}$, and $0 < x_2/3 < x_1 < x_2$, there is a constant $C_{b,\phi}$ so that, for $t \in S_\phi$, we have*

$$\int_0^{|t|}\left|(\partial_y y - b)k^b_{se^{i\theta}}(x_2,\alpha) - (\partial_y y - b)k^b_{se^{i\theta}}(x_2,\beta)\right| ds \leq C_{b,\phi}, \tag{A.178}$$

where α and β are defined in (6.35).

PROOF. We let $t = |t|e^{i\theta}$, where $|\theta| < \frac{\pi}{2} - \phi$, and use I to denote the quantity on the left in (A.178). Using (4.5), we see that, for t in the right half plane,

$$(\partial_y y - b)k^b_t(x,y) = \frac{1}{t}\left(\frac{y}{t}\right)^b e^{-\frac{(x+y)}{t}}\left[\left(\frac{x}{t}\right)\psi_b'\left(\frac{xy}{t^2}\right) - \psi_b\left(\frac{xy}{t^2}\right)\right], \tag{A.179}$$

and therefore:

$$|I| \leq \int_0^t \frac{e^{-\cos\theta\frac{x_2}{s}}}{s}$$
$$\left|\left(\frac{\alpha}{s}\right)^b e^{-\frac{\alpha e_\theta}{s}}\left[\left(\frac{\alpha e_\theta}{s}\right)\psi_b'\left(\frac{x_2\alpha e_{2\theta}}{s^2}\right) - \psi_b\left(\frac{x_2\alpha e_{2\theta}}{s^2}\right)\right] - \right.$$
$$\left.\left(\frac{\beta}{s}\right)^b e^{-\frac{\beta e_\theta}{s}}\left[\left(\frac{\beta e_\theta}{s}\right)\psi_b'\left(\frac{x_2\beta e_{2\theta}}{s^2}\right) - \psi_b\left(\frac{x_2\beta e_{2\theta}}{s^2}\right)\right]\right| ds. \tag{A.180}$$

As

$$\frac{1}{3} \leq \frac{x_1}{x_2} < 1, \tag{A.181}$$

the numbers

$$\frac{\alpha}{s} < \frac{x_1}{s} < \frac{x_2}{s} < \frac{\beta}{s}$$

are all comparable. If $s < x_1$, then we can use the asymptotic expansion to estimate the integrand by

$$\frac{C_b}{s} e^{-\cos\theta \frac{(\sqrt{x_2}-\sqrt{x_1})^2}{4s}} \left[\frac{(\sqrt{x_2}-\sqrt{x_1})}{2\sqrt{s}} + O\left(\sqrt{\frac{s}{\alpha}}\right) \right]. \tag{A.182}$$

Changing variables with

$$\sigma = \frac{(\sqrt{x_2}-\sqrt{x_1})^2}{s}, \tag{A.183}$$

the principal term in the integral from 0 to x_1 becomes:

$$C_b \int\limits_{\frac{(\sqrt{x_2}-\sqrt{x_1})^2}{x_1}}^{\infty} e^{-\cos\theta \frac{x}{4}} \frac{dx}{\sqrt{x}}. \tag{A.184}$$

This is uniformly bounded. The integral of the error term is bounded by

$$\int\limits_0^{x_1} \frac{ds}{\sqrt{\alpha s}} = \frac{1}{2}\sqrt{\frac{x_1}{\alpha}}. \tag{A.185}$$

As $x_1/x_2 > 1/3$ this is bounded by 1, completing the estimate of this part of the s-integral.

If $|t| > x_1$, then we also need to estimate the s-integral over $[x_1, t]$. If we let

$$F_\mu(z) = z^b e^{-ze_\theta}[ze_\theta \psi_b'(\mu z e_{2\theta}) - \psi_b(\mu z e_{2\theta})], \tag{A.186}$$

then the remaining part of the s-integral can be written:

$$\int\limits_{x_1}^{t} \frac{e^{-\cos\theta \frac{x_2}{s}}}{s} \left| F_{\frac{x_2}{s}}\left(\frac{\alpha}{s}\right) - F_{\frac{x_2}{s}}\left(\frac{\beta}{s}\right) \right| ds. \tag{A.187}$$

Lemma A.0.4 shows that this is estimated by

$$C \int\limits_{x_1}^{t} \frac{e^{-\cos\theta \frac{x_2}{s}}}{s} |F'_{\frac{x_2}{s}}(\xi)| \left(\frac{\beta-\alpha}{s}\right) ds. \tag{A.188}$$

Here $\xi \in [\frac{\alpha}{s}, \frac{\beta}{s}] \subset (0, \frac{\beta}{x_1}]$ and $\mu \in [\frac{x_2}{t}, \frac{x_2}{x_1}]$. The z^b-term in $F_\mu(z)$ is the only term which may contribute something unbounded to $F'_\mu(z)$, and this occurs only if $b < 1$.

The remaining terms are easily seen to contribute a term bounded by $C_b(1 - x_1/x_2)$. The z^{b-1}-term is bounded by

$$K = bC_b \int_{x_1}^{t} \frac{e^{-\cos\theta \frac{x_2}{s}}}{s} \left(\frac{\alpha}{s}\right)^{b-1} \left(\frac{\beta-\alpha}{s}\right) ds. \tag{A.189}$$

We let $w = x_2/s$ to obtain

$$K = bC_b \left(\frac{\beta-\alpha}{x_2}\right)\left(\frac{\alpha}{x_2}\right)^{b-1} \int_{\frac{x_2}{|t|}}^{\frac{x_2}{x_1}} w^{b-1} e^{-\cos\theta w} dw \leq C_{b,\theta} b\Gamma(b)\left(1 - \frac{x_1}{x_2}\right). \tag{A.190}$$

This completes the proof that there is a constant $C_{b,\phi}$ uniformly bounded with b, so that

$$I \leq C_{b,\phi}. \tag{A.191}$$

□

LEMMA A.3.4. *[*LEMMA *6.1.17] For $b > 0$, $0 < \gamma < 1$, $\phi < \frac{\pi}{2}$, and $0 < x_2/3 < x_1 < x_2$, if $J = [\alpha, \beta]$, with the endpoints given by* (6.35)*, there is a constant $C_{b,\phi}$ so that if $|\theta| < \frac{\pi}{2} - \phi$, then*

$$\begin{aligned} I_1 &= \int_0^{|t|}\int_\alpha^\beta |L_b k^b_{se^{i\theta}}(x_2, y)||\sqrt{y} - \sqrt{x_2}|^\gamma \, dy ds \leq C_{b,\phi}|\sqrt{x_2} - \sqrt{x_1}|^\gamma \\ I_2 &= \int_0^{|t|}\int_\alpha^\beta |L_b k^b_{se^{i\theta}}(x_1, y)||\sqrt{y} - \sqrt{x_1}|^\gamma \, dy ds \leq C_{b,\phi}|\sqrt{x_2} - \sqrt{x_1}|^\gamma . \end{aligned} \tag{A.192}$$

PROOF. Throughout these calculations we use the formula, valid for t in the right half plane:

$$L_b k^b_t(x, y) = \partial_t k^b_t(x, y) = \frac{1}{yt}\left(\frac{y}{t}\right)^b e^{-\frac{(x+y)}{t}}\left[\left(\frac{x+y}{t} - b\right)\psi_b\left(\frac{xy}{t^2}\right) - \left(\frac{2xy}{t^2}\right)\psi_b'\left(\frac{xy}{t^2}\right)\right]. \tag{A.193}$$

We give the argument for I_1; the argument for I_2 is essentially identical.

If we let

$$w = \frac{y}{s} \quad \lambda = \frac{x_2}{s}, \tag{A.194}$$

and

$$R_{\alpha,\beta,t} = \{(w, \lambda) : \frac{\alpha}{x_2}\lambda \leq w \leq \frac{\beta}{x_2}\lambda \text{ and } \frac{x_2}{|t|} \leq \lambda\}, \tag{A.195}$$

then I_1 becomes:

$$|I_1| \leq x_2^{\frac{\gamma}{2}} \iint\limits_{R_{\alpha,\beta,t}} w^{b-1} e^{-\cos\theta(w+\lambda)} |[(w+\lambda)e_\theta - b]\psi_b(w\lambda e_{2\theta}) - 2w\lambda e_{2\theta}\psi_b'(w\lambda e_{2\theta})| \times \left|1-\sqrt{\frac{w}{\lambda}}\right|^\gamma \frac{dw d\lambda}{\lambda}. \quad (A.196)$$

As in the previous cases we estimate ψ_b and ψ_b' using the Taylor expansion where $w\lambda < 1$ and using the asymptotic expansion where $w\lambda \geq 1$. In the present instance this divides the argument into two cases: (1) $\frac{x_1}{|t|} \geq 1$ and (2) $\frac{x_1}{|t|} < 1$. In case 1 we only need to use the asymptotic expansions, whereas in case 2 we also have to consider another term, where we estimate ψ_b and ψ_b' using the Taylor expansion. We begin with case 1.

The asymptotic expansion gives the estimate

$$|I_1| \leq C_b x_2^{\frac{\gamma}{2}} \iint\limits_{R_{\alpha,\beta,t}} w^{b-1}(w\lambda)^{\frac{1}{4}-\frac{b}{2}} e^{-\cos\theta(\sqrt{w}-\sqrt{\lambda})^2} \times \left|(\sqrt{w}-\sqrt{\lambda})^2 + b\left[1+O\left(\frac{1}{\sqrt{w\lambda}}\right)\right]\right| \left|1-\sqrt{\frac{w}{\lambda}}\right|^\gamma \frac{dw d\lambda}{\lambda}. \quad (A.197)$$

As w/λ is bounded above and below, this satisfies

$$|I_1| \leq C_b x_2^{\frac{\gamma}{2}} \iint\limits_{R_{\alpha,\beta,t}} \left(\frac{w}{\lambda}\right)^{\frac{b}{2}-\frac{1}{4}} e^{-\lambda\cos\theta\left(1-\sqrt{\frac{w}{\lambda}}\right)^2} \times \left[\lambda\left(1-\sqrt{\frac{w}{\lambda}}\right)^2+1\right] \left|1-\sqrt{\frac{w}{\lambda}}\right|^\gamma \frac{dw d\lambda}{\sqrt{w\lambda}}. \quad (A.198)$$

We let $z = \sqrt{w/\lambda} - 1$; taking into account that z is bounded we obtain:

$$|I_1| \leq C_b x_2^{\frac{\gamma}{2}} \int\limits_{\frac{x_2}{|t|}}^{\infty} \int\limits_{\sqrt{\frac{\alpha}{x_2}}-1}^{\sqrt{\frac{\beta}{x_2}}-1} e^{-\lambda\cos\theta z^2}\left[\lambda z^2+1\right]|z|^\gamma \frac{dz d\lambda}{\sqrt{\lambda}}. \quad (A.199)$$

We interchange the order of the integrations and set $x = \lambda z^2$, in the λ-integral, to see that

$$|I_1| \leq C_b x_2^{\frac{\gamma}{2}} \int\limits_{\sqrt{\frac{\alpha}{x_2}}-1}^{\sqrt{\frac{\beta}{x_2}}-1} \int\limits_{\frac{x_2 z^2}{|t|}}^{\infty} e^{-\cos\theta x}\left(\frac{1}{\sqrt{x}}+\sqrt{x}\right) dx |z|^{\gamma-1} dz. \quad (A.200)$$

The x-integral is bounded by a constant depending only on θ, and this shows that there is a constant $C_{b,\theta}$, which is bounded for $0 < b \leq B$, so that

$$|I_1| \leq C_{b,\theta}\|g\|_{WF,0,\gamma}|\sqrt{x_2}-\sqrt{x_1}|^\gamma. \quad (A.201)$$

Now we turn to case 2. The foregoing analysis is used to estimate the part of the integral where $w\lambda > 1$, by using 1 as the lower limit of integration in (A.199) instead of $x_2/|t|$. This leaves the part of the integral in (A.196) over the set

$$R_{\alpha,\beta,t} \cap \{(w,\lambda) : w\lambda < 1\}. \tag{A.202}$$

We replace this set, with the slightly larger set

$$R'_{\alpha,\beta,t} = \{(w,\lambda) : \frac{\alpha}{x_2}\lambda < w < \frac{\beta}{x_2}\lambda \text{ and } \frac{x_2}{|t|} \leq \lambda \leq \frac{x_2}{\alpha}\}. \tag{A.203}$$

Using the Taylor series, we see that this term is bounded by

$$Cx_2^{\frac{\gamma}{2}} \int\limits_{\frac{x_2}{|t|}}^{\frac{x_2}{\alpha}} \int\limits_{\lambda\frac{\alpha}{x_2}}^{\lambda\frac{\beta}{x_2}} w^{b-1}\left[(w+\lambda+b)\left(\frac{1}{\Gamma(b)}+w\lambda\right)+w\lambda\right]\times \left|1-\sqrt{\frac{w}{\lambda}}\right|^{\gamma}\frac{dwd\lambda}{\lambda}. \tag{A.204}$$

In the w-integral we let $\sigma = w/\lambda$ to see that this is bounded by

$$Cx_2^{\frac{\gamma}{2}} \int\limits_{0}^{\frac{x_2}{\alpha}} \int\limits_{\frac{\alpha}{x_2}}^{\frac{\beta}{x_2}} (\sigma\lambda)^{b-1}\left[(\sigma\lambda+\lambda+b)\left(\frac{1}{\Gamma(b)}+\sigma\lambda^2\right)+\sigma\lambda^2\right]\times \left|1-\sqrt{\sigma}\right|^{\gamma} d\sigma d\lambda. \tag{A.205}$$

As c in (A.181) is at least $1/3$, we know that range of the σ-integral satisfies

$$\frac{\sqrt{3}-1}{2} \leq \sqrt{\sigma} \leq 1. \tag{A.206}$$

In the domain of the σ-integral, the quantity $x_2^{\frac{\gamma}{2}}\left|1-\sqrt{\sigma}\right|^{\gamma}$ is bounded by a constant multiple of $|\sqrt{x_2}-\sqrt{x_1}|^{\gamma}$. As σ is bounded above and below, all that remains is the λ-integral. An elementary calculation shows that it remains bounded, even as $b \to 0$. This completes the proof, in all cases, that there is a constant C_b, bounded with b, so that (A.192) holds for I_1. The estimate for I_2 is essentially the same. □

LEMMA A.3.5. *[LEMMA 6.1.18] For* $b > 0$, $0 < \gamma < 1$, $0 < \phi < \frac{\pi}{2}$, *and* $0 < x_2/3 < x_1 < x_2$, *if* $J = [\alpha, \beta]$, *with the endpoints given by* (6.35), *there is a constant* $C_{b,\phi}$ *so that if* $|\theta| < \frac{\pi}{2} - \phi$, *then*

$$\int\limits_0^t \int\limits_{J^c} |L_b k^b_{se^{i\theta}}(x_2,y) - L_b k^b_{se^{i\theta}}(x_1,y)||\sqrt{y}-\sqrt{x_1}|^{\gamma}\,dyds \leq C_{b,\phi}|\sqrt{x_2}-\sqrt{x_1}|^{\gamma}. \tag{A.207}$$

PROOF. We use the formula for $L_b k_t^b$, given in (A.193), hence:

$$I^+ = \int_0^{|t|}\int_\beta^{\infty} \left(\frac{y}{s}\right)^{b-1} e^{-\cos\theta\frac{y}{s}} \times$$
$$\left| e^{-\frac{x_2 e_\theta}{s}} \left[\left(\frac{(x_2+y)e_\theta}{s} - b\right)\psi_b\left(\frac{x_2 y e_{2\theta}}{s^2}\right) - 2\left(\frac{x_2 y e_{2\theta}}{s^2}\right)\psi_b'\left(\frac{x_2 y e_{2\theta}}{s^2}\right)\right] - e^{-\frac{x_1 e_\theta}{s}} \times \right.$$
$$\left.\left[\left(\frac{(x_1+y)e_\theta}{s} - b\right)\psi_b\left(\frac{x_1 y e_{2\theta}}{s^2}\right) - 2\left(\frac{x_1 y e_{2\theta}}{s^2}\right)\psi_b'\left(\frac{x_1 y e_{2\theta}}{s^2}\right)\right]\right| |\sqrt{y}-\sqrt{x_1}|^\gamma \frac{dwds}{s^2}, \tag{A.208}$$

and

$$I^- = \int_0^{|t|}\int_0^{\alpha} \left(\frac{y}{s}\right)^{b-1} e^{-\cos\theta\frac{y}{s}} \times$$
$$\left| e^{-\frac{x_2 e_\theta}{s}} \left[\left(\frac{(x_2+y)e_\theta}{s} - b\right)\psi_b\left(\frac{x_2 y e_{2\theta}}{s^2}\right) - 2\left(\frac{x_2 y e_{2\theta}}{s^2}\right)\psi_b'\left(\frac{x_2 y e_{2\theta}}{s^2}\right)\right] - e^{-\frac{x_1 e_\theta}{s}} \times \right.$$
$$\left.\left[\left(\frac{(x_1+y)e_\theta}{s} - b\right)\psi_b\left(\frac{x_1 y e_{2\theta}}{s^2}\right) - 2\left(\frac{x_1 y e_{2\theta}}{s^2}\right)\psi_b'\left(\frac{x_1 y e_{2\theta}}{s^2}\right)\right]\right| |\sqrt{y}-\sqrt{x_1}|^\gamma \frac{dwds}{s^2}. \tag{A.209}$$

For this case we give the details for I^-, and leave I^+ to the interested reader.

We change variables, setting

$$w = \frac{y}{s} \quad \lambda = \frac{x_1}{s} \text{ so that } \frac{dyds}{s^2} = \frac{dwd\lambda}{\lambda}; \tag{A.210}$$

we also let $\mu = x_2/x_1$. The integral now satisfies:

$$|I^-| \leq \int_{\frac{x_1}{|t|}}^{\infty}\int_0^{\frac{\alpha\lambda}{x_1}} w^{b-1} e^{-\cos\theta w} \times$$
$$\left| e^{-\mu\lambda e_\theta}\left[[(\mu\lambda + w)e_\theta - b]\psi_b(\mu\lambda w e_{2\theta}) - 2(\mu\lambda w e_{2\theta})\psi_b'(\mu\lambda w e_{2\theta})\right] - e^{-\lambda e_\theta} \times \right.$$
$$\left.\left[[(\lambda + w)e_\theta - b]\psi_b(\lambda w e_{2\theta}) - 2(\lambda w e_{2\theta})\psi_b'(\lambda w e_{2\theta})\right]\right| |\sqrt{\lambda} - \sqrt{w}|^\gamma \frac{x_1^{\frac{\gamma}{2}} dwd\lambda}{\lambda^{1+\frac{\gamma}{2}}}. \tag{A.211}$$

We split the w-integral into the part, $I^{--}(\lambda)$ with $w \in [0, \frac{1}{\lambda}]$, and the rest, $I^{-+}(\lambda)$, which only arises when $\lambda^2 > x_1/\alpha$.

To estimate $I^{--}(\lambda)$ we let

$$F(\mu,\lambda,w)=e^{-\mu\lambda e_\theta}\left[[(\mu\lambda+w)e_\theta-b]\psi_b(\mu\lambda we_{2\theta})-2(\mu\lambda we_{2\theta})\psi_b'(\mu\lambda we_{2\theta})\right]. \tag{A.212}$$

It follows from Lemma A.0.4 that for some $\xi\in(1,\mu)$ we have that

$$\left|F(\mu,\lambda,w)-F(1,\lambda,w)\right|\leq(\mu-1)|\partial_\mu F(\xi,\lambda,w)|. \tag{A.213}$$

In the set $\{w\lambda<1\}$, we have the bound (see (A.220)):

$$|\partial_\mu F(\xi,\lambda,w)|\leq C_b e^{-\cos\theta\xi\lambda}\lambda(1+\lambda+w)\left[\frac{1}{\Gamma(b)}+(1+\lambda)w\right]. \tag{A.214}$$

In this case the w-integral is bounded by

$$C_b x_1^{\frac{\gamma}{2}}(\mu-1)\int_0^{\frac{1}{\lambda}} w^{b-1}e^{-\cos\theta(w+\xi\lambda)}\left[\frac{(1+\lambda)}{\Gamma(b)}+w(1+\lambda)^2+w^2(1+\lambda)\right]\times |\sqrt{\lambda}-\sqrt{w}|^\gamma\frac{dwd\lambda}{\lambda^{\frac{\gamma}{2}}}, \tag{A.215}$$

and therefore

$$I^{--}(\lambda)\leq C_{b,\theta}\|g\|_{WF,0,\gamma}x_1^{\frac{\gamma}{2}}(\mu-1)\frac{e^{-\cos\theta\lambda}(1+\lambda^{1+\frac{\gamma}{2}})}{\lambda^{\frac{\gamma}{2}}(1+\lambda^b)}. \tag{A.216}$$

As this is integrable from 0 to ∞, we see that

$$I^{--}\leq C_{b,\theta}\frac{x_2-x_1}{x_1^{1-\frac{\gamma}{2}}}. \tag{A.217}$$

As x_2/x_1 is bounded from above, it follows immediately that

$$I^{--}\leq C_{b,\theta}|\sqrt{x_2}-\sqrt{x_1}|^\gamma. \tag{A.218}$$

This leaves only I^{-+}, which is estimated by

$$|I^{-+}|\leq x_1^{\frac{\gamma}{2}}(\mu-1)\int\limits_{\max\{\frac{x_1}{|t|},\sqrt{\frac{x_1}{\alpha}}\}}^{\infty}\int\limits_{\frac{1}{\lambda}}^{\frac{\alpha\lambda}{x_1}} w^{b-1}e^{-\cos\theta w}|\sqrt{\lambda}-\sqrt{w}|^\gamma\times |F_\mu(\xi,\lambda,w)|\frac{dwd\lambda}{\lambda^{1+\frac{\gamma}{2}}}. \tag{A.219}$$

As before we apply Lemma A.0.4 as in (A.213) to see that we need to estimate:

$$|\partial_\mu F(\xi,\lambda,w)| = e^{-\cos\theta\xi\lambda}\big|\lambda e_\theta \psi_b(\xi\lambda w e_{2\theta})(1-(\xi\lambda+w)e_\theta+b)+ \\ \lambda w e_{2\theta}\psi_b'(\xi\lambda w e_{2\theta})[(3\xi\lambda+w)e_\theta+b-2]-2\xi(\lambda w e_{2\theta})^2\psi_b''(\xi\lambda w e_{2\theta})\big|, \quad \xi\in[1,\frac{x_2}{x_1}]. \tag{A.220}$$

To get a controllable error term, we must use the asymptotic expansions for ψ_b, $\psi_b' = \psi_{b+1}$ through second order:

$$\begin{aligned} \psi_b(z) &= \frac{z^{\frac{1}{4}-\frac{b}{2}}e^{2\sqrt{z}}}{\sqrt{4\pi}}\left[1-\frac{(2b-1)(2b-3)}{16\sqrt{z}}+O(\frac{1}{z})\right] \\ \psi_b'(z) &= \frac{z^{-\frac{1}{4}-\frac{b}{2}}e^{2\sqrt{z}}}{\sqrt{4\pi}}\left[1-\frac{(2b+1)(2b-1)}{16\sqrt{z}}+O(\frac{1}{z})\right]. \end{aligned} \tag{A.221}$$

Using the equation $z\psi_b'' = \psi_b - b\psi_b'$, and inserting these relations into (A.220), gives

$$|\partial_\mu F(\xi,\lambda,w)| \le \lambda e^{\cos\theta(2\sqrt{\xi\lambda w}-\xi\lambda)}\frac{(\xi\lambda w)^{\frac{1}{4}-\frac{b}{2}}}{\sqrt{4\pi}}\times \\ \left[\xi\lambda\left(\sqrt{\frac{w}{\lambda\xi}}-1\right)^3+a_1(b)\left(\sqrt{\frac{w}{\lambda\xi}}-1\right)+\right. \\ \left. a_2(b)\sqrt{\frac{w}{\lambda\xi}}\left(\sqrt{\frac{w}{\lambda\xi}}-1\right)+a_3(b)\left(\sqrt{\frac{w}{\lambda\xi}}-\sqrt{\frac{\lambda\xi}{w}}\right)+O\left(\frac{1}{\lambda}+\frac{1}{w}+\frac{1}{\sqrt{\lambda w}}\right)\right]. \tag{A.222}$$

Here $a_1(b)$, $a_2(b)$ and $a_3(b)$ are polynomials in b. We denote the contributions of these terms by M_0, M_1, M_2, M_3, M_e.

We first consider M_0 :

$$M_0 \le C_b x_1^{\frac{\gamma}{2}}(\mu-1)\int\limits_{\max\{\frac{x_1}{|t|},\sqrt{\frac{x_1}{\alpha}}\}}^{\infty}\int\limits_{\frac{1}{\lambda}}^{\frac{\alpha\lambda}{x_1}}\left(\frac{w}{\lambda}\right)^{\frac{b}{2}-\frac{1}{4}}e^{-\cos\theta(\sqrt{w}-\sqrt{\xi\lambda})^2}\times \\ |\sqrt{\lambda}-\sqrt{w}|^{\gamma}|\sqrt{\xi\lambda}-\sqrt{w}|^3\frac{dwd\lambda}{\sqrt{w}\lambda^{\frac{1+\gamma}{2}}}. \tag{A.223}$$

In this integral $w<\lambda$ and $1\le\xi\le x_2/x_1$, and therefore this is bounded by

$$M_0 \le C_b x_1^{\frac{\gamma}{2}}(\mu-1)\int\limits_{\max\{\frac{x_1}{|t|},\sqrt{\frac{x_1}{\alpha}}\}}^{\infty}\int\limits_{\frac{1}{\lambda}}^{\frac{\alpha\lambda}{x_1}}\left(\frac{w}{\lambda}\right)^{\frac{b}{2}-\frac{1}{4}}e^{-\cos\theta(\sqrt{w}-\sqrt{\xi\lambda})^2}\times \\ |\sqrt{\xi\lambda}-\sqrt{w}|^{3+\gamma}\frac{dwd\lambda}{\sqrt{w}\lambda^{\frac{1+\gamma}{2}}}. \tag{A.224}$$

We now let $z = \sqrt{w/\lambda} - \sqrt{\xi}$ to obtain that

$$M_0 \leq C_b x_1^{\frac{\gamma}{2}}(\mu - 1) \int\limits_{\max\{\frac{x_1}{|t|},\sqrt{\frac{x_1}{\alpha}}\}}^{\infty} \int\limits_{\frac{1}{\lambda}-\sqrt{\xi}}^{\sqrt{\frac{\alpha}{x_1}}-\sqrt{\xi}} (\sqrt{\xi}+z)^{b-\frac{1}{2}} e^{-\cos\theta\lambda z^2} \times$$
$$|z|^{3+\gamma}\lambda^{\frac{3}{2}} dz d\lambda. \quad \text{(A.225)}$$

As λ is bounded from below by $\sqrt{x_1/\alpha}$, we need to estimate the z-integral as $\lambda \to \infty$. We apply Lemma 6.1.24 to see that

$$\int\limits_{\frac{1}{\lambda}-\sqrt{\xi}}^{\sqrt{\frac{\alpha}{x_1}}-\sqrt{\xi}} (\sqrt{\xi}+z)^{b-\frac{1}{2}} e^{-\cos\theta\lambda z^2} |z|^{3+\gamma} dz \leq$$
$$\begin{cases} \frac{C_{b,\theta}}{\lambda} e^{-\cos\theta\frac{\lambda(\sqrt{x_2}-\sqrt{x_1})^2}{4x_1}} \left(\frac{\sqrt{x_2}-\sqrt{x_1}}{\sqrt{x_1}}\right)^{2+\gamma} & \text{if } \sqrt{\lambda}\left(\sqrt{\frac{x_2}{x_1}}-1\right) > 1 \\ \frac{C_{b,\theta}}{\lambda^{\frac{4+\gamma}{2}}} \text{ if } \sqrt{\lambda}\left(\sqrt{\frac{x_2}{x_1}}-1\right) \leq 1. \end{cases} \quad \text{(A.226)}$$

The large λ contribution (the first estimate in (A.226)) leads to terms of the form

$$C_{b,\theta}|\sqrt{x_2}-\sqrt{x_1}|^{\gamma} \frac{x_2 - x_1}{x_1} \frac{\sqrt{x_1}}{\sqrt{x_2}-\sqrt{x_1}}$$
$$\leq C_{b,\theta}\left(\frac{\sqrt{x_2}+\sqrt{x_1}}{\sqrt{x_1}}\right)|\sqrt{x_2}-\sqrt{x_1}|^{\gamma} \quad \text{(A.227)}$$

as above (see (A.217)–(A.218)). Integrating the second estimate in (A.226) over

$$\lambda \in \left[\max\left\{\frac{x_1}{|t|}, \sqrt{\frac{x_1}{\alpha}}\right\}, \frac{x_1}{(\sqrt{x_2}-\sqrt{x_1})^2}\right],$$

gives a term bounded by

$$C_{b,\theta}\left(\frac{\sqrt{x_2}+\sqrt{x_1}}{\sqrt{x_1}}\right)|\sqrt{x_2}-\sqrt{x_1}|^{\gamma}. \quad \text{(A.228)}$$

This completes the proof that M_0 satisfies the desired bound.

Using the same change of variables, we see that M_1 and M_2 are also bounded by the quantity in (A.228). To treat M_3 we let $z = \sqrt{w/\lambda}$; this gives the bound:

$$|M_3| \leq C_b x_1^{\frac{\gamma}{2}}(\mu - 1) \int\limits_{\max\{\frac{x_1}{|t|},\sqrt{\frac{x_1}{\alpha}}\}}^{\infty} \int\limits_{\frac{1}{\lambda}}^{\sqrt{\frac{\alpha}{x_1}}} z^{b-\frac{1}{2}} e^{-\cos\theta\lambda(\sqrt{\xi}-z)^2} \times$$
$$|\sqrt{\xi}-z|^{\gamma} \frac{|z^2-\xi|}{z\sqrt{\xi}} \sqrt{\lambda} dz d\lambda. \quad \text{(A.229)}$$

As $\lambda \to \infty$, the part of the z-integral from $1/\lambda$ to $1/2$ (e.g.,) is bounded by a constant multiple of $\lambda^{1-b}e^{-\cos\theta\frac{\lambda}{4}}$, and so contributes a term to M_3 that satisfies the desired estimate.

We are left to estimate the contribution from near the diagonal, i.e., for $\lambda \in [1/2, \sqrt{\alpha/x_1}]$. If $\sqrt{\lambda}(\sqrt{x_2/x_1} - 1) > 1/2$, then the z-integral is bounded by

$$C_{b,\theta}\left|\frac{\sqrt{x_2}-\sqrt{x_1}}{2\sqrt{x_1}}\right|^{\gamma}\frac{e^{-\cos\theta\lambda\left(\frac{\sqrt{x_2}-\sqrt{x_1}}{2\sqrt{x_1}}\right)^2}}{\lambda}; \tag{A.230}$$

the contribution of this term satisfies the desired bound. If $\sqrt{\lambda}(\sqrt{x_2/x_1} - 1) < 1/2$, then the z-integral is bounded by $\frac{C_{b,\theta}}{\lambda^{\frac{2+\gamma}{2}}}$. Integrating in λ completes the proof that

$$|M_3| \leq C_{b,\theta}\left(\frac{\sqrt{x_2}+\sqrt{x_1}}{\sqrt{x_1}}\right)|\sqrt{x_2}-\sqrt{x_1}|^{\gamma}. \tag{A.231}$$

To complete the estimate of I^{-+}, and thereby of I^{-}, we only need to show that the error terms satisfy the desired bound. To that end we let $z = \sqrt{w/\lambda}$; the contribution of the error terms is bounded by

$$|M_e| \leq C_b x_1^{\frac{\gamma}{2}}(\mu - 1)\int\limits_{\max\{\frac{x_1}{|t|},\sqrt{\frac{x_1}{\alpha}}\}}^{\infty}\int\limits_{\frac{1}{\lambda}}^{\sqrt{\frac{\alpha}{x_1}}} z^{b-\frac{1}{2}}e^{-\cos\theta\lambda(\sqrt{\xi}-z)^2}\times$$
$$|\sqrt{\xi}-z|^{\gamma}\left(\frac{1}{\sqrt{\lambda}}+\frac{1}{z}+\frac{1}{\sqrt{\lambda}z}\right)dzd\lambda. \tag{A.232}$$

Arguing as above, we see that these terms all satisfy the desired bound. As noted, the estimate of I^{+} is quite similar and is left to the reader. □

LEMMA A.3.6. *[LEMMA 6.1.19] For $b > 0$, $0 < \gamma < 1$, and $t_1 < t_2 < 2t_1$ there is a constant C_b so that*

$$\int\limits_{t_2-t_1}^{t_1}\int\limits_{0}^{\infty}|L_b k^b_{t_2-t_1+s}(x,y) - L_b k^b_s(x,y)||\sqrt{x}-\sqrt{y}|^{\gamma}dyds \leq C_b|t_2-t_1|^{\frac{\gamma}{2}}. \tag{A.233}$$

This lemma follows from the more basic:

LEMMA A.3.7. *[LEMMA 6.1.20] For $b > 0$, $0 < \gamma < 1$, and $t_1 < t_2 < 2t_1$ and $s > t_2 - t_1$, there is a constant C_b so that*

$$\int\limits_{0}^{\infty}|L_b k^b_{t_2-t_1+s}(x,y) - L_b k^b_s(x,y)||\sqrt{x}-\sqrt{y}|^{\gamma}dy \leq C_b(t_2-t_1)s^{\frac{\gamma}{2}-2}. \tag{A.234}$$

PROOF OF LEMMA 6.1.19. The derivation of (A.233) from (A.234) is quite easy:

$$\int_{t_2-t_1}^{t_1}\int_0^{\infty} |L_b k^b_{t_2-t_1+s}(x,y) - L_b k^b_s(x,y)||\sqrt{x}-\sqrt{y}|^{\gamma}\,dyds \le C_b(t_2-t_1)\int_{t_2-t_1}^{t_1} s^{\frac{\gamma}{2}-2}ds$$
$$\le C_b|t_2-t_1|^{\frac{\gamma}{2}}. \tag{A.235}$$

□

PROOF OF LEMMA 6.1.20. To prove (6.49) we need to apply Taylor's formula to estimate the difference $L_b k^b_{t_2-t_1+s}(x,y) - L_b k^b_s(x,y)$. To that end, we let $F(\tau,s,x,y) = L_b k^b_{\tau+s}(x,y)$; we denote the left-hand side in (6.49) as I, which we can rewrite as

$$I = \int_0^{\infty} [F(\tau,s,x,y) - F(0,s,x,y)]|\sqrt{x}-\sqrt{y}|^{\gamma}\,dy; \tag{A.236}$$

here $\tau = t_2 - t_1$. From the mean value theorem, we get the estimate

$$|I| \le \tau \int_0^{\infty} |\partial_\tau F(\xi,s,x,y)||\sqrt{y}-\sqrt{x}|^{\gamma}\,dy. \tag{A.237}$$

Using the differential equation satisfied by ψ_b we can show that

$$\partial_\tau F(\xi,s,x,y) =$$
$$\frac{y^{b-1}e^{-\left(\frac{x+y}{s+\xi}\right)}}{(s+\xi)^{b+2}}\left\{\psi_b\left(\frac{xy}{(s+\xi)^2}\right)\left[\left(\frac{x+y}{s+\xi}-b\right)\left(\frac{x+y}{s+\xi}-(b+1)\right)\right.\right.$$
$$\left.\left.-\left(\frac{x+y}{s+\xi}\right)+\frac{4xy}{(s+\xi)^2}\right]+\frac{2xy}{(s+\xi)^2}\psi_b'\left(\frac{xy}{(s+\xi)^2}\right)\left[3-2\left(\frac{x+y}{s+\xi}\right)\right]\right\}. \tag{A.238}$$

Since $s \in [t_2-t_1, t_1]$ and $\xi \in [0, t_2-t_1]$, we see that $s < s+\xi < 2s$, and therefore

$$\frac{xy}{4s^2} \le \frac{xy}{(s+\xi)^2} \le \frac{xy}{s^2}. \tag{A.239}$$

We can therefore split the y-integral into a compact part with $y \in [0, \frac{4s^2}{x}]$, I^- and the remaining non-compact part I^+. In the compact part we use the usual estimates for ψ_b and ψ_b'. Setting $w = y/s$, $\lambda = x/s$, we obtain that

$$|I^-| \le s^{\frac{\gamma}{2}-2}\tau \int_0^{\frac{4}{\lambda}} w^{b-1}e^{-\frac{1}{2}(w+\lambda)}\times$$
$$\left[(\lambda+w+1)^2\left(\frac{1}{\Gamma(b)}+w\lambda\right)+w\lambda+w+\lambda\right]\left|\sqrt{\lambda}-\sqrt{w}\right|^{\gamma}\,dw. \tag{A.240}$$

If λ is bounded then we easily see that this satisfies:

$$|I^-| \leq C_b \tau s^{\frac{\gamma}{2}-2}. \tag{A.241}$$

In the case that λ is large, then we see that

$$|I^-| \leq C s^{\frac{\gamma}{2}-2} \tau e^{-\frac{\lambda}{2}}, \tag{A.242}$$

which therefore applies for $\lambda \in [0, \infty)$.

To estimate I^+, we use the second order asymptotic expansions for ψ_b and ψ_b' to see that

$$|\partial_\tau F(\xi, s, x, y)| = \left(\frac{w}{\mu}\right)^{\frac{b}{2}-\frac{1}{4}} \frac{e^{-(\sqrt{\mu}-\sqrt{w})^2}}{\sqrt{w}(s+\xi)^3} \Big[(b-1)^2(\sqrt{w}-\sqrt{\mu})^4 + \\ \frac{9}{4}(\sqrt{w}-\sqrt{\mu})^2 + O(1+\sqrt{w/\mu}+\sqrt{\mu/w})\Big], \tag{A.243}$$

here

$$w = \frac{y}{s+\xi} \quad \text{and } \mu = \frac{x}{s+\xi}. \tag{A.244}$$

From this expansion, and the fact that $s \leq s+\xi \leq 2s$, it follows that

$$|I^+| \leq C_b \frac{\tau}{s^3} \int\limits_{\frac{4s^2}{x}}^{\infty} \left(\frac{y}{x}\right)^{\frac{b}{2}-\frac{1}{4}} \sqrt{\frac{s}{y}} e^{-\frac{x}{2s}\left(1-\sqrt{\frac{y}{x}}\right)^2} |\sqrt{y}-\sqrt{x}|^\gamma \times \\ \left[\left(\sqrt{\frac{y}{s}}-\sqrt{\frac{x}{s}}\right)^4 + \left(\sqrt{\frac{y}{s}}-\sqrt{\frac{x}{s}}\right)^2 + O\left(1+\sqrt{\frac{x}{y}}+\sqrt{\frac{y}{x}}\right)\right] dy. \tag{A.245}$$

To estimate this integral, we let $z = \sqrt{y/x}-1$, and $\lambda = x/s$, obtaining:

$$|I^+| \leq C_b \frac{\tau x^{\frac{\gamma}{2}}\sqrt{\lambda}}{s^2} \int\limits_{\frac{2}{\lambda}-1}^{\infty} (z+1)^{b-\frac{1}{2}} e^{-\frac{\lambda}{2}z^2} |z|^\gamma \times \\ \left[\lambda^2 z^4 + \lambda z^2 + O\left(1+z+\frac{1}{1+z}\right)\right] dz. \tag{A.246}$$

If $\lambda \to 0$, then the integral behaves like $e^{-\frac{1}{4\lambda}}$. As $\lambda \to \infty$, an application of Laplace's method shows that the z-integral behaves like $\lambda^{-\frac{1+\gamma}{2}}$, which, in turn, establishes (6.49).

□

A.4 OFF-DIAGONAL AND LARGE-T BEHAVIOR

We close this section with estimates valid for t, with positive real part, which do not use an assumption about the Hölder continuity of the data.

LEMMA A.4.1. *[LEMMA 6.1.21] For $0 < b < B, 0 < \phi < \frac{\pi}{2}$, and $j \in \mathbb{N}$ there is a constant $C_{j,B,\phi}$ so that if $t \in S_\phi$, then*

$$\int_0^\infty |\partial_x^j k_t^b(x,y)|dy \leq \frac{C_{j,B,\phi}}{|t|^j}, \tag{A.247}$$

and

$$\int_0^\infty |x^{\frac{j}{2}} \partial_x^j k_t^b(x,y)|dy \leq \frac{C_{j,B}}{|t|^{\frac{j}{2}}}. \tag{A.248}$$

PROOF. The proof of this lemma is easier than results proved above for small t behavior. We observe that for $0 < b < B$ we can write

$$k_t^b(x,y) = \frac{1}{y} F\left(\frac{x}{t}, \frac{y}{t}\right), \tag{A.249}$$

where, for z and ζ in the right half plane,

$$F(\zeta, z) = z^b e^{-(z+\zeta)} \psi_b(z\zeta). \tag{A.250}$$

This expression easily implies that

$$\partial_x^j k_b^t(x,y) = \frac{1}{yt^j} \partial_\zeta^j F\left(\frac{x}{t}, \frac{y}{t}\right). \tag{A.251}$$

From the form of F and the fact that $\partial_z \psi_b(z) = \psi_{b+1}(z)$, we see that a simple induction establishes:

$$\partial_\zeta^j F(\zeta, z) = z^b e^{-(z+\zeta)} \sum_{l=0}^{j} \binom{j}{l} (-1)^{j-l} z^l \psi_{b+l}(z\zeta). \tag{A.252}$$

We let $t = |t|e^{i\theta}$, with $|\theta| < \frac{\pi}{2} - \phi$, and set $w = y/|t|$, $\lambda = x/|t|$. To complete the proof of (A.247) it suffices to show that there are constants $C_{l,B,\phi}$ so that

$$\int_0^\infty w^{b+l} e^{-\cos\theta(w+\lambda)} |\psi_{b+l}(w\lambda e_{2\theta})| \frac{dw}{w} \leq C_{l,B,\phi}. \tag{A.253}$$

When $l = 0$ the integral is bounded in Lemma 6.1.3, so we can assume that $l \geq 1$.

We need to estimate

$$I_{j,b+l} = \frac{1}{|t|^j} \int_0^\infty w^{b+l} e^{-\cos\theta(w+\lambda)} |\psi_{b+l}(w\lambda e_{2\theta})| \frac{dw}{w}. \tag{A.254}$$

We split the integral into the part from $[0, 1/\lambda]$ and the rest; applying the asymptotic formula we obtain that

$$I_{j,b+l} \leq \frac{C_{b+l}}{|t|^j}\left[e^{-\cos\theta\lambda}\int_0^{\frac{1}{\lambda}} w^{b+l}e^{-\cos\theta w}\frac{dw}{w} + \int_{\frac{1}{\lambda}}^{\infty}\left(\frac{w}{\lambda}\right)^{\frac{b+l}{2}-\frac{1}{4}} e^{-\cos\theta(\sqrt{w}-\sqrt{\lambda})^2}\frac{dw}{\sqrt{w}}\right]. \tag{A.255}$$

The first term in the brackets is bounded by $\Gamma(b+l)e^{-\cos\theta\lambda}$, and the second term is rapidly decaying as $\lambda \to 0$. To study the second term as $\lambda \to \infty$, we let $z = \sqrt{w}-\sqrt{\lambda}$, to see that the second integral is bounded by

$$C_{b+l}\int_{\frac{1}{\sqrt{\lambda}}-\sqrt{\lambda}}^{\infty}\left(1+\frac{z}{\sqrt{\lambda}}\right)^{b+l-\frac{1}{2}} e^{-\cos\theta z^2}dz. \tag{A.256}$$

As $b+l-\frac{1}{2} > 0$, it follows easily that this integral is bounded as $\lambda \to \infty$, which completes the proof of (A.247).

The estimate in (A.248) for $x/|t| < 1$ follows immediately from this formula. To prove (A.248) for $x/|t| \geq 1$ requires more careful consideration. Using (A.252) and the asymptotic expansions for ψ_{b+l} we see that

$$|\partial_x^j k_t(x,y)| = \frac{(-1)^j}{t^j}\left(\frac{z}{\zeta}\right)^{\frac{b}{2}-\frac{1}{4}} e^{-(\sqrt{z}-\sqrt{\zeta})^2}\times$$
$$\left[\sum_{l=0}^{j}\binom{j}{l}(-1)^l\left(\frac{z}{\zeta}\right)^{\frac{l}{2}}\left[\sum_{k=0}^{[\frac{j}{2}]}\frac{(-1)^k\Gamma(b+l+k-\frac{1}{2})}{4^k(z\zeta)^{\frac{k}{2}}\Gamma(b+l-k-\frac{1}{2})} + O\left(\frac{1}{(z\zeta)^{\frac{j}{4}}}\right)\right]\right]. \tag{A.257}$$

We observe that the ratios of Γ-functions are polynomials in l, which can be expressed as

$$\frac{\Gamma(b+l+k-\frac{1}{2})}{\Gamma(b+l-k-\frac{1}{2})} = p_{k,0}(b) + \sum_{m=1}^{2k} p_{k,m}(b)l(l-1)\cdots(l-m+1). \tag{A.258}$$

The coefficients $\{p_{k,m}(b)\}$ are polynomials in b. Putting this expression into the previous formula and using the fact that

$$(1-u)^j = \sum_{l=0}^{j}\binom{j}{l}(-1)^l u^l, \tag{A.259}$$

we see there are polynomials, $P_{j,k}(b,u)$ in (b,u), so that

$$\partial_x^j k_t(x,y)dy = \frac{(-1)^j}{t^j}\left(\frac{z}{\zeta}\right)^{\frac{b}{2}-\frac{1}{4}} e^{-(\sqrt{z}-\sqrt{\zeta})^2}\times \left[\sum_{k=0}^{[\frac{j}{2}]}\frac{\left(1-\sqrt{\frac{z}{\zeta}}\right)^{j-2k}}{(z\zeta)^{\frac{k}{2}}} P_{j,k}\left(b,\sqrt{\frac{z}{\zeta}}\right)+ \left[\sum_{l=0}^{j}\binom{j}{l}(-1)^l\left(\frac{z}{\zeta}\right)^{\frac{l}{2}}\right]\cdot O\left(\frac{1}{(z\zeta)^{\frac{j}{4}}}\right)\right]. \tag{A.260}$$

Using this expression and the analysis from the previous case we easily show that

$$\int_0^\infty |x^{\frac{j}{2}}\partial_x^j k_t^b(x,y)|dy \leq \frac{C_{b,j,\phi}}{|t|^{\frac{j}{2}}}, \tag{A.261}$$

which completes the proof of the lemma. □

LEMMA A.4.2. *[LEMMA 8.2.15] For $j\in\mathbb{N}$ and $0<\phi<\frac{\pi}{2}$ there is a constant $C_{j,\phi}$ so that if $t\in S_\phi$, then*

$$\int_{-\infty}^\infty |\partial_x^j k_t^e(x,y)|dy \leq \frac{C_{j,\phi}}{|t|^{\frac{j}{2}}}. \tag{A.262}$$

PROOF. These estimates, which are classical, follow easily from homogeneity considerations, and the formula

$$\partial_x^j k_t^e(x,y) = \frac{1}{t^{\frac{j}{2}}}\sum_{l=0}^{j} c_{j,l}\left(\frac{x-y}{2\sqrt{t}}\right)^l k_t^e(x,y). \tag{A.263}$$

□

We consider the off-diagonal behavior.

LEMMA A.4.3. *[LEMMA 9.3.2] Let $b>0$, $\eta>0$ and for $x\in\mathbb{R}_+$ define the set*

$$J_{x,\eta} = \{y\in\mathbb{R}_+ : |\sqrt{x}-\sqrt{y}|\geq\eta\}. \tag{A.264}$$

For $0\leq b<B$, $0<\phi<\frac{\pi}{2}$, and $j\in\mathbb{N}_0$ there is a constant $C_{\eta,j,B,\phi}$ so that if $t=|t|e^{i\theta}$, with $|\theta|\leq\frac{\pi}{2}-\phi$, then

$$\int_{J_{x,\eta}} |\partial_x^j k_t^b(x,y)|dy \leq C_{\eta,j,B,\phi}\frac{e^{-\cos\theta\frac{\eta^2}{2|t|}}}{|t|^j}. \tag{A.265}$$

For the Euclidean models we have

LEMMA A.4.4. *[LEMMA 9.3.3] Let $\eta > 0$ and for $x \in \mathbb{R}$ define the set*

$$J_{x,\eta} = \{y \in \mathbb{R} : |x - y| \geq \eta\}. \tag{A.266}$$

For $j \in \mathbb{N}_0$, $0 < \phi < \frac{\pi}{2}$, there is a constant $C_{\eta,j,\phi}$ so that if $t = |t|e^{i\theta}$, with $|\theta| \leq \frac{\pi}{2} - \phi$, then

$$\int\limits_{J_{x,\eta}} |\partial_x^j k_t^e(x, y)|dy \leq C_{\eta,j,\phi} \frac{e^{-\cos\theta \frac{\eta^2}{8t}}}{|t|^{\frac{j}{2}}}. \tag{A.267}$$

PROOF OF LEMMA 9.3.2. Recall that for $0 < b$, the kernel is given by

$$k_t^b(x, y) = \frac{1}{y}\left(\frac{y}{t}\right)^b e^{-\frac{x+y}{t}} \psi_b\left(\frac{xy}{t^2}\right), \tag{A.268}$$

where

$$\psi_b(z) = \sum_{j=0}^{\infty} \frac{z^j}{j!\Gamma(j + b)}. \tag{A.269}$$

Using a simple inductive argument, and the fact that

$$\partial_z^l \psi_b(z) = \psi_{b+l}(z), \tag{A.270}$$

we can show that there are constants $\{c_{j,l}\}$ so that

$$\partial_x^j k_t^b(x, y) = \frac{1}{yt^j}\left(\frac{y}{t}\right)^b e^{-\frac{x+y}{t}} \sum_{l=0}^{j} c_{j,l} \left(\frac{y}{t}\right)^l \psi_{b+l}\left(\frac{xy}{t^2}\right). \tag{A.271}$$

To prove the assertion of the lemma, it therefore suffices to prove it for each function,

$$\frac{1}{yt^j}\left(\frac{y}{t}\right)^b e^{-\frac{x+y}{t}} \left(\frac{y}{t}\right)^l \psi_{b+l}\left(\frac{xy}{t^2}\right) \tag{A.272}$$

where $0 \leq l \leq j$.

Letting $w = y/|t|$ and $\lambda = x/|t|$, we see that we must estimate the integrals

$$I_{l,j}(x, t, \eta) = \frac{1}{|t|^j} \int\limits_{|t|^{-1}J_{x,\eta}} w^{b+l} e^{-\cos\theta(w+\lambda)} |\psi_{b+l}(w\lambda e_{2\theta})| \frac{dw}{w}. \tag{A.273}$$

There are two cases: if $x < \eta^2$, then

$$|t|^{-1}J_{x,\eta} = \left[\frac{(\sqrt{x} + \eta)^2}{|t|}, \infty\right); \tag{A.274}$$

otherwise,

$$|t|^{-1}J_{x,\eta} = \left[0, \frac{(\sqrt{x}-\eta)^2}{|t|}\right] \bigcup \left[\frac{(\sqrt{x}+\eta)^2}{|t|}, \infty\right). \tag{A.275}$$

Without loss of generality, we can assume that $|t| < \eta^2$.

We first consider the case where $x < \eta^2$. Here again there are two cases to examine: if $|t|^2 < x(\sqrt{x}+\eta)^2$ ("small $|t|$ case"), then we only need to use the asymptotic expansion for ψ_{b+l}; otherwise ("large $|t|$ case") we also need to separately estimate the integral over $\left[\frac{(\sqrt{x}+\eta)^2}{|t|}, \lambda^{-1}\right)$. We begin with the small $|t|$ case. The product $w\lambda > 1$ and we can use the asymptotic expansion

$$\psi_{b+l}(z) \sim \frac{z^{\frac{1}{4}-\frac{b+l}{2}}}{\sqrt{4\pi}} e^{2\sqrt{z}}. \tag{A.276}$$

There is a constant $C_{b,l}$ so that

$$\begin{aligned} I_{l,j}(x,t,\eta) \leq & C_{b,l}\frac{1}{|t|^j}\int\limits_{\frac{(\sqrt{x}+\eta)^2}{|t|}}^{\infty}\left(\frac{w}{\lambda}\right)^{\frac{b+l}{2}-\frac{1}{4}} e^{-\cos\theta(\sqrt{w}-\sqrt{\lambda})^2}\frac{dw}{\sqrt{w}} \\ \leq & C_{b,l}\frac{1}{|t|^j}\int\limits_{\frac{\eta}{\sqrt{|t|}}}^{\infty}\left(1+\frac{z}{\sqrt{\lambda}}\right)^{b+l-\frac{1}{2}} e^{-\cos\theta z^2} bz \end{aligned} \tag{A.277}$$

where we set $z = \sqrt{w}-\sqrt{\lambda}$ in the second line. Since $\eta/\sqrt{|t|\lambda} \leq 2\eta^2/|t|$, an elementary integration by parts argument shows that

$$I_{l,j}(x,t,\eta) \leq C_{b,l,\theta}\left(\frac{\eta^2}{|t|}\right)^{b+l}\frac{e^{-\cos\theta\frac{\eta^2}{|t|}}}{|t|^j}. \tag{A.278}$$

For the large $|t|$ case we need to consider

$$I'_{l,j}(x,t,\eta) = \frac{1}{|t|^j}\int\limits_{\frac{(\sqrt{x}+\eta)^2}{|t|}}^{\frac{1}{\lambda}} w^{b+l}e^{-\cos\theta(w+\lambda)}\psi_{b+l}(w\lambda e_{2\theta})\frac{dw}{w}. \tag{A.279}$$

In this case we approximate $\psi_{b+l}(z)$ by a constant to obtain

$$I'_{l,j}(x,t,\eta) \leq \frac{C_{b,l}e^{-\cos\theta\lambda}}{|t|^j}\int\limits_{\frac{(\sqrt{x}+\eta)^2}{|t|}}^{\frac{1}{\lambda}} w^{b+l}e^{-\cos\theta w}\frac{dw}{w}. \tag{A.280}$$

If $b+l \geq 1$, then this is estimated by

$$\frac{C_{b,l,\theta}}{|t|^j}\left(\frac{\eta^2}{|t|}\right)^{b+l-1} e^{-\cos\theta\frac{\eta^2}{|t|}} \leq \frac{C_{b,l,\theta}}{|t|^j} e^{-\cos\theta\frac{\eta^2}{2|t|}}. \tag{A.281}$$

If $0 < b+l < 1$, then because $\eta^2/|t| > 1$, we have that

$$I'_{l,j}(x,t,\eta) \leq \frac{C_{b,l,\theta} e^{-\cos\theta\frac{\eta^2}{|t|}}}{|t|^j}. \tag{A.282}$$

The other part of $I_{l,j}(x,t,\eta)$ is bounded by

$$I''_{l,j}(x,t,\eta) \leq C_{b,l}\frac{1}{|t|^j}\int\limits_{\frac{1}{\lambda}}^{\infty}\left(\frac{w}{\lambda}\right)^{\frac{b+l}{2}-\frac{1}{4}} e^{-\cos\theta(\sqrt{w}-\sqrt{\lambda})^2}\frac{dw}{\sqrt{w}}. \tag{A.283}$$

Estimating the integral shows that

$$I''_{l,j}(x,t,\eta) \leq C_{b,l,\theta}\frac{\lambda^{-(b+l)}e^{-\frac{\cos\theta}{\lambda}}}{|t|^j}. \tag{A.284}$$

Since $1/\lambda > \frac{(\sqrt{x}+\eta)^2}{|t|}$, this is again easily seen to satisfy

$$I''_{l,j}(x,t,\eta) \leq C_{b,l,\theta}\frac{e^{-\cos\theta\frac{\eta^2}{2|t|}}}{|t|^j}. \tag{A.285}$$

This establishes the estimates

$$I_{l,j}(x,t,\eta) \leq C_{b,l,j,\theta}\frac{e^{-\cos\theta\frac{\eta^2}{2|t|}}}{|t|^j}, \text{ when } x \leq \eta^2. \tag{A.286}$$

The constants $C_{b,l,j,\theta}$ are uniformly bounded for $0 < b < B$, and $|\theta| < \frac{\pi}{2} - \phi$.

We now consider $x \geq \eta^2$; as before we assume that $|t| < \eta^2$, so that $1/\lambda < (\sqrt{x}+\eta)^2/|t|$. We first estimate the non-compact part of the integral:

$$\begin{aligned} I''_{l,k,j}(x,t,\eta) \leq & C_{b,l}\frac{1}{|t|^j}\int\limits_{\frac{\eta}{\sqrt{|t|}}}^{\infty}\left(1+\frac{z}{\sqrt{\lambda}}\right)^{b+l-\frac{1}{2}} e^{-\cos\theta z^2} bz \\ \leq & C_{b,l,\theta}\frac{e^{-\cos\theta\frac{\eta^2}{|t|}}}{|t|^j}. \end{aligned} \tag{A.287}$$

This leaves only

$$I'_{l,j}(x,t,\eta) = \frac{1}{|t|^j}\int\limits_0^{\frac{(\sqrt{x}-\eta)^2}{|t|}} w^{b+l}e^{-\cos\theta(w+\lambda)}|\psi_{b+l}(w\lambda e_{2\theta})|\frac{dw}{w}. \tag{A.288}$$

If $|t|^2 < x(\sqrt{x}-\eta)^2$, then we need to split this integral into two parts: from 0 to $1/\lambda$ and the rest. We first assume that there is just one part. If $b+l \geq 1$, then we have the estimate

$$I'_{l,j}(x,t,\eta) \leq \frac{C_{b,l}}{|t|^j} \int\limits_0^{\frac{(\sqrt{x}-\eta)^2}{|t|}} w^{b+l} e^{-\cos\theta(w+\lambda)} \frac{dw}{w} \leq \frac{C_{b,l,\theta} e^{-\cos\theta\lambda}}{|t|^j}. \tag{A.289}$$

In this case the fact that $x \geq \eta^2$, shows that there is a constant $C_{b,l,\theta}$ so that

$$I'_{l,j}(x,t,\eta) \leq C_{b,l,\theta} \frac{e^{-\cos\theta\frac{\eta^2}{|t|}}}{|t|^j}. \tag{A.290}$$

If $l=0$ and $b<1$, then we need to use the approximation

$$\psi_b(z) = \frac{1}{\Gamma(b)} + O(z) \tag{A.291}$$

to see that

$$I'_{0,j}(x,t,\eta) \leq \frac{C_{b,0,\theta} e^{-\cos\theta\lambda}}{|t|^j} \int\limits_0^{\frac{(\sqrt{x}-\eta)^2}{|t|}} w^{b-1} e^{-\cos\theta w} \left[\frac{1}{\Gamma(b)} + O(w\lambda)\right] dw, \tag{A.292}$$

which again implies that

$$I'_{l,j}(x,t,\eta) \leq C_{b,0,\theta} \frac{\lambda e^{-\cos\theta\lambda}}{|t|^j} \leq C'_{b,0,\theta} \frac{e^{-\cos\theta\frac{\eta^2}{2|t|}}}{|t|^j}. \tag{A.293}$$

Here $C'_{b,0,\theta}$ is bounded as $b \to 0$.

The only case that remains is when $|t|^2 < x(\sqrt{x}-\eta)^2$, wherein

$$I'_{l,j}(x,t,\eta) \leq \frac{C_{b,l}}{|t|^j}\Bigg[\int\limits_0^{\frac{1}{\lambda}} w^{b+l} e^{-\cos\theta(w+\lambda)} |\psi_{b+l}(w\lambda e_{2\theta})| \frac{dw}{w} +$$

$$\int\limits_{\frac{1}{\lambda}}^{\frac{(\sqrt{x}-\eta)^2}{|t|}} \left(\frac{w}{\lambda}\right)^{\frac{b+l}{2}-\frac{1}{4}} e^{-\cos\theta(\sqrt{w}-\sqrt{\lambda})^2} \frac{dw}{\sqrt{w}}\Bigg]. \tag{A.294}$$

If $b+l \geq 1$, then we can estimate ψ_{b+l} by a constant to see that the first term is bounded by

$$\frac{C_{b,l,\theta} e^{-\cos\theta\lambda}}{|t|^j} \leq \frac{C_{b,l,\theta} e^{-\cos\theta\frac{\eta^2}{|t|}}}{|t|^j}. \tag{A.295}$$

If $l = 0$ and $b < 1$, then, as before, we need to use (A.291) to see that this term is bounded by

$$\frac{C_{b,0,\theta}\lambda e^{-\cos\theta\lambda}}{|t|^j} \leq \frac{C'_{b,0,\theta}e^{-\cos\theta\frac{\eta^2}{2|t|}}}{|t|^j}, \tag{A.296}$$

where again $C'_{b,0,\theta}$ is bounded for $b < B$. This leaves only

$$\frac{C_{b,l}}{|t|^j}\int\limits_{\frac{1}{\lambda}}^{\frac{(\sqrt{x}-\eta)^2}{|t|}}\left(\frac{w}{\lambda}\right)^{\frac{b+l}{2}-\frac{1}{4}} e^{-\cos\theta(\sqrt{w}-\sqrt{\lambda})^2}\frac{dw}{\sqrt{w}} =$$

$$\frac{C_{b,l}}{|t|^j}\int\limits_{\frac{\eta}{\sqrt{|t|}}}^{\sqrt{\lambda}-\frac{1}{\sqrt{\lambda}}}\left(1-\frac{z}{\sqrt{\lambda}}\right)^{b+l-\frac{1}{2}} e^{-\cos\theta z^2}dz. \tag{A.297}$$

If $b + l - \frac{1}{2} > 0$, then this is bounded by

$$\frac{C_{b,l}}{|t|^j}\int\limits_{\frac{\eta}{\sqrt{|t|}}}^{\sqrt{\lambda}-\frac{1}{\sqrt{\lambda}}} e^{-\cos\theta z^2}dz \leq \frac{C_{b,l,\theta}e^{-\cos\theta\frac{\eta^2}{|t|}}}{|t|^j}. \tag{A.298}$$

This leaves only the case $l = 0$, $b < \frac{1}{2}$. To obtain a good estimate in this case, as $\lambda \to \infty$, we split the integral into two parts:

$$\int\limits_{\frac{\eta}{\sqrt{|t|}}}^{\sqrt{\frac{2\lambda}{3}}}\left(1-\frac{z}{\sqrt{\lambda}}\right)^{b-\frac{1}{2}} e^{-\cos\theta z^2}dz + \int\limits_{\sqrt{\frac{2\lambda}{3}}}^{\sqrt{\lambda}-\frac{1}{\sqrt{\lambda}}}\left(1-\frac{z}{\sqrt{\lambda}}\right)^{b-\frac{1}{2}} e^{-\cos\theta z^2}dz. \tag{A.299}$$

The first term is bounded by $C_{b,0,\theta}e^{-\cos\theta\frac{\eta^2}{t}}$; and the second by

$$C_{b,0,\theta}\sqrt{\lambda}e^{-\cos\theta\frac{2\lambda}{3}} \leq C_{b,0,\theta}e^{-\cos\theta\frac{\eta^2}{2|t|}}.$$

Altogether we have shown that

$$I'_{l,j}(x,t,\eta) \leq \frac{C_{b,l,\theta}e^{-\cos\theta\frac{\eta^2}{2|t|}}}{|t|^j}. \tag{A.300}$$

□

The proof of Lemma 9.3.3 is similar, but easier. It follows from the formula, valid for $t \in S_0$:

$$\partial_x^j k_t^e(x,y) = \frac{1}{t^{\frac{j}{2}}} \sum_{l=0}^{j} c_{j,l} \left(\frac{(x-y)}{2\sqrt{t}} \right)^l k_t^e(x,y). \tag{A.301}$$

The details of the proof are left to the reader.

Bibliography

[1] S. R. Athreya, M. T. Barlow, R. F. Bass, and E. A. Perkins, *Degenerate stochastic differential equations and super-Markov chains*, Probab. Theory Related Fields, 123 (2002), pp. 484–520.

[2] A. D. Barbour, S. N. Ethier, and R. C. Griffiths, *A transition function expansion for a diffusion model with selection*, The Annals of Applied Probability, 10 (2000), pp. 123–162.

[3] R. F. Bass and A. Lavrentiev, *The submartingale problem for a class of degenerate elliptic operators*, Probab. Theory Related Fields, 139 (2007), pp. 415–449.

[4] R. F. Bass and E. A. Perkins, *Degenerate stochastic differential equations with Hölder continuous coefficients and super-Markov chains*, Transactions of the American Math Society, 355 (2002), pp. 373–405.

[5] A. J. Castro and T. Z. Szarek, *Calderón-Zygmund operators in the Bessel setting for all possible type indices*, ArXiv e-prints, 1110.5659 (2011).

[6] S. Cerrai and P. Clément, *Schauder estimates for a class of second order elliptic operators on a cube*, Bull. Sci. Math., 127 (2003), pp. 669–688.

[7] ——, *Well-posedness of the martingale problem for some degenerate diffusion processes occurring in dynamics of populations*, Bull. Sci. Math., 128 (2004), pp. 355–389.

[8] ——, *Schauder estimates for a degenerate second order elliptic operator on a cube*, J. Differential Equations, 242 (2007), pp. 287–321.

[9] F. A. C. C. Chalub and M. O. Souza, *The frequency-dependent Wright-Fisher model: diffusive and non-diffusive approximations*, ArXiv:, 1107.1549v1 (2011).

[10] L. Chen and D. Stroock, *The fundamental solution to the Wright-Fisher equation*, SIAM J. Math. Anal., 42 (2010), pp. 539–567.

[11] P. Daskalopoulos and R. Hamilton, *Regularity of the free boundary for the porous medium equation*, Jour. of the AMS, 11 (1998), pp. 899–965.

[12] ——, *The free boundary on the gauss curvature flow with flat sides*, J. Reine Angew. Math., 510 (1999), pp. 187–227.

[13] E. B. DAVIES, *One-parameter Semigroups*, Academic Press, New York, 1980.

[14] C. L. EPSTEIN, *Convergence of the Neumann series in higher norms*, Comm. in PDE, 29 (2004), pp. 1429–1436.

[15] C. L. EPSTEIN AND R. MAZZEO, *Wright-Fisher diffusion in one dimension*, SIAM J. Math. Anal., 42 (2010), pp. 568–608.

[16] A. ETHERIDGE AND R. GRIFFITHS, *A coalescent dual process in a Moran model with genic selection*, Theoretical Population Biology, 75 (2009), pp. 320–330.

[17] S. ETHIER AND R. GRIFFITHS, *The transition function of a Fleming-Viot process*, The Annals of Prob., 21 (1993), pp. 1571–1590.

[18] S. N. ETHIER, *A class of degenerate diffusion processes occurring in population genetics*, CPAM, 29 (1976), pp. 417–472.

[19] W. EWENS, *Mathematical Population Genetics, I, 2nd edition*, vol. 27 of Interdisciplinary Applied Mathematics, Springer Verlag, Berlin and New York, 2004.

[20] W. FELLER, *The parabolic differential equations and the associated semi-groups of transformations*, Ann. of Math., 55 (1952), pp. 468–519.

[21] D. FERNHOLZ AND I. KARATZAS, *On optimal arbitrage*, Ann. Appl. Probab., 20 (2010), pp. 1179–1204.

[22] C. GOULAOUIC AND N. SHIMAKURA, *Regularité Hölderienne de certains problèmes aux limites elliptiques dégénérés*, Ann. Scuola Norm. Sup. Pisa Cl. Sci. (4), 10 (1983), pp. 79–108.

[23] C. R. GRAHAM, *The Dirichlet problem for the Bergman Laplacian. I*, Comm. Partial Differential Equations, 8 (1983), pp. 433–476.

[24] ——, *The Dirichlet problem for the Bergman Laplacian. II*, Comm. Partial Differential Equations, 8 (1983), pp. 563–641.

[25] R. GRIFFITHS, *Stochastic processes with orthogonal polynomial eigenfunctions*, Journal of Computational and Applied Mathematics, 233 (2009), pp. 739–744.

[26] E. HILLE AND R. PHILLIPS, *Functional Analysis and Semi-groups, revised edition*, vol. XXXI of AMS Colloquium Publications, American Mathematical Society, Providence, RI, 1957.

[27] S. KARLIN AND H. M. TAYLOR, *A second course in stochastic processes*, Academic Press [Harcourt Brace Jovanovich Publishers], New York, 1981.

[28] H. KOCH, *Non-Euclidean singular integrals and the porous medium equation*, Habilitation thesis, University of Heidelberg, 1999.

[29] N. KRYLOV, *Lectures on Elliptic and Parabolic Equations in Hölder Spaces*, vol. 12 of Graduate Studies in Mathematics, American Mathematical Society, Providence, RI, 1996.

[30] G. LUMER AND R. S. PHILLIPS, *Dissipative operators in a Banach space*, Pacific J. Math., 11 (1961), pp. 679–698.

[31] A. LUNARDI, *Analytic semigroups and optimal regularity in parabolic problems*, Progress in Nonlinear Differential Equations and their Applications, 16, Birkhäuser Verlag, Basel, 1995.

[32] R. MAZZEO, *Elliptic theory of differential edge operators. I*, Comm. Partial Differential Equations, 16 (1991), pp. 1615–1664.

[33] R. MAZZEO AND A. VASY, *Resolvents and Martin boundaries of product spaces*, Geom. Funct. Anal., 12 (2002), pp. 1018–1079.

[34] R. MAZZEO AND A. VASY, *Analytic continuation of the resolvent of the Laplacian on* $\mathrm{SL}(3)/\mathrm{SO}(3)$, Amer. J. Math., 126 (2004), pp. 821–844.

[35] R. B. MELROSE, *Calculus of conormal distributions on manifolds with corners*, International Mathematics Research Notices, (1992), pp. 51–61.

[36] ——, *The Atiyah-Patodi-Singer index theorem*, vol. 4 of Research Notes in Mathematics, A K Peters Ltd., Wellesley, MA, 1993.

[37] N. SHIMAKURA, *Équations différentielles provenant de la génétique des populations*, Tôhoku Math. J., 29 (1977), pp. 287–318.

[38] ——, *Formulas for diffusion approximations of some gene frequency models*, J. Math. Kyoto Univ., 21 (1981), pp. 19–45.

[39] D. STROOCK AND S. VARADHAN, *Multidimensional Diffusion Processes*, vol. 233 of Grundlehren Series, Springer-Verlag, Berlin, Heidelberg, 1979.

[40] D. W. STROOCK, *Partial Differential Equations for Probabilists*, vol. 112 of Cambridge Studies in Advanced Mathematics, Cambridge University Press, Cambridge, 2008.

Index